FOUNDATION FOR COLLEGE MATHEMATICS

Enhanced STEM Edition

By James P. Fulton, Ph.D.
Adrienne Chu, Ph.D.
Elizabeth Chu, M.S.
Rachael Millings, M.S.

Suffolk County Community College

Preface to the Student:

Mathematics is a language by which we try and make sense out of the world around us. As a student you have probably studied mathematics for the better part of the last 13 years, if you just came out of high school.

Simply stated, Algebra is a language that allows us to release the power of numbers by creating things called **models.** Models are the mathematical representations of quantifiable phenomena that allow us to calculate and/or predict useful outcomes.

For example, algebra can be used to calculate the amount of a loan repayment in business; the estimated time of death in forensics; the age of ancient artifacts in archaeology; the discounted price in consumer math; the profit of a business in accounting.

The only way to learn a language is to practice it and use it over-and-over again. You will not be able to watch the instructor and then flawlessly perform the same steps, just like you would not be able to play a piece of music; paint a picture; perform surgery; drive a car; or play a particular sport, by simply watching a person and then repeating what you saw them do. You must try it yourself, again-and-again-and-again …

One thing Algebra is not, and that is easy for most people. In this book, we try to make it as sensible and as understandable as possible, but there will be times that you will become confused and frustrated. This is normal! It happens to everyone. Expect it, and when it inevitably does happen, seek out help, so that you can move on to the next concept. The concepts build upon one another, and failure to understand one critical idea could cause problems later on. Also, you will find that Algebra has a lot of rules to remember, which is why you need to speak it often. Get used to it!

As we all know, mathematics is a very precise language, therefore, we need an element of precision in how we approach it. However, when new mathematics is created or discovered, the process by which this happens is almost always informal, and with a lot of trial and error. When the subject matter is taught, though, the material is presented in the formal context first, and then informal examples are used to highlight the theory after the fact. This is quite contrary to how it was created in the first place. We will aim for more of a mixed, informal-approach first, and then summarize the formal results and definitions at the end of each section, not at the beginning.

This course is a prerequisite for intermediate algebra, statistics, and laboratory science courses, therefore, we will show some examples of concepts and ideas you will see in future courses.

Preface to the Instructor:

This book is intended for students that need a refresher in the mathematics needed to prepare them for either a course in liberal arts mathematics, introductory statistics, intermediate algebra, or any of the basic laboratory science classes, that don't require intermediate algebra or higher as a prerequisite.

You will notice that this book does not follow any standard outline of topics. Topics were chosen after an extensive survey was given to faculty in the science departments, that used this course as a prerequisite. In addition, we included topics that would also prepare students to take the next course in the STEM math sequence; intermediate algebra. It is a tall order, so some difficult choices had to be made regarding how in-depth we should go into any one topic. The list of topics presented in the Table of Contents is a compromise, and we hope it adequately prepares students to go down the different paths as they start their college experiences. Although some topics had to be omitted, we still feel confident that a student will be adequately prepared to take any of the follow-up courses that this course feeds into.

The course that this book was written for is extremely challenging. For the most part these are at risk students, and after 12 or 13 years of secondary school mathematics training, they still do not understand basic algebra, even though virtually all of these students have taken and passed an intermediate algebra course, or higher. Before modifying the course, we shared our course outline with Dr. Hung Hsi Wu (Professor Emeritus, University of California at Berkeley) and received this reply. "I looked at the description of your course and told myself that you had been put in an impossible situation. For students who are at risk, they are probably wary of math. They underachieve in math most likely because they have been mistreated mathematically, and their knowledge base is therefore shaky from the ground up. Your course has to first restore their trust in math. Yet, in one semester you are called upon to cover a whole lot of ground, and I myself don't see how to do that while still managing to win their trust. You have an unenviable task alright, one that I do not know how to do."

Obviously, we will not be able to correct for many years of misunderstanding in a single semester course. Our hope, however, is to give the students enough understanding for them to be successful in their follow-up college courses that require a certain level of mathematical understanding. Now we understand that there are many instructors that would like to see a much higher and more rigorous level of mathematics achieved in this course, but to do so would probably confuse many of these students even further, because the pace would have to be extremely fast, with so much to cover. The goal of this course is not simply to prepare all our students for intermediate algebra, but instead to prepare them for their next college course, which requires math as a prerequisite. That is why we have chosen this approach, and these set of topics.

Acknowledgments:

We'd like to thank members of Suffolk County Community College Science Departments for taking the time to sit with us, and go over many of the ideas behind the topics in this book. They were also very generous with their lesson plans and exams, and are responsible for a number of the questions in the exercises. The faculty that assisted us are:

Jean Anastasia (Marine Biology), Jing Yi Chin (Chemistry), Amy Czura (Biology), Rosa Gambier (Biology), Todd Gardner (Marine Biology), Scott Gianelli (Automotive Physics), Janet Haff (Natural Sciences), Michael Inglis (Astronomy), Scott Mandia (Meteorology), Joseph Napolitano (Biology), Matt Pappas (Astronomy), Richa Prakash (Chemistry), Kevin Reilly (Academic Skills Center), Lou Roccanova (Biology), Hope Sasway (Natural Sciences), Peter Smith (Biology), Phoebe Smith (Biology), and Sean Tvelia (Earth Sciences),

We would also like to thank the College Administration for giving us the time to pursue this effort and in allowing us to present this new approach to our students. In particular, we'd like to thank President Shaun McKay and Associate Vice President for Academic Affairs Paul Beaudin for all their support and encouragement. Furthermore, the chairs of our departments on the three campuses of this college were extremely supportive. A special thanks to the chairs: Dennis Reissig, Ted Koukounas, and John Jerome for their support.

TABLE OF CONTENTS

Chapter 7: Ratios, Percents, Rates, and Proportions

Chapter 8: Linear Equations with Two Variables

Chapter 9: Solving Systems of Linear in Two Variables

Chapter 10: Basic Math for Lab Science Courses

INTRODUCTION

Measurement – Measuring Your Way Through Mathematics

Students always ask – 'Why do I have to study math?,' or 'When will I ever use this?' The answer quite frankly is simple. Math is everything! It is everywhere around us, and yes, it is true that we may not need the specific math skill we are learning at a particular time, but if we would only open up ourselves to the possibilities, we might notice that it could be useful in some other setting or other time in the future. For you see, math is really all around us. It truly is the language of the universe. It is a language that allows us to describe all the attributes of this thing we call the universe, in a very definable and precise way. Mathematics is the language of patterns, and patterns are what make up the universe!

The world is full of patterns. The sun "rises" and "sets," the seasons come and go, and cicadas breed in unique and consistent time delays, all forming cyclical patterns. Birds fly in formation, spiders spin webs, trees grow branches, flowers form seed packs and petals, and bees build hives, all forming rich geometric patterns. The weather dramatically changes, and the stock market seems to erratically rise and fall, all giving rise to seemingly chaotic patterns.

When Aristotle sought to define logic, the essence of human thought and reasoning, as a way to make language more precise, little did he foresee that it would be mathematics that would some day add true precision to it, and clear up the ambiguities he sought to dismiss. He never saw the insights of Leibniz, Boole, Frege and Russell that would show that even the idea of human thought could potentially be understood by and reduced to mathematics – the mathematical patterns associated with human reason.

Mathematics has been called the Queen of all the Sciences, essentially because; mathematics is the science of recognizing, analyzing, and deciphering patterns. Patterns make the world the way it is and understanding the code needed to decipher them gives us power to predict and alter their effects. That code is written in the language of mathematics. The only way we can grasp these patterns is to use our senses to observe, characterize, and quantify their unique attributes. We accomplish the important act of quantification through measuring. Our understanding hinges on our ability to reproducibly measure the attributes we observe.

To measure means we determine either the amount, size, or degree of something, by either counting, using a device calibrated with some set of standardized attribute units, or by comparing it to an object with a known attribute value. Measuring is the most important tool of any mathematician. Without it, we would have only vague words to try and describe what we "see," but no concrete reproducible way to determine if the pattern persists in a precise unchanging way. We would have no way to determine if the pattern is caused by a fundamental law of nature, or simply by the appearances of what we think we "see".

In this textbook we will embark on the journey of gaining an understanding of the art of measuring. It is a valuable skill for all to have, not just for mathematicians and scientists, but for anyone to survive in this world. Measuring is how we make sense and navigate our way through this complex world and life.

To be able to master the art of measuring means we need to be proficient with several concepts. The first concept is that of **number**. Now by number we shall mean, whole numbers, fractions, decimals (decimal fractions), irrational numbers, exponential numbers, and even negative numbers. We furthermore need to be aware of the idea of accuracy and precision of our measurements related to these numbers. This leads to the idea of obtaining approximate values for our measurements, and also being able to effectively reproduce our results.

The next important concept related to measuring is **geometry**. Even the name geometry; (from **geo** meaning 'earth', and **metria** meaning 'measure'), or earth–measure, involves the concept of measurement in its name. The shapes of objects and their relationships are important tools we use to characterize objects and things around us. How and what we measure geometrically is very important.

We also need to understand how measuring different attributes of things are related to each other. Thus we need to see how our measured data are related to each other, and so we need to know how to create and interpret relations between measure values. We do this in the form of graphs of data, and this inevitably leads to the idea of **algebra** and its relation to measuring.

This textbook will use the concept of measuring, which we have just stated is critical to our ability to live in this world, to unify several branches of mathematics related to; numbers, geometry, and algebra. Along with whatever we are introducing or defining, we will also consider; how does this affect how we measure? Furthermore, the idea of measuring is essential to any scientific investigation. Thus, it is central in any laboratory science class, as well as any mathematics or quantification related course. Hence, another purpose of this textbook is to equip students with the necessary skills to succeed in a college level science course as well as a college level mathematics course.

MEASUREMENT

One of the most important aspects of mathematics has to do with solving problems. Not the concocted word-problems we have all become frustrated by, but problems of real consequence. We frequently need to know "how much?", and "of what?" How much time do we have? How far do we have to travel? How high up is it? How cold is it? How much flour do I add? How much rain did we get? How long does the board need to be? How much fuel do I need? The list can go on.

To be able to answer the **how-much** question, we need to know how to measure. Whether it is with a ruler, a weight-scale, or a thermometer, we need to know about measuring. It is how we accurately compare quantities to see which is larger, or how we determine how much of a certain quantity is present relative to another quantity, or how we properly mix ingredients, or how we determine if things fit together properly or not. It is a vital and necessary skill.

All measurements are relative and are made up of two distinct parts. The first part answers the "how-much?" question. It is the numerical characterization of the measurement, such as 1, 2, 3, 6.5, or 10.75. The second part answers the "of what?" question. It identifies what we are measuring — it is the **units** part of our measurement, which also has two aspects, its type or attribute, and its scale. Any measurement must contain both a number and a unit (attribute), otherwise it is useless and meaningless. When measuring we are really asking "How much", and "of what, are we measuring?" The number expresses the magnitude, and the unit expresses the type of quantity we are considering, but given in a specific scale. See the figure below:

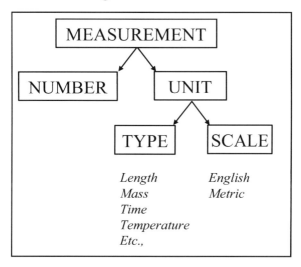

Examples of measurements and units: 25 mm, 16 in^2, 35.7 g, 2.5 gal., 64 km, 153 miles, 36°F, 5 lbs, and 12 oz., 1.3 kg, 75 mph, 16 Amps., 39 cm^3, etc.

In mathematics, we often disregard the units, which does a great disservice to problem solving. Units, however, are key to every problem, so in this textbook we will focus on, and pay close attention to units.

In the first chapter we begin with the idea of number. A concept you have seen many times before. It is our hope that if you have had problems in the past grasping some of the concepts related to numbers, that we can help you to master these concepts. For others it will be a review of something you may not have seen for a while. For us, the authors, it is an opportunity to make sure that all students are starting with the same foundation that we can begin to build upon.

4

CHAPTER 1

Introduction to Numbers –

Their Properties and Operations Performed With Them

Numbers are the fundamental building blocks on which all of mathematics is formed and based. You have worked with numbers for most of your life, and have studied them and their properties extensively in school. In this chapter we provide a review of numbers and their properties, which are essential for a thorough understanding of the mathematical concepts required in many of your technical college courses. This is prerequisite material that you should already be familiar with, but you may either be a little rusty, or it never really made sense the first time you saw it, or out of lack of interest you never really paid much attention to it.

1.1 TYPES OF NUMBERS AND THE NUMBER LINE

As you probably recall from early lessons in mathematics, there are many different types of numbers. You need to be well versed with them, and be able to distinguish between the different types. There are essentially six groupings of numbers involved: they are called Natural, Whole, Integer, Rational, Irrational and Real. What they are and how they are related to each other is the subject of this section. We will begin with a brief overview of the different types, and then later on define them more precisely. We do this only because you are already familiar with them, otherwise we would take a step-by-step development process to define them more concretely at the beginning.

Natural Numbers

The number system we use today was developed over many centuries. As with all developments, its introduction was largely based upon its necessity to solve problems. Counting or natural numbers were the first numbers to be developed. The Natural Numbers are represented by the bold capital letter **N**, and can be written as a list of numbers such as 1, 2, 3, 4, 5 and 6. Obviously the list goes on and on without end. We represent this list with the following mathematical notation:

$$\mathbf{N} = \{1, 2, 3, 4, 5, 6, \ldots\}$$

The braces, { }, represent the fact that we are talking about a **set** or a group of numbers. For now, we simply take a set to mean a collection of numbers. The "…" are called ellipses and simply mean that the list goes on and does not have an end.

A very useful aid in understanding our system of numbers is through the use of a number line. A number line is simply a horizontal line that continues endlessly in both directions as represented by the arrows in the figure below. Throughout our early introduction of numbers we will use the concept of the number line to understand all types of numbers. Thus, we need to move away from thinking about numbers as simply counting objects, but more like measuring distances. If you can make this first abstract distinction, you are on your way towards a better understanding of our modern number system. In the graph below, we have placed our natural numbers on the number line:

Whole Numbers

Adding zero to the set of natural numbers gives the set of Whole Numbers, **W**.

$$\mathbf{W} = \{0, 1, 2, 3, 4, 5, \ldots\}$$

It took hundreds of years before people were ready to accept zero as a number. It was not an easy concept to understand. Why do we need a symbol, which is something, to represent nothing? Also, zero doesn't behave quite the same as other numbers. When you multiply by it, it annihilates everything, and you can't divide by it. It was, and is, a very strange number, indeed.

Adding zero to the number line we have:

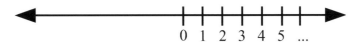

Now zero plays a unique role in our number system. It is the number that "centers" our number system or our number line. You can think of it as the origin, from which the distances of other numbers are measured. If you do, then the next set of numbers will make more sense to you.

Integers

After zero comes another bizarre concept, the negative number. For many brilliant minds throughout history, the concept of negative numbers made no sense. When the need for them came up in the course of solving mathematical problems, they were thrown out or ignored, and originally called false solutions. This was because at this stage in their development, numbers only represented the counting of objects, not points on a number line. In fact, in 16th century France, they were cited as the main reason for the problem with the French educational system. How could we have a number that was less than nothing? Mathematicians didn't like or understand it, but the business and trade world needed it. Negative numbers could be used to represent a debt. The concept was needed for trade! However, along with this need came many conceptual problems, as we shall see later.

Now the negative of the natural numbers, along with zero and the set of natural numbers, make up the set of Integers, **Z**.

$$\mathbf{Z} = \{\ldots, {}^-5, {}^-4, {}^-3, {}^-2, {}^-1, 0, 1, 2, 3, 4, 5, \ldots\}$$

Now you may wonder why we choose the letter **Z** to represent the set of integers and not the letter I. The reason is that many of the early mathematicians that explored properties of numbers were

German, and they used the German word for number, which is zhalen, which they abbreviated with the letter **Z** to represent the set of all the whole numbers and their negatives, and it stuck.

We should also point out that we have also chosen to use a unique way of representing negative numbers (It's not really unique because it was actually done this way many years ago, and then abandoned for some unknown reason.) We will use a superscript minus sign before the number, so that we know we are talking about a negative number. This will help us later on as we try to avoid the confusion between a negative number and the operation of subtraction. This allows us to more clearly distinguish between the name of an object (noun) and an arithmetic operation on that object (verb). Hopefully, in this presentation it will clear up some of the confusion introduced by using the same symbol to represent two very different things. In later math courses, as well as later on in this book, we will not highlight this distinction and instead will use the same symbol to represent both ideas. However, when we first present the two distinct concepts, we find it less confusing to show that they are really two different things, and need to be considered as such, to help avoid confusion early on.

Negative numbers are used to represent distances to the left of zero on the number line. We have placed the negative integers onto the number line a unit distance apart. By convention, we put the **larger** numbers to the right of the **smaller** numbers.

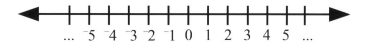

This is where the concept of negative numbers starts to get a bit confusing. When we talk about larger and smaller numbers in mathematics, we are really talking about their order on the number line. Thus, if a number comes before another number as we traverse the number line from left to right it is considered smaller. We use this convention even though a smaller negative number is actually a larger distance away from zero on our number line. This is precisely what can cause confusion in the thought process. For example ⁻10 is actually "smaller" than ⁻3, although ⁻10 is a "larger" distance from zero on the number line than ⁻3.

Later on we'll define this more precisely, but for now, just note that this is another area of potential confusion as we introduce these new types of numbers, and it will need careful consideration when we are working problems involving these types of numbers.

The number system can be extended further if we continue with the distance analogy. In between the integers on the number line there are gaps. We can begin to fill in these gaps by introducing a new set of numbers, called Rational Numbers.

Rational Numbers

The set of all numbers that can be formed as a quotient (ratio) of two integers, where the denominator can never be zero, are defined as Rational Numbers, **Q**. This set contains all those dreaded fractions. Again, another bold and confusing concept emerges. We now must accept that a ratio of two numbers, can now be considered to be a single number! The ancient Greeks didn't accept this, and the early Egyptians only considered and worked with something called "unit-

8

fractions," fractions with a one on top (this is almost correct, since they did have a few non-unit-fractions for special calculations, but these were exceptions). For Pythagoras and his religious sect the Pythagorean's, the rationals were not numbers, but were only thought of as ratios of two distinct numbers. Unsurprisingly, it was centuries later before the single-number concept took hold, and this only begins to make sense when you think of numbers as representing distances on a number line, and not as a way to count individual objects.

The set of rational numbers contains the set of integers, since every integer can be written as a ratio over one. And since the integers contain the whole numbers, which contain the natural numbers, the rational numbers contain ALL of them!

Now, just like the Egyptians and the Pythagorean's, most students don't fully understand fractions. It is hard for many to see them as numbers in the same way they see numbers such as 1, 2, or 3. It is not a realization that comes easily for many students, though. It takes quite a bit of time to properly develop the concept of seeing a fraction as a number.

Examples of Rational Numbers

$\dfrac{3}{4}$ 3 and 4 are both integers

$\dfrac{1}{3}$ 1 and 3 are both integers

$\dfrac{23}{17}$ 23 and 17 are both integers

$\dfrac{15}{3} = 5$ 15 and 3 are both integers, but 15 divided by 3 simplifies to 5

$\dfrac{^-21}{2} = {}^-21/2 = {}^-10\dfrac{1}{2}$ $^-21$ and 2 are both integers, the numbers can also be written as a mixed number

7   This number can be written as the quotient of two integers 7 and 1, $\dfrac{7}{1}$ so 7 is also considered to be a rational number

Rational numbers can be represented either as a ratio of two numbers, as shown above, or as a decimal number (also called decimal fractions). A decimal number is what results when we take the quotient (divide the two numbers). We have all used a calculator and should be very familiar with this. We will explore decimal numbers in greater detail later in this chapter, but for now we need to introduce them to identify a key property of rational numbers. Decimals are also more closely recognizable as distances when we think about numbers.

All rational numbers can be represented as either terminating or repeating decimals. For the examples above we see that:

$\dfrac{3}{4} = 0.75$ This is a terminating decimal since the division ends.

$\dfrac{1}{3} = 0.333333.....$ This is a repeating decimal since the digit 3 is repeating.

$\dfrac{23}{17} = 1.3529411764705882352941176470 5882...$ This is a repeating decimal since the digits 3529411764705882 repeat. There are 16 digits in this pattern!

$\dfrac{15}{3} = 5$ This terminates.

$-\dfrac{21}{2} = {}^-21/2 = {}^-10.5$ This is a terminating decimal.

More examples of terminating and repeating decimals:

$\dfrac{3}{5} = 0.6$ Terminating decimal

$\dfrac{17}{99} = 0.171717....$ Repeating decimal

Irrational Numbers

Surprisingly, the set of Rational Numbers does not fill up all of the spaces on the number line. It turns out that there are more gaps, and these gaps are filled by a new set of numbers that are not rational. They are the Irrational Numbers. The prefix "ir" in irrational is from the Latin, meaning "not". Thus, it means **not-rational**, i.e. not a ratio of integers. These are very strange numbers indeed. In fact, nothing in real life that we can measure is an irrational number. They exist only in our mathematical world, but we can't get rid of them. The important number π is irrational, and we all know how useful this number is in calculating circular areas and volumes. Irrational numbers are quite troublesome, though. Believe it or not, they involve another incomprehensible concept, that of infinity. We'll have to leave this for another discussion, though.

If we consider the decimal representation of a number, it turns out that an irrational number cannot be represented using either a terminating or a repeating decimal number.

For example:

$$\dfrac{3}{5} = 0.6 \qquad\qquad \dfrac{3}{25} = 0.12 \qquad\qquad \dfrac{1}{3} = 0.33333... \qquad\qquad \dfrac{17}{99} = 0.171717...$$

All of these are rational numbers and they can be represented by a decimal that either terminates, as in the first two examples above, or repeats, as in the last two examples.

On the other hand, the decimal representation of an irrational number will never terminate nor will it display a pattern of numbers that repeats.

Examples of irrational numbers and their decimal equivalents are:

$$\sqrt{2} = 1.41421356237...,$$
$$\pi = 3.1415926535...,$$
$$-\sqrt{13} = {}^{-}3.6055512754...,$$
$$\sqrt[4]{3} = 1.3160740129...$$

All these numbers are irrational, since a terminating or repeating decimal pattern is not possible. Thus, the exact representation of irrational numbers can only come from their symbolic form, i.e. $\sqrt{2}$ instead of 1.4142. Because of this property, we can never exactly represent an irrational number on a calculator or a computer. We can only use its **rational approximation**. This is another reason why the concept of approximating an answer becomes important.

Real Numbers

The set of irrational numbers combined with the set of rational numbers does fill in every space on the number line, and this combined set of numbers is called the set of real numbers.

The relationship between the various sets of numbers is illustrated in the figure below. We see that the Naturals are a subset of the Wholes, which are a subset of the Integers, which are a subset of the Rationals. The Irrationals are a set with nothing in common with the Rationals, and together the Rationals and the Irrationals make up the set of Real Numbers.

If the natural numbers were a fish, then the whole numbers would be a slightly larger fish that would swallow up the Natural fish. The Integers would be a larger fish to swallow the Whole, and the largest, Rational fish, would swallow the Integer fish. This Rational fish swims in the water, which is made up of the Irrational numbers. Together, the fish and the water make up the ocean of Real numbers.

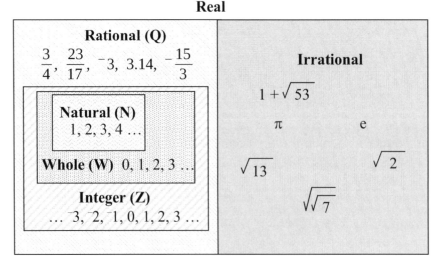

Quite a confusing concept that has caused one famous mathematician, Leopold Kronecker, to comment, "God created the natural numbers; all the rest is the work of man."

The main take-away from this section is to know that the number concept is quite broad. If you are going to be successful in this course, you furthermore need to understand that when we say number, we mean something we can measure. Thus, the idea of a number line is very important in helping us to see that all these different types of numbers, can be viewed simply as points on a number line that starts at zero and proceeds in two opposite directions at a distance equivalent to the value of the number. If the numbers are negative, they are to the left of zero, and if positive, they are to the right. The different types of numbers allow us to identify any location (point) on the number line, as shown below with some specific examples of different types of numbers; $^-15/4$, $^-2$, 0, $\sqrt{2}$, π, and $16/3$.

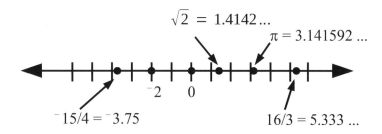

In the remainder of this chapter we review some of the more important aspects of numbers and operations on them that will be required for your college math and laboratory science classes. We will not review everything, but focus more on problem areas and topics you will need.

1.2 A FRESH LOOK AT NUMBERS: Understanding Numbers, Place–Value, and more

In the first section of this book we reviewed all the basic types of numbers you have seen in the past and will see throughout this course. In this section, however, we wish to revisit the introduction of these numbers, and focus more on the number concept. We realize that you have seen this material many times before, but it is our hope that we can give you some new tools and perspectives to better understand the associated concepts related to the different types of numbers. This will be especially true when we introduce fractions and negative numbers later on. These are two very fundamental and difficult concepts that students need to fully comprehend.

<u>Defining the Number Concept and Axiomatic Systems</u>
Let's begin again. We start with a conceptual definition of number. We should point out that this is not an easy task mathematically. What seems so obvious and simple is much more involved than we can get into at this level. Strangely it brings deep philosophical issues to light that will not be discussed in this book.

A mathematical description of number introduces numbers, along with the operations we define on them, using something called an axiomatic system. This is a formal system based upon principles of logic that allows us to avoid mistakes and paradoxical results as we develop the concepts and their relationships.

An axiomatic system consists of an undefined term/terms, along with statements called axioms that are assumed to be true. The axioms are the rules of our axiomatic game. Definitions, in terms of the undefined term/terms, of objects are then created so that properties of and relationships between these newly defined objects can be proven. A complete mathematical theory of numbers and their properties and relationships is obtained by proving new statements, called theorems, using only the axioms, logic, and previous theorems.

Parts of an Axiomatic System:
1. Undefined terms/primitive terms
2. Defined terms
3. Axioms/postulates-accepted unproved statements
4. Theorems-proved statements

In the theory of numbers, our undefined object is the concept of what we mean by the word number. We will start with pseudo–definition where we instead describe what we'd like our numbers object to do.

Pseudo–Definition: A **number** is a mathematical object used to count, measure, and label.

Now we could start out in a very formal way by listing all the axioms and definitions and begin to prove new theorems, but we will not. We did, however, want to introduce the idea of an axiomatic system, so that you would know there is a logically consistent approach for developing each concept and idea we come across. We will, from time–to–time, refer to this implied connection as we develop the subject further, but for now we'll proceed less formally.

We take a less formal/rigorous approach because the concept of number, believe it or not, is quite hard to define in a logically consistent mathematical way, that allows all the types of numbers we discussed earlier to be considered as numbers. The above definition is not a true definition of number because the concepts of count, measure, and label all require that we already have the concept of number defined to be able to define them. Thus, our definition above is circular. As a result we simply take this as a pseudo–definition and use our intuition about what numbers are, and proceed from there. This is not rigorous, but necessary if we are going to move forward without losing all readers as we proceed.

From our working definition we see that a number can be used to count individual objects, but that it is also used to measure quantities. The quality of being able to measure is what requires "in–between" values for our numbers. We could be asked to measure distances, weights, volumes, etc., all things which may not come in simple distinct units of measure, which our simple counting numbers would give us. Thus, we need to extend the number concept beyond simply adding another unit and counting higher. For us the key feature in extending the number concept beyond counting, will be through geometry and the concept of the number line. We will think of a number as a point on a number line whose value is the distance our point is away from a fixed point of our choosing on this number line. We call this point the origin and define its location to be the number zero. It is this number line idea, of number being a point on this line whose distance/location from zero is its value, that is required for us to understand all the concepts we will introduce in this book.

If you want to truly grasp the number ideas we present in this course, instead of just memorizing names and facts, then you have to shift away from thinking of numbers as units for counting, and more like distances for measuring. If you can do this, then what we show in this course will make more sense to you. The counting numbers are still there, they are now just points a fixed number of unit distances away from the origin (zero), but now numbers such as fractions, decimals, negative numbers, and irrational numbers will be easier to make sense out of. That being said, we still will use the idea of counting "objects," but now these objects may simply be parts of a whole. This will preserve the idea of number regardless of their type.

The Hindu–Arabic System of Numerals

Now that we have our number concept defined as points on a number line, we next move towards how to easily communicate these values to other people. This requires that we create symbols to represent these numbers. Now this is not as easy as you might think. We have many different types of numbers, but we'd like a single simple system to be able to represent ALL the different types of numbers. Over time the system that was chosen over all other possible ways to do this was something called the Hindu–Arabic Number System. Throughout history there were many different systems developed to represent numbers such as; Egyptian Hieroglyphs, Roman Numerals, Greek Numbers, Mayan Numbers, and Babylonian Numbers, to name a few. However, the one system that won out to be the standard that the entire world adopted, is the Hindu–Arabic Number System.

Our Hindu–Arabic number system got its name due to its creation in India (by the Hindu's) and preservation and improvement by Arab scholars. The Hindu–Arabic Number System is what we call a positional number system. This means that we have symbols we call numerals (or digits) to represent the first nine counting numbers and zero, as well as place–values in our number representation. That means when a numeral (zero through nine) is in what we call our number, where this numeral occurs is important. Its "place" in our number representation, has a value. Since we have ten fingers, it is easy for humans to count in groups of ten, so the Hindu–Arabic Number System counts in groups of ten, and uses place–values using these groups of ten.

Let us begin by reintroducing how we represent these numbers. It is what is known as a base–ten positional number system. This is because we count in groups of ten, and the location of a numeral within a number is important. Its position is called its place, and that place, we will show, represents a specific value. It is a way to represent all types of numbers very concisely, and has been adopted around the entire world, with only a few minor variations, such as the use of commas and periods within a number representation.

Numerals and Place–Values

First we need our numerals (digits) to represent values of numbers from zero to nine. We are all familiar with the symbols chosen to do so; 0, 1, 2, 3, 4, 5, 6, 7, 8, 9. Next, we define the value of the place (location) our numeral falls in. We have also chosen to write our number representations from left to right, with the leftmost numeral being in the largest place–value and the rightmost numeral being in the smallest (for now) one's location. After the ones, to the left, we have the tens' location. Then we have the hundreds' (or ten groups of ten) location. Next we have the thousands', etc..

For example, to represent the first ten numbers from zero to nine we simply write a single symbol for that number: 0, 1, 2, 3, 4, 5, 6, 7, 8, 9. To write the next largest number ten, however, we use two of our base symbols, a one and a zero (1, 0). By putting the one and zero together as 10 (one ten and zero ones) we have two symbols whose position (place) matters in our representation of the number ten. The position of the 1 means we have one group of ten and the position of the 0 means we have no ones in our number. This is what we mean by place–value. The place where our number symbol is located has a designated group value.

Thus, the number 26 means that we have two groups of ten and six ones, which is quite different from 62 which has six groups of ten and two ones. Even though the symbols are the same (2,6), the order in which they appear matters.

We can continue this process to represent even larger numbers so the next place –value is defined as ten groups of ten or one hundred, written as 100. That's a one in the hundreds place, a zero in the tens place, and a zero in the ones place. Continuing in this way we have the place–value chart below:

...	Ten Thousands	Thousands	Hundreds	Tens	Ones
...	Ten groups of a thousand	Ten groups of a hundred	Ten groups of ten	Ten groups of one	One group of one.
...	1,0000	1,000	100	10	1

This is a topic students frequently have trouble with, but is extremely important. You need to understand the place–value concept, otherwise numbers will not make sense to you.

You should also know that it is possible to represent a number in several distinct ways. The first is called **standard form**, where we write the number as you are used to seeing it. For example the number 247.

Now, we can also write this number using words, or in **word-form** as, two–hundred and forty–seven. Notice how we use the place–value information in the words.

Alternatively, the **place–value** form of the number becomes; two groups of a hundred, four groups of ten, and seven ones, or simply as, 2 hundreds, 4 tens, and 7 ones.

The number can also be represented in **chart-form** both with and without our numeral symbols for numbers as,

Hundreds	Tens	Ones
2	4	7
• •	• • • •	• • • • • • •

Where each disk in the chart represents one group of the named amount in each of the place–values.

Finally, we can write this number in something called **expanded form**. Expanded form simply writes the full place–value with the appropriate values and expands the number out using addition, for example:

$$247 = 200 + 40 + 7$$
$$\text{or}$$
$$= 2\times100 + 4\times10 + 7\times1$$

Being able to understand and use the various representations of a number, means you have a full understanding of the number concept and how it is symbolized in our Hindu–Arabic number system. You should take the time to understand the different representations of whole numbers before proceeding to arithmetic. It will help you better understand the more difficult concepts we will encounter later on.

Consider the following examples:

EXAMPLES:

1. The number 87 has eight groups of ten and seven ones, or 8 tens, and 7 ones.

    We write this number in words as; <u>eighty–seven</u>.

    In chart form we can represent the number as follows:

Tens	Ones
8	7
• • • • • • • •	• • • • • • •

    In expanded form this is written as:
$$87 = 80 + 7$$
$$\text{or}$$
$$= 8\times10 + 7\times1$$

2. The number 32,459 has three groups of ten thousand, two groups of a thousand, four groups of a hundred, five groups of ten, and nine ones, or 3 ten-thousands, 2 thousands, 4 hundreds, 5 tens, and 9 ones.

We write this number in words as; <u>thirty–two thousand four hundred fifty–nine</u>.

In chart form we can represent the number as follows:

Ten Thousands	Thousands	Hundreds	Tens	Ones
3	2	4	5	9
• • •	• •	• • • •	• • • • •	• • • • • • • • •

Again, each disk represents one group of the named amount in each of the place–values.

In expanded form this is written as:

$$32,459 = 30,000 + 2,000 + 400 + 50 + 9$$
$$\text{or}$$
$$= 3 \times 10,000 + 2 \times 1,000 + 4 \times 100 + 5 \times 10 + 9 \times 1$$

3. The number 109,386 has one group of a hundred thousand, zero groups of ten thousand, nine groups of a thousand, three groups of a hundred, eight groups of ten, and six ones, or 1 hundred–thousands, 9 thousands, 3 hundreds, 8 tens, and 6 ones.

We write this number in words as; <u>one hundred nine thousand three hundred eighty–six</u>.

In chart form we can represent the number as follows:

Hundred Thousands	Ten Thousands	Thousands	Hundreds	Tens	Ones
1	0	9	3	8	6
•		• • • • • • • • •	• • •	• • • • • • • •	• • • • • •

Again, each disk represents one group of the named amount in each of the place–values.

In expanded form this is written as:

$$109,386 = 100,000 + 9,000 + 300 + 80 + 6$$
$$\text{or}$$
$$= 1 \times 100,000 + 9 \times 1,000 + 3 \times 100 + 8 \times 10 + 6 \times 1$$

4. The number 7,326,243 has seven groups of a million, three groups of a hundred thousand, two groups of ten thousand, six groups of a thousand, two groups of a hundred, four groups of ten, and three ones, or 7 million, 3 hundred–thousands, 2 ten–thousands, 6 thousands, 2 hundreds, 4 tens, and 3 ones.

We write this number in words as; <u>seven million three hundred twenty–six thousand two hundred forty–three</u>.

In chart form we can represent the number as follows:

Millions	Hundred Thousands	Ten Thousands	Thousands	Hundreds	Tens	Ones
7	3	2	6	2	4	3
• • • • • • • •	• • •	• •	• • • • • •	• •	• • • •	• • •

Again, each disk represents one group of the named amount in each of the place–values.

In expanded form this is written as:

$$7,326,243 = 7,000,000 + 300,000 + 20,000 + 6,000 + 200 + 40 + 3$$

or
$$= 7 \times 1,000,000 + 3 \times 100,000 + 2 \times 10,000$$
$$+ 6 \times 1,000 + 2 \times 100 + 4 \times 10 + 3 \times 1$$

As we begin to introduce the arithmetic of numbers, we will find the different forms of writing numbers to be useful in understanding the various arithmetic operations. This is why it is important that you understand the concept of place–value along with not only its standard form, but its word, place–value, chart, and expanded form too.

EXERCISES 1.2

Understanding Numbers, Place–Value, and more

Write each number in words, expanded form, and using a place–value chart.

1.	29	4.	237	7.	13,452
2.	36	5.	3,458	8.	79,872
3.	125	6.	7,239	9.	326,913

Given each of the following numbers, identify the place value of the underlined digit.

10.	2,32<u>6</u>,900	12.	1,023,<u>6</u>77	14.	82<u>1</u>,000,352
11.	<u>6</u>5,280	13.	4<u>9</u>2,381	15.	123,4<u>5</u>6,789

1.3 INTEGERS AND OPERATIONS ON INTEGERS

Having defined the set of real numbers, we now need to be able to use them to solve problems. Before we can begin solving problems with any real number, we will start with operations on the set of integers first. This will allow us to focus on the specific challenges of working with and understanding negative numbers.

Arithmetic consists of the process of learning how to perform four important operations on numbers: addition, subtraction, multiplication and division.

Adding Integers

Adding (or finding the sum of) two numbers simply means we are finding the final location on the number line after performing our addition operation on the two numbers. For example, to add the two numbers 2 and 3 together is equivalent to first moving 2 units from zero to the right on the number line and then moving an additional 3 units to the right. This takes us to the final location of 5 units to the right of zero on the number line.

Or similarly, if we are given 2 dollars and then given an additional 3 dollars we would have a total of 5 dollars.

Mathematically, we express this using the addition operator, $+$, and write this as $2+3=5$.

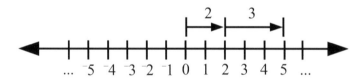

Now add together the numbers $^-2$ and $^-3$. The $^-2$ takes us two units to the left of zero and then the $^-3$ takes us an additional three units further to the left (we moved left since both of the numbers were negative). The net result is that we are now five units to the left of zero, or we are at the $^-5$ location on the number line.

We can also think of this using the following equivalent representation: if we are 2 dollars in debt ($^-2$) and then add another debt of 3 dollars ($^-3$), we are now in debt 5 dollars ($^-5$).

Mathematically, we express this as: $^-2+ {}^-3 = {}^-5$

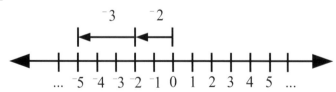

To add two numbers of different signs, we follow the same procedure as above. For example, adding a negative three and a positive four together, we first move three units to the left of zero putting us at the $^-3$ location on the number line, and then move four units to the right (since the 4 is positive). This puts us at the final location of a positive one (1) on the number line.

This is equivalent to being in debt 3 dollars ($^-3$) and then getting 4 dollars (4) to leave us with a net gain of 1 dollar (1).

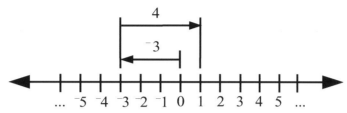

Mathematically, we have that, $^-3 + 4 = 1$

We should note that whenever we add two real numbers together, we have one of two possibilities. Either the signs of the two numbers are the same or they are different. We can show that the above procedure is equivalent to applying the following rules:

Like Signs. Add two numbers with the same sign by adding their values without the sign and taking the sign of the original numbers.

Unlike Signs. Add two numbers with different signs by subtracting the smaller value of the two numbers without the signs from the larger and taking the sign of the number with the largest value without the sign.

EXAMPLE 1: Adding numbers with like signs:

$$(a) \quad 5 + 7$$
$$(b) \quad ^-5 + ^-7$$

Solution: (a) $\quad 5 + 7 = ^+(5 + 7) = ^+12 = 12$
                 (b) $\quad ^-5 + ^-7 = ^-(5 + 7) = ^-12$

EXAMPLE 2: Adding numbers with unlike signs:

$$(a) \quad 5 + ^-7$$
$$(b) \quad ^-5 + 7$$

Solution: (a) $\quad 5 + ^-7 = ^-(7 - 5) = ^-(2) = ^-2$
                 (b) $\quad ^-5 + 7 = ^+(7 - 5) = ^+(2) = ^+2 = 2$

Subtracting Integers

Subtracting (or finding the difference of) any two real numbers is expressed mathematically by using the subtraction operator, $^-$. Subtraction is very similar to addition. In fact, one approach is to transform our subtraction problem into addition and then perform the addition operation. To change subtraction into addition, we simply change the sign of the term we are subtracting and then add the resulting two numbers together. We can think of subtraction as adding the opposite signed number, or simply stated as, adding the opposite.

Thus,

$$8 - 3 \quad \text{becomes:} \quad 8 + {}^-3 = 5$$

and

$$^-7 - 5 \quad \text{becomes:} \quad ^-7 + {}^-5 = {}^-12$$

In other words, if we initially have 8 dollars and subtract 3 dollars from it, it is equivalent to adding a debt of 3 dollars to the initial 8 dollars, giving us a net of 5 dollars. Or in the second example, if we are in debt 7 dollars and subtract 5, it is equivalent to being in debt 7 dollars and adding an additional debt of 5 dollars.

EXAMPLES: Calculate the subtractions below.

1. Subtract: $7 - {}^-8$ **Solution:** $7 - {}^-8 = 7 + {}^+8 = 7 + 8 = 15$

2. Subtract: $2 - 6$ **Solution:** $2 - 6 = 2 + {}^-6 = {}^-(6 - 2) = {}^-(4) = {}^-4$

3. Subtract: $^-6 - {}^-9$ **Solution:** $^-6 - {}^-9 = {}^-6 + {}^+9 = {}^+(9 - 6) = {}^+(3) = {}^+3 = 3$

4. Subtract: $^-7 - {}^-3$ **Solution:** $^-7 - {}^-3 = {}^-7 + {}^+3 = {}^-(7 - 3) = {}^-(4) = {}^-4$

Multiplying Integers

Multiplying or finding the product of two real numbers can sometimes be interpreted as performing addition a multiple number of times. The multiplication operator is represented by one of three different symbols, or no symbol at all. Multiplication is identified either by placing a *, $\times$ or $\cdot$ between the two numbers. Sometimes we do not write a multiplication symbol. When an operation symbol is missing, we assume multiplication. This approach can cause confusion, and we'll try and point it out in sections ahead.

As we just said, multiplication is equivalent to repeated additions. So, if I had the number 3 and I wanted to multiply it by 4, that would be equivalent to taking the number 3 and adding it to itself 3 times $3 + 3 + 3 + 3$, which would yield the quantity 12. Written in operator notation, we could either write this as:

$4*3 = 12,$ $4 \cdot 3 = 12,$ $4 \times 3 = 12,$ or $4(3) = 12$ Note: no operation symbol between 3 and 4, just parentheses

We already know how to multiply two positive integers together, as it is just the repeated additions of the numbers. The real question we have to answer is; what happens if one or both of the numbers are negative?

Let's start with multiplying a positive number times a negative number, for example:

$$3 \times {}^-4$$

Now we don't know what to do with a positive times a negative, but we do have rules and definitions on how to proceed. There is one rule, or axiom, and one definition we need to move forward. The first rule is called the distributive property, which tells us how to multiply a number times a sum of two numbers. Note: we will talk more about the distributive property in section 1.5, but we need it here, so we'll introduce it now. However, since it is very important we'll review it again in greater detail later in the chapter.

For example consider: $3 \times (5+2)$ The distributive property tells us we get the same result whether we add the 5 and the 2 and then multiply, or if we distribute the 3 to the 5 and then the 2 and then add the two products together, i.e.

$$3\times(5+2) = 3 \times 5 + 3 \times 2 = 6 + 15 = 21$$
$$= 3\times7 = 21$$

That is the distributive property.

The other thing we use is a definition which says that every number has something defined to be its additive inverse. An additive inverse is a number that when added to your number gives zero as a result. For example the additive inverse of 4 is ${}^-4$, since $4 + {}^-4 = 0$. We just did addition in the last section.

Using these two aspects of our number system we can now answer our original question: What is $3 \times {}^-4$?

First, let's start with another fact that any number times zero is equal to zero. That is how we define multiplication by zero, or more specifically:

$$3 \times 0 = 0$$

Next, we use our additive inverse to replace zero with what it's equivalent to, namely $4 + {}^-4 = 0$.

$$3 \times (4 + {}^-4) = 0$$

We now apply the distributive property to obtain:

$$3 \times 4 + 3 \times {}^-4 = 0$$

We can do the first product, since that is just two positive numbers.

$$12 + 3 \times {}^-4 = 0$$

We now have that 12 added to a number must equal zero. By definition, that number must be its additive inverse, ${}^-12$!

That means we can now answer the question of what is $3 \times {}^-4$ it is ${}^-12$, or

$$3 \times {}^-4 = {}^-12$$

So a positive number times a negative number is a negative number. Now, we could do a similar justification to show that a negative times a positive is also negative.

The next question we ask is what is a negative times a negative number, for example:

$${}^-6 \times {}^-9$$

Here we follow the same approach and start with

$${}^-6 \times (9 + {}^-9) = 0$$

Distribute the ${}^-6$ to obtain

$${}^-6 \times 9 + {}^-6 \times {}^-9 = 0$$

Since we now know that ${}^-6 \times 9 = {}^-54$, we can replace that value to obtain

$${}^-54 + {}^-6 \times {}^-9 = 0$$

This is our additive inverse statement again, so we are looking for the additive inverse of ${}^-54$, which is 54. This tells us that

$${}^-6 \times {}^-9 = 54$$

Thus, a negative times a negative is positive!

Now, this is not a formal proof, but we hope it provides some justification for this sometimes confusing fact in mathematics. It is not the consequence of any analogy we may try and conjure up, it is instead a fact based upon the rules of arithmetic that arise from our basic axioms and definitions. With these two rules we can now proceed in a very simple way with the multiplication of signed numbers.

> **Rule 1 of Multiplication (unlike signs):** A negative times a positive or a positive times a negative is equal to a negative of the product of the two numbers without their signs.
>
> **Rule 2 of Multiplication (like signs):** A negative times a negative is equal to a positive of the product of the two numbers without their negative signs. We already know that multiplying two positive integers gives a positive result.

EXAMPLES: Calculate the products below.

1. Multiply: ${}^-6 \times 3$

Since this is a product of a negative times a positive we use Rule 1 to compute this as

$${}^-(6 \times 3) = {}^-18$$

2. Multiply: $5 \times {}^-13$

Since this is a product of a negative times a positive we use Rule 1 to compute this as

$${}^-(5 \times 13) = {}^-65$$

3. Multiply: ${}^-5 \times {}^-7$

Since this is a product of a negative times a negative we use Rule 2 to compute this as

$${}^-5 \times {}^-7 = 5 \times 7 = 35$$

4. Multiply: ${}^-3 \times 5 \times {}^-4$

This is a product of three integers. We simply multiply left to right pairs of integers using the rules above. We start with ${}^-3 \times 5$ and obtain ${}^-15$ using Rule 1 above. We then replace the ${}^-3 \times 5$ with ${}^-15$, and use Rule 2 above to obtain:

$${}^-15 \times {}^-4 = 15 \times 4 = 60$$

We now move on to division.

Dividing Integers

Dividing or finding the quotient of two integers is represented mathematically with one of two symbols; /, or $\div$. Thus, 15 divided by 3 can be written as 15/3 (or $\frac{15}{3}$) or $15 \div 3$. Division is the "opposite" operation of multiplication. For example, when we multiply 3 times 5 and get 15, we know both the numbers 3 and 5 and we are asked to calculate (or find) the number 15. In division, we know the value of the "opposite" side (say 15 in this case), and we are asked to find one of the numbers used to multiply the other number by to give us 15, e.g.

$$15 \div 3 = 5 \text{ since } 3 \times 5 = 15.$$

Thus, we can view division as really a type of multiplication problem, in that $15 \div 3 = (?)$ is really asking $3 \times (?) = 15$. This implies that division should follow the same sign rules as multiplication.

Rule 1 of Division (unlike signs): A negative divided by a positive or a positive divided by a negative is equal to a negative of the quotient of the two numbers without their signs.

Rule 2 of Division (like signs): A negative divided by a negative is equal to a positive of the quotient of the two numbers without their negative signs.

We have one caveat when we divide with integers; that is, we can never divide by zero! Anything divided by zero is undefined. Why? Consider a division by zero and its related multiplication problem;

$$15 \div 0 = (?) \text{ or } 0 \times (?) = 15$$

We are asked to find a number that when multiplied by 0, gives us 15. This is impossible, which is why division by zero cannot be defined!

EXAMPLE 1: Divide: $\dfrac{^-10}{^-5}$

Solution: $\dfrac{^-10}{^-5}=(?)$ or $^-5\times(?)=^-10$, so the answer is 2. We also see that by Rule 2 for Division, a negative divided by a negative is positive.

EXAMPLE 2: Divide: $\dfrac{^-15}{3}$

Solution: $\dfrac{^-15}{3}=(?)$ or $3\times(?)=^-15$, so the answer is $^-5$. We also see that by Rule 1 for Division, a negative divided by a positive is negative.

If we begin to randomly divide different integers we will see that division of integers leads us to having to define a new type of number. For example, what is $10\div3=(?)$ or $3\times(?)=10$. Try as much as you want, but you won't find an integer that, when multiplied by 3, gives 10. For this type of problem, we have to introduce the idea of a fraction and expand our number concept to include rational numbers. In the next section we review the arithmetic of fractions.

EXERCISES 1.3

Adding Integers
Evaluate each expression

1. $^-3+4$
2. $^-7+^-4$
3. $7+13$
4. $^-6+^-6$
5. $^-33+33$
6. $^-26+^-16$
7. $^-6+^-39$
8. $^-56+^-44$
9. $^-3+4+^-4+7$
10. $5+^-7+2+^-10$
11. $^-42+36+^-8$
12. $73+^-90+13$

Subtracting Integers
Evaluate each expression

13. $3-9$
14. $^-6-^-3$
15. $^-9-^-13$
16. $^-23-^-32$
17. $^-67-^-21$
18. $14-^-21$
19. $32-^-78$
20. $^-3-8$
21. $5-1$
22. $^-3-^-5$
23. $4-21$
24. $^-38-^-32$

Adding and Subtracting Three or More Integers

Evaluate each expression

25. $23 - 18 + {}^-5$

26. $14 - 23 + {}^-6 - {}^-10$

27. ${}^-10 + {}^-10 - {}^-14 + 6$

28. ${}^-6 - 5 + {}^-3$

29. ${}^-8 + 7 - {}^-12$

30. ${}^-11 + {}^-6 + 7 - 9$

31. $7 + {}^-3 - 4 - {}^-6$

32. ${}^-9 - {}^-5 + 7 - 6$

33. ${}^-15 + 13 - 6 - {}^-2$

34. $12 - {}^-3 - 8$

35. ${}^-3 - {}^-5 - {}^-7$

36. ${}^-2 - 3 - 5 - 1$

Multiplying Integers

Find each product

37. ${}^-7 \cdot 8$

38. ${}^-6 \cdot {}^-4$

39. $13({}^-8)$

40. $25({}^-7)$

41. ${}^-16 \cdot {}^-3$

42. $11 \cdot {}^-9$

43. ${}^-12 \times {}^-12$

44. $10 \times {}^-8$

45. ${}^-21 \cdot {}^-6$

46. ${}^-3 \cdot 15$

47. $({}^-6)({}^-23)$

48. $({}^-43)({}^-7)$

49. $14({}^-14)$

50. ${}^-5 \times {}^-13$

51. ${}^-7 \cdot 8 \cdot {}^-2$

52. $3 \cdot {}^-4 \cdot {}^-6$

53. ${}^-5 \times 8 \times {}^-6$

54. ${}^-7 \times {}^-3 \times {}^-2$

55. $(9)(4)({}^-3)$

56. $(3)({}^-3)({}^-2)(3)$

Dividing Integers

Find each quotient and write the equivalent multiplication sentence.

57. $\dfrac{12}{6}$

58. $\dfrac{26}{13}$

59. $\dfrac{{}^-72}{{}^-9}$

60. $\dfrac{125}{{}^-5}$

61. $\dfrac{{}^-44}{4}$

62. $\dfrac{{}^-72}{{}^-4}$

63. $\dfrac{{}^-40}{{}^-4}$

64. $\dfrac{35}{{}^-7}$

65. $\dfrac{75}{{}^-15}$

66. $\dfrac{{}^-56}{{}^-14}$

67. ${}^-65 \div 5$

68. $72 \div {}^-8$

69. $138 \div {}^-23$

70. ${}^-85 \div {}^-17$

1.4 FRACTIONS AND OPERATIONS ON FRACTIONS

As we saw in the previous section, dividing integers can lead to a new type of number. This new number represents parts of a whole and can allow us to identify values in–between the integers on the number line. These new numbers are known as fractions. If you do not have a solid foundation concerning fractions, we suggest you visit the Appendix at the end of the textbook to get a better understanding of them. Fractions are key to understanding many concepts we will study in this book as well as subsequent mathematics courses. Our focus here will be to provide a brief review on the arithmetic of fractions.

Adding Fractions

Adding wholes is much easier, since we can simply count how many we have. Fractions, however, are like being given a bag of different length sticks and then asked to determine how long it would be if they were laid end to end. The only way to be able to measure how long the entire bag is in this way, is to have to measure each piece with a common measuring device, and then add all these different lengths of measure together. In fractions, this common device is called the unit fraction. We have to transform each fraction into a sum of equivalent unit fractions, and then we can "count" how many equivalent unit fractions we have. Thus, we try and turn adding fractions into a process similar to adding whole numbers, but instead of counting wholes, we count unit fractions. We can only add two quantities if they have the same units (unit fraction). This makes sense since you wouldn't add feet and inches directly, as if they were the same unit, you would instead first convert inches or feet to the other unit and then add. The same idea is true for fractions. Let's illustrate this with a few examples.

In our first example we start with a problem that is already set up for us, with each fraction based upon a common unit fraction. This means that they have the same denominator.

EXAMPLE 1: Add the following fractions: $\frac{3}{7} + \frac{2}{7}$

Since each fraction is based upon the common unit fraction of $\frac{1}{7}$, we can add all the sevenths together.

$$\frac{3}{7} = \frac{1}{7} + \frac{1}{7} + \frac{1}{7} \text{ and } \frac{2}{7} = \frac{1}{7} + \frac{1}{7}$$

so

$$\frac{3}{7} + \frac{2}{7} = \left(\frac{1}{7} + \frac{1}{7} + \frac{1}{7}\right) + \left(\frac{1}{7} + \frac{1}{7}\right) = \frac{5}{7}$$

EXAMPLE 2: Add the following fractions: $\frac{9}{16} + \frac{15}{16}$

Since each fraction is based upon the common unit fraction of $\frac{1}{16}$, we can add all the sixteenths together.

$$\frac{9}{16} + \frac{15}{16} = \left(\frac{1}{16} + \frac{1}{16} + \frac{1}{16} + \frac{1}{16} + \frac{1}{16} + \frac{1}{16} + \frac{1}{16} + \frac{1}{16} + \frac{1}{16}\right) +$$
$$\left(\frac{1}{16} + \frac{1}{16} + \frac{1}{16} + \frac{1}{16} + \frac{1}{16} + \frac{1}{16} + \frac{1}{16} + \frac{1}{16} + \frac{1}{16} + \frac{1}{16} + \frac{1}{16} + \frac{1}{16} + \frac{1}{16} + \frac{1}{16} + \frac{1}{16}\right)$$
$$= \frac{24}{16}$$

EXAMPLE 3: Add the following fractions: $\frac{3}{4} + \frac{-5}{8}$

Since one fraction is based upon the unit fraction of $\frac{1}{4}$, and the other is based on the unit fraction of $\frac{1}{8}$ we have to find a common equivalent fraction before we can add them together.

In this problem we see that 4 and 8 are related, since $8 = 2 \times 4$. This means we only have to change one of the fractions into its equivalent form, and that is the $\frac{3}{4}$, which becomes,

$$\frac{3}{4} = \frac{3 \times 2}{4 \times 2} = \frac{6}{8}$$

Then we can write

$$\frac{3}{4} + \frac{-5}{8} = \frac{6}{8} + \frac{-5}{8} = \frac{6 + {}^{-}5}{8} = \frac{1}{8}$$

EXAMPLE 4: Add the following fractions: $\frac{3}{7} + \frac{5}{6}$

In this problem we see that 7 and 6 are not "related," meaning that they have no factors in common.

We need to find the common equivalent unit fraction, so that we can find out how many equivalent unit fractions they have between the two of them.

> If the denominators do not have any factors in common, we simply multiply the denominators together to get the smallest common equivalent denominator.

In this case it is

$$7 \times 6 = 42$$

This means we have to change both of the fractions into their equivalent forms with 42 in the denominator.

$$\frac{3}{7} \text{ becomes } \frac{3}{7} = \frac{3 \times 6}{7 \times 6} = \frac{18}{42}$$

Notice, for this fraction we simply multiply by the opposite factor of 6.

$$\frac{5}{6} \text{ becomes } \frac{5}{6} = \frac{5 \times 7}{6 \times 7} = \frac{35}{42}$$

In this fraction we multiplied by the opposite factor of 7 to get our equivalent fraction.

Then we can write

$$\frac{3}{7} + \frac{5}{6} = \frac{18}{42} + \frac{35}{42} = \frac{53}{42}$$

Our answer is the improper fraction, $\frac{53}{42}$

EXAMPLE 5: Add the following fractions: $\frac{-7}{8} + \frac{11}{12}$

We first observe that 8 and 12 have a common factor of 4, since

$$8 = 2 \times 4 \qquad \text{and} \qquad 12 = 3 \times 4$$

We see that the uncommon factors are 2 and 3. The smallest common equivalent denominator is obtained by multiplying by the uncommon factors (2 and 3) and the common factor (4), or

$$2 \times 3 \times 4 = 24$$

> If the denominators do have factors in common, we simply multiply the uncommon factors by the common factors to get our common denominator.

This means that we now have to change both of the fractions into their equivalent forms with 24 in the denominator.

$$-\frac{7}{8} \text{ becomes } \frac{^-7}{8} = \frac{^-7 \times 3}{8 \times 3} = \frac{^-21}{24} = \frac{-21}{24}$$

Notice, for this fraction, we simply multiply by the opposite uncommon factor of 3.

$$\frac{11}{12} \text{ becomes } \frac{11}{12} = \frac{11 \times 2}{12 \times 2} = \frac{22}{24}$$

In this fraction, we multiplied by the opposite uncommon factor of 2 to get our equivalent fraction.

Then we can write:

$$\frac{^-7}{8} + \frac{11}{12} = \frac{-21}{24} + \frac{22}{24} = \frac{^-21 + 22}{24} = \frac{1}{24}$$

Our answer is, $\frac{1}{24}$.

Subtracting Fractions

We have spent a great deal of time showing how to do addition of fractions. This should now make the concept of subtraction easier. Subtraction of fractions is very similar to addition. Only now we remove common unit fractions instead of adding them.

Let's see this through a few examples.

EXAMPLE 1: Subtract the following fractions: $\frac{6}{7} - \frac{4}{7}$

In this case, the denominators are the same, so we already know our unit fraction is 1/7.

So we can simply subtract the numerators.

$$\frac{6}{7} - \frac{4}{7} = \frac{6-4}{7} = \frac{2}{7}$$

So our final answer is $\frac{2}{7}$.

EXAMPLE 2: Subtract the following fractions: $\frac{3}{4} - \frac{^-5}{8}$

In this case, the denominators are not the same, so we must obtain a common denominator (equivalent unit fraction) for both. We notice that we can simply change 3/4 and rewrite it as the equivalent fraction of $\frac{6}{8}$.

Thus, we can obtain.

$$\frac{3}{4} - {}^{-}\frac{5}{8} = \frac{6}{8} + \frac{5}{8} = \frac{6+5}{8} = \frac{11}{8}$$

In the first step, we turned subtraction into adding the opposite sign, just as we did for integers.

Multiplying Fractions

Multiplication is the process of repeated additions. It is stated as $(\text{Number } 1) \times (\text{Number } 2)$ and asks the question;

What number is obtained when (Number 2) is added to itself (Number 1) times?

Multiplying Whole Numbers Times Fractions

Let's start with the simplest form of multiplication, that of a whole number times a fraction. Now, we have already seen this in our multiplication statement for adding unit fractions. For example, we saw that

$$9 \times \frac{1}{16} = \frac{9}{16}$$

We essentially multiplied the 9 times the 1 in the numerator of the unit fraction. This was because 9/16 was just 9 copies of 1/16. The same holds true if the term we are multiplying is not a unit fraction. Let's illustrate this with some examples.

EXAMPLE 1: Multiply the following: $4 \times \frac{3}{5}$

Arithmetically, we have the standard algorithm which says we just multiply the numerator by the whole number we are multiplying by.

$$4 \times \frac{3}{5} = \frac{4 \times 3}{5} = \frac{12}{5}$$

We have left our final answer as an improper fraction. We could change it to a mixed number and get $2\frac{2}{5}$.

We should point out that the operation of multiplication is what we call commutative. That means $3 \times 9 = 9 \times 3$. We can reverse the order in which we multiply and get the same answer. The reason why we mention this now is because if we see the multiplication problem $\frac{5}{7} \times 6$, we can always rewrite it as $6 \times \frac{5}{7}$ and use the same interpretation and approach for multiplying that we just introduced. Both expressions mean that we have 6 copies of $\frac{5}{7}$.

Furthermore, we should also point out that if we don't use the above analogy, there is another analogy we will use later on in interpreting $\frac{5}{7} \times 6$. This expression also means that we are trying to find out what number is $\frac{5}{7}$ of 6. Depending upon the type of problem we are working on, one analogy or the other will be easier, so we need to understand both. We will explain it in detail in what follows.

Multiplying Fractions Times Whole Numbers

EXAMPLE 2: As we just said, multiplying a whole number by a fraction, such as $\frac{5}{7} \times 14$, is the same as asking the question: $\frac{5}{7}$ of 14 is equal to what number?

To compute using the standard algorithm we simply multiply the 5 times the 14 and divide by 7,

$$\frac{5}{7} \times 14 = \frac{5 \times 14}{7} = \frac{70}{7} = \frac{10 \times 7}{7} = 10$$

EXAMPLE 3: $\frac{3}{5}$ of 35 is what number?

Again, if we use the standard algorithm we get:

$$\frac{3}{5} \times 35 = \frac{3 \times 35}{5} = \frac{115}{5} = \frac{21 \times 5}{5} = 21$$

In these examples, the answer worked out to be a whole number. Most of the time the result is a fraction, though.

EXAMPLE 4: $\frac{2}{7}$ of 9 is what number?

For this problem, our answer will not be a whole number.

$$\frac{2}{7} \times 9 = \frac{2 \times 9}{7} = \frac{18}{7}$$

The final answer is the fraction $\frac{18}{7}$.

Multiplying Fractions Times Fractions

We now consider multiplying a fraction by another fraction, such as $\frac{4}{5} \times \frac{2}{3}$. Again, this is still the multiplication operation, so the same ideas and concepts we've been discussing in this section still apply. Multiplication is still a repeated addition. What we are adding, and how many times, however, is not as obvious. If you want or need a more detailed explanation of this concept, please look at section A4 in the Appendix of this book. For now we look at examples using the standard algorithm.

EXAMPLE 5: $\frac{4}{5}$ of $\frac{2}{3}$ is what number? Or $\frac{4}{5} \times \frac{2}{3}$ equals what number?

Let's now see how this works for the standard algorithm.

The standard algorithm tells us to multiply the numerators (tops) of each fraction by the denominators (bottoms) of each fraction.

$$\frac{4}{5} \times \frac{2}{3} = \frac{4 \times 2}{5 \times 3} = \frac{8}{15}$$

Dividing Fractions

Division is the process of finding the number used in the repeated additions of our related multiplication problem. Division is sometimes called the reciprocal relation of multiplication, and we will show why.

The division problem:

$$(\text{Number 1}) \div (\text{Number 2}) = (\text{Number 3})$$

is essentially asking the question; What number (Number 3) added to itself (Number 2) times equals (Number 1)?

$$(\text{Number 2}) \times (\text{Number 3}) = (\text{Number 1})$$

Sometimes we rephrase this question and ask: How many (Number 2's) are there in (Number 1)?

If you think about this carefully, you will see that both questions are actually asking you to find the same thing, but only in a different way. Why have two ways to ask the question? Because sometimes one question gives you a better grasp on understanding what is being asked for in a particular problem than the other question. We'll show this through examples below.

We'd also like to point out that these questions are addressing what the concept of division really is. Eventually we will end up with a standard algorithm to use, but we don't want to lose sight of the concept of what we are actually doing. The algorithm is not division. It is just a set of steps used to quickly find our answer, with no obvious connection to the concept of division. We stress this point so that you don't get confused later on. Don't mistake the standard algorithm for division. Division is a concept, the standard algorithm is not.

We'll start by showing examples of division, for different types of division problems involving fractions.

Dividing a Fraction by a Whole Number

Let's start with the simplest form of division, that of dividing a fraction by a whole number. For this type of problem, the question:

What number (Number 3) added to itself (Number 2) times equals (Number 1)?

makes more sense intuitively. Anytime the fraction is the first number and a whole number is the second number, you should use this conceptual interpretation of the division problem.

EXAMPLE 1: Perform the division $\frac{3}{4} \div 6$

We are actually asking: What number added to itself 6 times is equal to $\frac{3}{4}$? This question makes more sense than asking: How many 6's are in $\frac{3}{4}$? They both, however, are asking the same question.

The standard algorithm is as follows. We start by changing all division problems to a multiplication by the reciprocal of the term we are dividing by. This makes sense for the problem above says that $\frac{3}{4} \div 6$ is equivalent to finding 3/4 of 1/6, or

$$\frac{3}{4} \div 6 = \frac{3}{4} \times \frac{1}{6}$$

Then we use the multiplication algorithm we developed earlier to obtain:

$$\frac{3}{4} \times \frac{1}{6} = \frac{3 \times 1}{4 \times 6} = \frac{3}{24} \cdot$$

Which, when we simplify (reduce), is equivalent to $\frac{1}{8}$.

$$\frac{3}{24} = \frac{3 \div 3}{24 \div 3} = \frac{1}{8}$$

If we think about this, we can see that we were also looking for 3/4 of 1/6. This really is our multiplication problem in reverse. We want to know what number is 3/4 of 1/6 and that number is 1/8. This number is what we must add 6 times to get 3/4.

EXAMPLE 2:

Let's look at a slightly different example. Imagine that you have $\frac{2}{5}$ of a cup of frosting to share equally among three desserts. How would we write this as a division question? It is asking us to divide $\frac{2}{5}$ into 3 equal parts, or mathematically:

$$\frac{2}{5} \div 3$$

Again, we are asking; What number added to itself 3 times is equal to $\frac{2}{5}$?

$$\frac{2}{5} \div 3 = \frac{2}{5} \times \frac{1}{3} = \frac{2 \times 1}{5 \times 3} = \frac{2}{15}$$

Dividing a Whole Number by a Fraction

For these type of problems we are asking the question; How many (Number 2's) are in (Number 1)? Anytime a whole number comes first it may make sense to phrase the division problem in this way. Again, it makes more sense intuitively for certain types of division problems. This is due to our intuitive bias to think of a number as being a whole number and not a fraction. Sometimes the second question: "What number added to itself (Number 2) times equals (Number 1)?", will make more sense, though. You should look at both so that you better understand what you are being asked to find.

EXAMPLE 3: Perform the division $5 \div \frac{1}{2}$

This problem asks the question: How many $\frac{1}{2}$'s are there in 5?

Again, we just multiply 5 by the reciprocal of 1/2, which is 2.

$$5 \div \frac{1}{2} \;=\; 5 \times \frac{2}{1} \;=\; 10$$

EXAMPLE 4: Perform the division $3 \div \frac{2}{5}$

This problem asks the question; How many $\frac{2}{5}$'s are there in 3? or What number added to itself $\frac{2}{5}$ times, equals 3. As you can see the second question is less intuitive.

$$3 \div \frac{2}{5} \;=\; 3 \times \frac{5}{2} \;=\; \frac{3 \times 5}{2} \;=\; \frac{15}{2}$$

Dividing a Fraction by a Fraction

This is the final type of division problem we can expect. Since the second number is a fraction it makes more sense to ask the question:

How many (Number 2)'s are there in (Number 1)?

We start with an example where we divide a fraction by a fraction with the same denominator. We do this to develop an intuitive understanding before it becomes more abstract.

EXAMPLE 5: Perform the division $\frac{8}{9} \div \frac{2}{9}$

How many $\frac{2}{9}$'s are there in $\frac{8}{9}$?

$$\frac{8}{9} \div \frac{2}{9} \;=\; \frac{8}{9} \times \frac{9}{2} \;=\; \frac{8 \times 9}{9 \times 2} \;=\; \frac{8}{2} \;=\; 4$$

EXAMPLE 6: Perform the division $\frac{9}{12} \div \frac{3}{12}$

How many $\frac{3}{12}$'s are there in $\frac{9}{12}$?

$$\frac{9}{12} \div \frac{3}{12} \;=\; \frac{9}{12} \times \frac{12}{3} \;=\; \frac{9 \times \cancel{12}}{\cancel{12} \times 3} \;=\; \frac{9}{3} \;=\; 3$$

We now do an example where the denominators are not the same.

EXAMPLE 7: Perform the division $\dfrac{5}{6} \div \dfrac{3}{8}$

How many $\dfrac{3}{8}$'s are there in $\dfrac{5}{6}$?

$$\frac{5}{6} \div \frac{3}{8} = \frac{5}{6} \times \frac{8}{3} = \frac{5 \times 8}{6 \times 3} = \frac{40}{18} = \frac{20}{9}, \text{ or } 2\,\frac{4}{18} = 2\,\frac{2}{9}$$

Multiplying and Dividing Negative Fractions

Fractions are just numbers, like the integers. Thus, they follow the same rules when multiplying and dividing them. If they are of opposite sign, the result is a negative number. If they are the same sign, the result is a positive number.

We illustrate this with a few examples.

EXAMPLE 8: Multiply the following: $^-4 \times \dfrac{3}{5}$

First, we note that we are multiplying a negative number times a positive number. The result will be a negative number. Thus, we can take care of this first, and proceed accordingly. Arithmetically, we have the standard algorithm which says we just multiply the numerator by the integer we are multiplying by, and keep the sign of the number negative.

$$^-4 \times \frac{3}{5} = \,^-\!\left[\frac{4 \times 3}{5}\right] = \,^-\frac{12}{5}$$

EXAMPLE 9: Multiply the following: $\dfrac{^-4}{7} \times \dfrac{^-3}{5}$

First, we note that we are multiplying a negative by a negative, which gives a positive value. Thus, we take care of this first and then proceed accordingly. Arithmetically, we have the standard algorithm which says we just multiply the numerator by the numerator and the denominator by the denominator.

$$\frac{^-4}{7} \times \frac{^-3}{5} = \frac{4 \times 3}{7 \times 5} = \frac{12}{35}$$

EXAMPLE 10: Perform the division $\dfrac{5}{3} \div \dfrac{^-3}{4}$

First we note that we are dividing a positive number by a negative number, and the result will be a negative number. Next we apply the standard algorithm, which says to invert and multiply the fraction we are dividing by.

$$\frac{5}{3} \div \frac{^-3}{4} = \,^-\!\left[\frac{5}{3} \times \frac{4}{3}\right] = \,^-\!\left[\frac{5 \times 4}{3 \times 3}\right] = \,^-\frac{20}{9}$$

EXERCISES 1.4

Adding and Subtracting Fractions

Add the following fractions

1. $\dfrac{2}{8} + \dfrac{5}{8}$

2. $\dfrac{8}{25} + \dfrac{9}{25}$

3. $\dfrac{^-4}{19} + \dfrac{11}{19}$

4. $\dfrac{7}{9} + \dfrac{^-2}{9}$

5. $\dfrac{3}{13} + \dfrac{5}{26}$

6. $\dfrac{5}{21} + \dfrac{1}{7}$

7. $\dfrac{7}{18} + \dfrac{8}{9}$

8. $\dfrac{3}{5} + \dfrac{^-4}{15}$

9. $\dfrac{^-9}{16} + \dfrac{7}{24}$

10. $\dfrac{5}{8} + \dfrac{7}{18}$

11. $\dfrac{3}{16} + \dfrac{^-5}{12}$

12. $\dfrac{9}{22} + \dfrac{8}{33}$

Subtract the following fractions

13. $\dfrac{9}{10} - \dfrac{7}{10}$

14. $\dfrac{3}{8} - \dfrac{^-1}{8}$

15. $\dfrac{13}{25} - \dfrac{5}{25}$

16. $\dfrac{11}{15} - \dfrac{8}{15}$

17. $\dfrac{3}{13} - \dfrac{^-5}{26}$

18. $\dfrac{^-10}{21} - \dfrac{1}{7}$

19. $\dfrac{7}{18} - \dfrac{2}{9}$

20. $\dfrac{3}{5} - \dfrac{4}{15}$

21. $\dfrac{9}{16} - \dfrac{^-5}{24}$

22. $\dfrac{5}{8} - \dfrac{5}{18}$

23. $\dfrac{7}{16} - \dfrac{5}{12}$

24. $\dfrac{^-9}{22} - \dfrac{5}{33}$

Add or Subtract fractions applications

25. Kaitlyn read $\dfrac{2}{5}$ of a book on Monday, then read $\dfrac{1}{4}$ of the same book on Tuesday, and the remainder of the book on Wednesday, What part of the book did she read on Wednesday?

26. Ethan ate $\dfrac{1}{4}$ of a huge cookie his mother made for him. Two hours later, he ate $\dfrac{1}{3}$ of the same cookie. What part of the cookie is left?

27. A student did $\dfrac{3}{5}$ of her homework on Saturday morning and $\dfrac{1}{4}$ of her homework in the afternoon. She said she would do the remaining homework in the evening. What part of her homework does she still have to do?

28. A farmer stated that he and his crew harvested $\dfrac{5}{8}$ of his strawberry field on day 1. They harvested $\dfrac{1}{5}$ of the field on day 2. How much of the whole field did the farmer and his crew harvest in the two days?

Multiplying and Dividing Fractions

Multiplication: Solve using the standard algorithm.

29. $\dfrac{1}{3}$ of 18

30. $\dfrac{1}{3}$ of 36

31. $\dfrac{^-3}{4} \times 24$

32. $\dfrac{3}{8} \times 24$

33. $\dfrac{4}{5} \times 25$

34. $\dfrac{1}{7} \times 140$

35. $\dfrac{1}{2} \times {}^-8$

36. $8 \times \dfrac{1}{2}$

37. $\dfrac{3}{5} \times 10$

41. $\dfrac{1}{2}$ of $\dfrac{2}{2}$

45. $\dfrac{1}{2} \times \dfrac{3}{5}$

49. $\dfrac{^-3}{4} \times \dfrac{5}{6}$

38. $10 \times \dfrac{3}{5}$

42. $\dfrac{2}{3}$ of $\dfrac{1}{2}$

46. $\dfrac{2}{3} \times \dfrac{^-1}{4}$

50. $\dfrac{2}{3} \times \dfrac{6}{7}$

39. $14 \times \dfrac{3}{7}$

43. $\dfrac{3}{4}$ of $\dfrac{^-4}{5}$

47. $\dfrac{3}{4} \times \dfrac{2}{3}$

51. $\dfrac{^-4}{9} \times \dfrac{^-3}{10}$

40. $\dfrac{3}{4} \times 36$

44. $\dfrac{2}{5}$ of $\dfrac{^-2}{3}$

48. $\dfrac{^-4}{5} \times \dfrac{^-5}{8}$

52. $\dfrac{3}{11} \times \dfrac{7}{9}$

Applications of Multiplying Fractions

53. There are 48 students going on a field trip. One-fourth are girls. How many boys are going on the trip?

54. Abbie spent $\dfrac{5}{8}$ of her money and saved the rest. If she spent $45, how much money did she have at first?

55. A marching band is rehearsing in rectangular formation. $\dfrac{1}{5}$ of the marching band members play percussion instruments. $\dfrac{1}{2}$ of the percussionists play the snare drum. What fraction of all the band members play the snare drum?

56. Phillip's family traveled $\dfrac{3}{10}$ of the distance to his grandmother's house on Saturday. They traveled $\dfrac{4}{7}$ of the remaining distance on Sunday. What fraction of the total distance to his grandmother's house was traveled on Sunday?

57. Santino bought a $\dfrac{3}{4}$ pound bag of chocolate chips. He used $\dfrac{2}{3}$ of the bag while baking. How many pounds of chocolate chips did he use while baking?

58. Farmer Dave harvested his corn. He stored $\dfrac{5}{9}$ of his corn in one large silo and $\dfrac{3}{4}$ of the remaining corn in a small silo. The rest was taken to market to be sold.
    a. What fraction of the corn was stored in the small silo?
    b. If he harvested 18 tons of corn, how many tons did he take to market?

59. $\dfrac{5}{8}$ of the songs on Harrison's music player are hip-hop. $\dfrac{1}{3}$ of the remaining songs are rhythm and blues. What fraction of all the songs are rhythm and blues?

60. Three-fifths of the students in a room are girls. One-third of the girls have blond hair. One-half of the boys have brown hair.
    a. What fraction of all the students are girls with blond hair?
    b. What fraction of all the students are boys without brown hair?

Division: Solve using the standard algorithm.

61. $\dfrac{7}{9} \div \dfrac{8}{11}$ 63. $\dfrac{3}{7} \div \dfrac{5}{16}$ 65. $\dfrac{9}{13} \div \dfrac{^-3}{10}$ 67. $\dfrac{1}{9} \div \dfrac{2}{5}$

62. $\dfrac{^-5}{11} \div \dfrac{3}{7}$ 64. $\dfrac{11}{2} \div \dfrac{4}{3}$ 66. $\dfrac{6}{5} \div \dfrac{1}{7}$ 68. $\dfrac{^-12}{5} \div \dfrac{^-7}{5}$

Applications for division of fractions

69. Mother has 20 pounds of salad. She wants each person to have $\dfrac{1}{4}$ pounds of salad. How many people can she feed with the 20 pounds of salad that she has?

70. I bought 10-lbs of sugar to make cakes. Each cake needs $\dfrac{3}{4}$ cups of sugar. Assuming that the 10-lbs of sugar is approximately 20 cups, how many cakes can I make with the 10 -lbs of sugar?

71. A piece of ribbon $\dfrac{18}{5}$ m long is cut into 12 shorter pieces of equal length. What is the length of each short piece?

1.5 DECIMAL FRACTIONS AND PERCENTS

In the previous section, we saw that fractions can be considered as numbers, since they enable us to measure in–between values. However, the notation that we used for fractions did not make them appear to be actual numbers, but rather pairs of numbers used to represent parts of a whole. Then add to that there were infinitely many ways to write the same numbers using equivalent fractions, and you are bound to be confused. Finally, the rules for arithmetic are really quite complicated, so that the standard algorithms can confuse even the brightest students. This is the major reason why fractions are so difficult for many students to understand well enough to be able to master.

In this section, we will show that fractions can be represented in an alternative way, so that they behave more like the natural and whole numbers we are familiar with. Instead of being written using the fraction notation of having a specific number for the numerator and one for the denominator, where we can have infinitely many different ways of writing the same value, we can standardize what we use for the denominators and just vary the numerator. This is really just an extension of the place–value numeral system we developed earlier in the chapter (section 1.2), and makes many of the operations on fractions easier to work with and understand. This new notation is called decimal fractions or decimals for short.

Decimal Notation

In decimals, the denominators are always powers of 10. for example:

$\dfrac{1}{10}$ or one-tenth, $\dfrac{1}{100}$ or one-hundredth, $\dfrac{1}{1000}$ or one-thousandth, etc.

We remove the denominator from our representation, since we will always use the same denominators with our decimal fractions, and just the numerator term will change.

For example, to write the combined sum of fractions we simply write this in the following way:

$$\frac{3}{10} + \frac{5}{100} + \frac{7}{1000}$$

First, we need to identify that the number we are writing is a fraction. For this, we introduce the decimal point notation. Anytime we see a period in a number, we are to interpret whatever comes after the period to the right as a fraction less than one. The period is called a decimal point.

Finally, we only write the numerator terms of the fractions, since the denominators are already known. Thus, we can write the sum of the three fractions above as

$$0.357$$

where it is to be understood that the period, or decimal point, means that the numbers following this point are all numerators for fractions with their denominators coming in increasing powers of 10 as shown above.

Now, for numbers that have a whole part greater than one, we simply write the whole number part of the number to the left of the decimal point, and the fraction part to the right of the decimal.

As an example, the number 524.75 is understood to be

$$500 + 20 + 4 + \frac{7}{10} + \frac{5}{100}$$

Using this new notation, we can also see the connection to the place–values we used earlier.

Alternatively, we could write this number as:

$$5 \times 100 + 2 \times 10 + 4 \times 1 + 7 \times \frac{1}{10} + 5 \times \frac{1}{100}$$

with the values after the multiplication symbols being the place–values.

This means that we can extend our place values to numbers to the right of the decimal point representing fractions with powers of 10 in their denominators. These are the place values to represent quantities less than one 1 in magnitude. The place–values are now:

...	Hundreds	Tens	Ones	.	Tenths	Hundredths	Thousandths	...
...	100	10	1	.	1/10	1/100	1/1000	...

We now provide a few examples of writing decimals in expanded notation.

EXAMPLE 1: Write the decimal in expanded form, 0.59

$$0.59 = 5 \times \frac{1}{10} + 9 \times \frac{1}{100}$$

EXAMPLE 2: Write the decimal in expanded form, 1.703

$$1.703 = 1 \times 1 + 7 \times \frac{1}{10} + 0 \times \frac{1}{100} + 3 \times \frac{1}{1000}$$

EXAMPLE 3: Write the decimal in expanded form, 36.34

$$36.34 = 3 \times 10 + 6 \times 1 + 3 \times \frac{1}{10} + 4 \times \frac{1}{100}$$

EXAMPLE 4: Write the decimal in expanded form, 237.4981

$$237.4981 = 2 \times 100 + 3 \times 10 + 7 \times 1 + 4 \times \frac{1}{10} + 9 \times \frac{1}{100} + 8 \times \frac{1}{1,000} + 1 \times \frac{1}{10,000}$$

We also need to be able to identify the place value of specific numbers in our decimal number. Consider the following examples.

EXAMPLE 5: Identify the place value of the underlined digit, 2.32<u>6</u>

The 6 digit is located in the one-thousandth's $\left(\frac{1}{1000} \right)$ place-value location

EXAMPLE 6: Identify the place value of the underlined digit, 0.0<u>2</u>79

The 2 digit is located in the one-hundredth's $\left(\frac{1}{100} \right)$ place-value location

Percents

In the previous section we were introduced to the concept of decimals or decimal fractions. Fractions allow us to distinguish between parts of a whole, but as we have shown in previous sections, comparing fractions of different sizes can be quite complicated. In ancient Rome, long before the creation of decimals, parts out of a total of 100 became common. It was an easy way to separate small parts from a whole, especially when taxes were levied. Taxes would typically be assigned in parts of a hundred, or in our modern fractional notation, $\frac{1}{100}$. In words this would be per one-hundred. In Latin one-hundred is a centum, so per one-hundred became percent (with centum shortened to cent and attached to the prefix per.) This new word was given a special symbol to identify it, and so we now use the % symbol.

So, when you take a test, and you receive a grade of 90%, what does it mean? It means that you scored 90 points out of 100 total points. The symbol % means out of 100 (per 100) and can also be written as $\frac{1}{100}$.

40

When we shop, we usually have to pay sales tax on the item that we purchased. For example, if the sales tax is 6%, then the question could be how much is the sales tax on a piece of furniture that I purchase. We see percents in many applications other than sales tax.

We can also illustrate this using the following visual model. The following is a 10 × 10 grid. To represent 90%, we can shade 90 of the unit boxes as shown to the right:

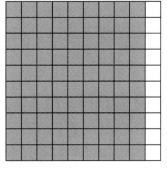

Suppose we have the following visual model to the right. What percent does it represent?

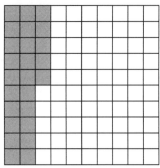

The shaded area is 25 boxes out of a total of 100 boxes, or $\frac{25}{100}$ which we call 25%. Here is another way of visually representing 25% as 1 out of 4 boxes, or $\frac{1}{4}$.

Percents can be written as a fraction or as a decimal. The example above represents 25% which can be written as either $\frac{25}{100}$ or $\frac{1}{4}$ or in decimal form as 0.25.

EXAMPLE 1: Write 63% in fraction form, in decimal form, and draw the visual model.

Fraction form: $63\% = 63 \times \frac{1}{100} = \frac{63}{100}$

Decimal form: $\frac{63}{100} = 0.63$

Visual model to the right: Shade 63 out of 100 squares.

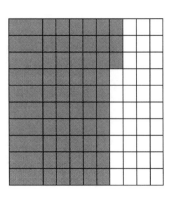

EXAMPLE 2: Write 22% as a fraction in reduced form.

$$22\% = \frac{22}{100} = \frac{2 \times 11}{2 \times 50} = \frac{\cancel{2} \times 11}{\cancel{2} \times 50} = \frac{11}{50}$$

EXAMPLE 3: Write 44.6% as a fraction in reduced form.

$$44.6\% = \frac{44.6}{100} = \frac{10 \times 44.6}{10 \times 100} = \frac{446}{1000} = \frac{2 \times 223}{2 \times 500} = \frac{\cancel{2} \times 223}{\cancel{2} \times 500} = \frac{223}{500}$$

EXAMPLE 4: Write $20\frac{1}{5}$ % as a fraction in reduced form.

$$20\frac{1}{5}\% = \left(20 + \frac{1}{5}\right) \div 100 = \frac{20}{100} + \frac{1}{5} \div 100 = \frac{20}{100} + \frac{1}{5} \times \frac{1}{100}$$

$$= \frac{20}{100} + \frac{1}{500} = \frac{5 \times 20}{5 \times 100} + \frac{1}{500} = \frac{100}{500} + \frac{1}{500}$$

$$= \frac{101}{500}$$

EXAMPLE 5: Express 15% as a decimal

$$15\% = \frac{15}{100} = 0.15$$

EXAMPLE 6: Express 29.75% as a decimal

$$29.75\% = \frac{29.75}{100} = \frac{100 \times 29.75}{100 \times 100} = \frac{2,975}{10,000} = 0.2975$$

EXAMPLE 7: Express the fraction $\frac{13}{20}$ as a decimal and then as a percent.

$$\frac{13}{20} = \frac{5 \times 13}{5 \times 20} = \frac{65}{100} = 0.65$$

$$= \left(\frac{65}{100}\right) \times 100\% = \left(\frac{65}{\cancel{100}}\right) \times \cancel{100}\% = 65\%$$

EXAMPLE 8: Express the decimal 0.45 as a percent then as a fraction in reduced form.

$$0.45 = 0.45 \times 100\% = 45\%$$

$$= \frac{45}{100} = \frac{5 \times 9}{5 \times 20} = \frac{\cancel{5} \times 9}{\cancel{5} \times 20} = \frac{9}{20}$$

EXAMPLE 9: Express the decimal 1.2 as a percent then as a fraction in reduced form.

$$1.2 = 1.2 \times 100\% = 120\%$$

$$= \frac{120}{100} = \frac{20 \times 6}{20 \times 5} = \frac{\cancel{20} \times 6}{\cancel{20} \times 5} = \frac{6}{5}$$

EXAMPLE 10: Sofia's take home salary is $3,200 each month. She spends $1,300 for rent and $150 for gasoline for her car. What percent of her salary is her rent? What percent of her salary is for the gasoline she uses each month?

For rent we use 1,300 and 3,200 to obtain: $\frac{1300}{3200} = 0.40625 = 40.625\% \approx 41\%$

For gasoline we use 150 and 3,200 to obtain:
$$\frac{150}{3200} = 0.046875 = 4.6875\% \approx 5\%$$

In addition to finding the percentage of a number given both the percentage and the number, we are sometimes given the percentage and the final value, and asked to find the number we are taking the percentage of only knowing the final value. Or given a part and the percent, can we find the whole. Sometimes we are given the part and the whole and are asked to find the associated percentage. Let's consider some examples to understand this better.

Let's consider another type of problem.

EXAMPLE 11: 60 out of 300 = what percent?

60 out of 300 can be written as $\frac{60}{300}$. Now divide both numerator and denominator by 3 to get 100 in the denominator:

$$\frac{60 \div 3}{300 \div 3} = \frac{20}{100}$$

So 60 out of 300 = 20%

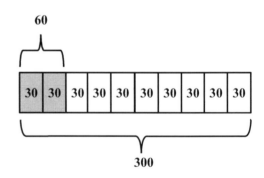

Using decimals, we will divide 60 by 300 and multiply by 100%.
$$\frac{60}{300} \times \frac{100}{100} = 0.2 \times 100 \times \frac{1}{100} = 20 \times \frac{1}{100} = \frac{20}{100} = 20\%$$

EXAMPLE 12: As an application of percentages, in chemistry class, you may be asked to determine percent composition of compounds (by mass). For example, if you want to determine the percent composition of the compound, water (H_2O), you want to determine what percentage of water is made up of hydrogen and what percentage of water is made up of oxygen, by mass (grams). Let's say you have a sample of 50 g of water, where 5.6 g is hydrogen and the remaining 44.4 g is oxygen.

This means then that $\dfrac{5.6}{50}\times100\%=11.2\%$ of the compound is made up of hydrogen and

$\dfrac{44.4}{50}\times100\%=88.8\%$ of the compound is made up of oxygen.

Thus, the percent composition of the compound is: 11.2% hydrogen; 88.8% oxygen

EXERCISES 1.5

Decimal Fractions

Write the decimals in expanded form

1. 0.035
2. 1.19
3. 16.0843
4. 25.0938
5. 11.002
6. 45.02023

Identify the place value of the underlined digit

7. 0.0$\underline{3}$5
8. 1.$\underline{1}$9
9. 16.084$\underline{3}$
10. 25.0$\underline{9}$38
11. 11.$\underline{0}$02
12. 45.020$\underline{2}$3

Percents

Express each percentage as fractions in simplest form:

13. 19%
14. 50%
15. 38%
16. 90%
17. 62.5%
18. 125%
19. $30\dfrac{1}{2}$ %
20. $66\dfrac{2}{3}$ %

Express each of the following percents as decimals:

21. 50%
22. 30%
23. 27.5%
24. 21.25%
25. 45%
26. 62.5%

Express each fraction as a decimal and then as a percent. If an exact answer is not possible, then round all answers to the hundredths place.

27. $\dfrac{2}{5}$
28. $\dfrac{7}{8}$
29. $\dfrac{3}{40}$
30. $\dfrac{12}{5}$

Express each decimal as a percent and then as a fraction in lowest terms:

31. 0.4
32. 0.38
33. 0.875
34. 1.92

Applications of percents: (Use calculators as needed, and round percentages to the nearest hundredhs place)

35. If 18 pieces of 50 pieces of candies are jolly ranchers, what percent does the 18 pieces represent?
36. Bernard's take home salary is $2000 each month. He spends $600 for rent and $100 for gasoline for his car. What percent of his salary is his rent? What percent of his salary is the gasoline he uses for each month?

37. In a class of 24 students, if 6 students received a grade of A, what percent of the class received a grade of A? What percent of the class did not receive a grade of A?

38. Aluminum chloride is formed by reacting 13.43 g aluminum with 53.18 g chlorine. What is the percent composition of the compound (i.e., % aluminum and % chlorine)?

39. A compound is found to consist of 1.37g of iron and 2.62g of chlorine. What is the percent composition of the compound (i.e. % iron and % chlorine)?

40. A 73.16 g sample of an interesting barium silicide reported to have superconducting properties was found to contain 33.62 g barium and the remainder silicon. Calculate the percent composition of the compound (i.e. % barium and % silicon).

41. A compound is found to consist of 4.87 g of silver, 3.5 g of nitrogen, and 8.92 g of oxygen. What is the percent composition of the compound (i.e. % silver, % nitrogen, % oxygen)?

42. What percent of 70 is 21?   43. 37 is what percent of 148?

44. There are 15 female and 25 male students in a class. What is the total number of students in the class? What percent of the class does the female students represent? What percent of the class does the male students represent?

45. You are performing a differential white blood cell (WBC) count. In a total of 100 WBCs counted, you observe that 57 of them are neutrophils. What percent of the WBCs are neutrophils?

46. You are performing a differential white blood cell (WBC) count. In a total of 100 WBCs counted, you observe that 28 of them are lymphocytes. What percent of the WBCs are lymphocytes?

47. You are determining the hematocrit of a patient's blood. If the amount of plasma is 36.3 ml and the amount of the whole blood (red blood cells, white blood cells, and plasma) is 66 ml, what percent of the patient's blood is plasma?

1.6 PROPERTIES OF REAL NUMBERS

The various types of numbers and arithmetic operators we defined above, are governed by the following properties and laws:

Associative Law

When adding three or more numbers, it makes no difference, which two pairs of numbers in left to right order we add first. This is called the associative law of addition.

$$(a+b)+c = a+(b+c)$$

For example:

$$(2+5)+7 = 7+7 = 14$$
$$2+(5+7) = 2+12 = 14 \qquad \text{so} \quad (2+5)+7 = 2+(5+7) = 14$$

Caution: The associative law is not true for subtraction:

$$(a-b)-c \neq a-(b-c)$$

To illustrate this:

$$(2-5)-7 = -3-7 = -10$$
$$2-(5-7) = 2-(-2) = 4 \qquad \text{so} \quad (2-5)-7 \neq 2-(5-7)$$

However, if we convert the subtraction to addition and then apply the Associative law

$$(2+(-5))+(-7) = (-3)+(-7) = -10$$
$$2+(-5+(-7)) = 2+(-12) = -10 \qquad \text{so} \quad (2+(-5))+(-7) = 2+(-5+(-7)) = -10$$

The associative law is true for multiplication:

$$(a \cdot b) \cdot c = a \cdot (b \cdot c)$$

For example:

$$(3 \cdot 4) \cdot 5 = 12 \cdot 5 = 60$$
$$3 \cdot (4 \cdot 5) = 3 \cdot 20 = 60 \qquad \text{so} \quad (3 \cdot 4) \cdot 5 = 3 \cdot (4 \cdot 5)$$

Commutative Law

Two numbers can be added in either order. This is called the Commutative Law of Addition.

$$a+b = b+a$$

For example:

$$2+5 = 7$$
$$5+2 = 7 \qquad \text{so} \quad 2+5 = 5+2 = 7$$

Two numbers can be multiplied in either order. This is called the Commutative Law of Multiplication.

$$a \cdot b = b \cdot a$$

For example:

$$5 \cdot 2 = 10$$
$$2 \cdot 5 = 10 \quad \text{so} \quad 5 \cdot 2 = 2 \cdot 5 = 10$$

Distributive Law

Multiplying the sum of two or more terms by some number n, is the same as the sum of the individual products.

$$n \cdot (a+b+c) = n \cdot a + n \cdot b + n \cdot c \quad \text{or}$$

$$(a+b+c) \cdot n = a \cdot n + b \cdot n + c \cdot n$$

This is called the distributive law of multiplication over addition.

For example:

$$(3+5+7) \cdot 2 = (15) \cdot 2 = 30$$

$$\text{so} \quad (3+5+7) \cdot 2 = 3 \cdot 2 + 5 \cdot 2 + 7 \cdot 2 = 30$$

$$3 \cdot 2 + 5 \cdot 2 + 7 \cdot 2 = 6 + 10 + 14 = 30$$

Additive Identity

An additive identity is a real number that when added to any other number, a, will not change the value of the original number. In the set of real numbers, zero is the additive identity and it satisfies the following property for an arbitrary real number, a:

$$a + 0 = 0 + a = a$$

For example: $7 + 0 = 7$

Multiplicative Identity

A multiplicative identity is a real number that when multiplied by any other number will not change its value. In the set of real numbers, one is the multiplicative identity and it satisfies the following property for an arbitrary real number, a:

$$1 \cdot a = a \cdot 1 = a$$

For example:

$$1 \cdot 3 = 3 \cdot 1 = 3$$
or
$$1 \cdot (-5) = (-5) \cdot 1 = -5$$

Additive Inverse

An additive inverse is a real number that when added to specific number makes the resulting sum equal to zero. In the set of real numbers, the additive inverse of the number "a" is " −a." The additive inverse satisfies the following property for an arbitrary real number, a:

$$a + (-a) = (-a) + a = 0$$

For example: $5 + (-5) = 0$

Multiplicative Inverse

A multiplicative inverse is a real number that when multiplied by a specific number makes the resulting product equal to one. In the set of real numbers, the multiplicative inverse of the number "a" is "$\frac{1}{a}$" or "a^{-1}." The multiplicative inverse satisfies the following property for an arbitrary real number, a:

$$a \cdot \frac{1}{a} = \frac{1}{a} \cdot a = 1 \qquad \text{or} \qquad a \cdot a^{-1} = a^{-1} \cdot a = 1$$

For example: $\frac{1}{3} \cdot 3 = 1$

EXAMPLES: State the property or properties used in each of the following exercises:

	Problem	Property
1	$(2+3)+4 = 2+(3+4)$	Associative law of addition
2	$(9 \cdot 3) \cdot 4 = 9 \cdot (3 \cdot 4)$	Associative law of multiplication
3	$4 \cdot (3+7+9) = 4 \cdot 3 + 4 \cdot 7 + 4 \cdot 9$	Distributive law of multiplication over addition
4	$55 + 74 = 74 + 55$	Commutative law of addition
5	$12 \cdot 55 = 55 \cdot 12$	Commutative law of multiplication
6	$99 + 34 = 34 + 99$	Commutative law of addition
7	$121 + (49 + 55) = (121 + 49) + 55$	Associative law of addition
8	$(34+15) \cdot 3 = 34 \cdot 3 + 15 \cdot 3$	Distributive law of multiplication over addition
9	$(-4) \cdot (-4)^{-1} = 1$	Multiplicative Inverse
10	$0 + (-8) = -8$	Additive Identity
11	$12 + (-12) = 0$	Additive Inverse
12	$13 \cdot 1 = 13$	Multiplicative Identity

48

Properties or Real Numbers

State the property or properties used in each of the following exercises

1. $(2+3)+4 = 2+(3+4)$

2. $(9\cdot3)\cdot4 = 9\cdot(3\cdot4)$

3. $15\cdot\dfrac{1}{15} = 1$

4. $1\cdot17 = 17$

5. $55+74 = 74+55$

6. $16+(-16) = 0$

7. $12\cdot55 = 55\cdot12$

8. $(-1)\cdot1 = -1$

9. $99+34 = 34+99$

10. $(-13)+13 = 0$

11. $(-6)^{-1}(-6) = 1$

12. $42+0 = 42$

13. $(34+13)\cdot3 = 34\cdot3+13\cdot3$

14. $5\cdot(2+11+8) = 5\cdot2+5\cdot11+5\cdot8$

15. $121+(49+55) = (121+49)+55$

1.7 APPLYING THE DISTRIBUTIVE PROPERTY

The commutative and the associative properties of addition and multiplication tell us how to manipulate an arithmetic expression with a single arithmetic operator, either addition or multiplication. What is missing is a property that links multiplication and addition. That property is called the distributive property, or more precisely the distributive property of multiplication over addition. It tells us how multiplication and addition are related when they occur in the same expression.

For example, if we have the expression $2 \cdot (3 + 5)$, we have a choice. We can either add the 3 and the 5 together, or we can distribute (multiply) the 2 times the 3 and the 5 separately and get the same result, i.e.

$$2 \cdot (3 + 5) = 2 \cdot 8 = 16$$

or

$$2 \cdot (3 + 5) = 2 \cdot 3 + 2 \cdot 5 = 6 + 10 = 16$$

Not only does the distributive property work by distributing left to right, but it also works by distributing right to left as shown below when considering $(3 + 5) \cdot 2$.

$$(3 + 5) \cdot 2 = 8 \cdot 2 = 16$$

or

$$(3 + 5) \cdot 2 = 3 \cdot 2 + 5 \cdot 2 = 6 + 10 = 16$$

In general, we have the **Distributive Law of Multiplication over Addition**

$$a \cdot (b + c) = a \cdot b + a \cdot c \quad \text{or} \quad (b + c) \cdot a = b \cdot a + c \cdot a,$$

where a, b, and c are any real numbers.

We illustrate the distributive property (law) through several examples:

EXAMPLES: Use the distributive property to simplify the expressions. Then show you get the same answer by calculating what is in the parentheses first and then multiplying.

1. $^-3 \cdot (4 + 3) = {}^-3 \cdot 4 + {}^-3 \cdot 3 = {}^-12 + {}^-9 = {}^-21$

    $^-3 \cdot (4 + 3) = {}^-3 \cdot 7 = = {}^-21$

2. $4 \cdot (2 - 4) = 4 \cdot (2 + {}^-4) = 4 \cdot 2 + 4 \cdot {}^-4 = 8 + {}^-16 = {}^-8$

    $4 \cdot (2 - 4) = 4 \cdot ({}^-2) = {}^-8$

50

3. $^-5\cdot(8-6) = {}^-5\cdot(8+{}^-6) = {}^-5\cdot8 + {}^-5\cdot{}^-6 = {}^-40 + 30 = {}^-10$

$^-5\cdot(8-6) = {}^-5\cdot(2) = {}^-10$

4. $\frac{1}{2}\cdot(8-2) = \frac{1}{2}\cdot(8+{}^-2) = \frac{1}{2}\cdot8 + \frac{1}{2}\cdot{}^-2 = 4 + {}^-1 = 3$

$\frac{1}{2}\cdot(8-2) = \frac{1}{2}\cdot(6) = \frac{6}{2} = 3$

5. $(9-6)\cdot\frac{2}{3} = (9+{}^-6)\cdot\frac{2}{3} = 9\cdot\frac{2}{3} + {}^-6\cdot\frac{2}{3} = \frac{18}{3} + \frac{{}^-12}{3} = 6 + {}^-4 = 2$

$(9-6)\cdot\frac{2}{3} = (3)\cdot\frac{2}{3} = \frac{6}{3} = 2$

6. $3\cdot(3-5+6) = 3\cdot(3+{}^-5+6) = 3\cdot3 + 3\cdot{}^-5 + 3\cdot6 = 9 + {}^-15 + 18$
$= {}^-6 + 18 = 12$

$3\cdot(3-5+6) = 3\cdot(4) = 12$

EXERCISES 1.7

Distributive Property

Use the distributive property to simplify the expressions. Then show you get the same answer by calculating what is in the parentheses first and then multiplying.

1. $^-6\cdot(3+8)$
2. $4\cdot(6-15)$
3. $^-4\cdot(3-8)$
4. $\frac{3}{5}\cdot(9-4)$
5. $\frac{4}{3}\cdot(5+10)$
6. $(8-13)\cdot\frac{5}{9}$
7. $(^-3-13)\cdot{}^-\frac{2}{3}$
8. $(7+5)\cdot{}^-5$
9. $(3+{}^-9)\cdot{}^-3$
10. $^-4\cdot(^-2-4)$
11. $^-7\cdot(^-3-1)$
12. $5\cdot(8-7+3)$
13. $^-8\cdot(^-2+5-3)$
14. $^-4\cdot(^-3-8+4)$
15. $5\cdot(4+{}^-5+7)$
16. $\frac{4}{7}\cdot(7+4+3)$
17. $\frac{3}{8}\cdot(2-16-10)$
18. $^-\frac{3}{4}\cdot(^-3+13+2)$
19. $^-\frac{4}{5}\cdot(6-12)$
20. $(^-4-8)\cdot{}^-\frac{5}{9}$
21. $(4+{}^-9)\cdot{}^-\frac{2}{9}$

1.8 ORDER OF OPERATIONS

In the earlier sections we defined two different operations and their inverses: addition/subtraction, and multiplication/division. We looked at each separately. However, most calculations involve many if not all of the operators in a single expression, and this can cause confusion as to which operation to do first, second, etc.

For example, consider the expression: $12 \div 3 \times 2 - 3 + 12/2 \times 6$. Now this looks simple enough, but it raises one important question: Which operation do we perform first, second, third, etc.? To avoid this confusion and to ensure that we all get the same result when evaluating this expression, an order of operations was established. It is very critical that one understands this, since the calculator is set up to use this protocol and, regardless of what we might intend, it will calculate a result based upon this set of rules. This is an area in which student mistakes are very common.

The order of operation can sometimes be remembered by the acronym **PEMDAS** or Please Excuse My Dear Aunt Sally. The order of operations states that we do all calculations inside Parentheses first. If we have nested parentheses (parentheses within parentheses), we do the innermost parentheses first and work outwards. Next we do all Exponent operations. Exponents have not been introduced yet, they will be discussed in detail later in the next chapter, so for now we skip this part. After the exponents, we do all Multiplication or Division simultaneously (whichever comes first) from left to right in the expression. Finally, we do the Addition or Subtraction simultaneously from left to right.

EXAMPLE 1: Perform the indicated operations using the order of operations:
$$(5)(5) - 3(3 - 6)$$
Solution: using the order of operations
$$(5)(5) - 3(3 - 6)$$
$$= (5)(5) - 3(^-3)$$
$$= 25 + 9$$
$$= 34$$

EXAMPLE 2: Perform the indicated operations using the order of operations:
$$(^-5)(^-2) - [3 + (8 - 10)]$$
Solution: using the order of operations
$$(^-5)(^-2) - [3 + (8 - 10)]$$
$$= (^-5)(^-2) - [3 + {}^-2]$$
$$= (^-5)(^-2) - [1]$$
$$= 10 - 1$$
$$= 9$$

EXAMPLE 3: Perform the indicated operations using the order of operations:

$$(^-8+3)-\frac{(5+10)}{7-2}$$

Solution: using the order of operations

$$(^-8+3)-\frac{(5+10)}{7-2}$$

$$= (^-5)-\frac{15}{5}$$

$$= ^-5-3$$

$$= ^-5+^-3$$

$$= ^-8$$

EXAMPLE 4: Perform the indicated operations using the order of operations:

$$[4-3(^-2)+25](^-2)$$

Solution: using the order of operations

$$[4-3(^-2)+25](^-2)$$

$$= [4+6+25](^-2)$$

$$= [10+25](^-2)$$

$$= [35](^-2)$$

$$= ^-70$$

EXERCISES 1.8

Order of Operations
Evaluate each expression
Basic Problems

1. $5(3+8)$

2. $63\div7+2$

3. $3+6\cdot4$

4. $5-(2-3)$

5. $(8+3)\cdot2-6$

6. $2\cdot4+3(6-3)$

7. $3(4-7)$

8. $15\div(3-(7-9))$

9. $5+7\cdot3+6$

10. $3+4(2-4)$

11. $7-5(2+3)$

12. $1+3(3-6)$

13. $4+5(2-4)$

14. $72\div(6+3)$

15. $24\cdot2-36$

16. $13-2(4+2)$

17. $14\div2+6$

18. $12\div4+(2+10)$

19. $2+3(7-11)$

20. $5-3+2(7-4+5)$

21. $2(3+1)-6\div3$

22. $2(3-6+4)-4+9$

23. $-2(5-7)-5+9$

24. $25\div5+10-4$

25. $-4+8\div2+4\times3$

26. $27\div3\times3-9$

27. $15-12\div(-4)\times(-3)-2$

28. $-32\div2\times2+32$

Intermediate to Advanced Problems

29. $\dfrac{(8+4)\cdot\dfrac{5}{6}-3}$

30. $5-3(5-7)$

31. $6-7(14-15)$

32. $3(7-7)-(4-3)$

33. $1-\dfrac{32}{2\cdot4+8}+(5-9)$

34. $3\cdot6-\dfrac{8+7}{15-10}$

35. $\dfrac{64}{8}\cdot\dfrac{75}{18-(1-8)}$

36. $\dfrac{3}{4-2}\cdot(1-3)\cdot5$

37. $6-\dfrac{27}{3}-(3+6)$

38. $(7+3)\div10\cdot3$

39. $(4+3\cdot2-9)\cdot3-1$

40. $(4+17)\div3-8$

41. $30\div3-(5-9)\div(1-3)$

42. $(8+4)\div(12-6)$

43. $(8+5-5)\cdot4-2$

44. $(6+36-2)\div8+2$

45. $\left[14+(6-10)\right]\div5-3$

46. $\dfrac{68}{(2+5)-3}$

47. $\left[3-(2+7)\right]\div3-2$

Chapter 1 Practice Test

Simplify each expression into a single number using standard algorithms.

1. $^-3-2$

2. $4-^-3$

3. $\dfrac{1}{2}+\dfrac{3}{4}$

4. $\dfrac{5}{8}-\dfrac{1}{16}$

5. $\dfrac{2}{5}-\dfrac{2}{3}$

6. $^-6\cdot7$

7. $^-5\cdot^-8$

8. $15\cdot^-6$

9. $25\cdot^-9$

10. $3\cdot\dfrac{2}{3}$

11. $^-\dfrac{4}{5}\left(^-\dfrac{25}{16}\right)$

12. $\dfrac{1}{2}\cdot\left(\dfrac{^-8}{3}\right)$

13. $\left(\dfrac{^-4}{9}\right)\div\left(\dfrac{2}{3}\right)$

14. $\dfrac{5}{7}\div10$

15. $7\div\left(\dfrac{^-14}{15}\right)$

16. $\dfrac{\left(\dfrac{^-2}{3}\right)}{\left(\dfrac{^-4}{9}\right)}$

Write two equivalent fractions for the following

17. $\dfrac{2}{3}$

18. $\dfrac{3}{8}$

Rewrite the decimals in expanded notation

19. 93,059

20. 21.325

21. 197.0134

Identify the place-value of the underlined digit

22. 17,038,769

23. 102.0347

24. 0.00395

Express each percentage as a fraction in simplest form:

25. 5%

26. 38%

27. 110%

Express each percent as a decimal:

28. 17%

29. 105%

30. 2%

Express each fraction as a decimal and then as a percent:

31. $\dfrac{4}{5}$

32. $\dfrac{3}{4}$

Express each decimal as a percent and then as a fraction in lowest terms:

33. 0.15

34. 0.04

35. 2.5

Evaluate each expression

36. $5(3+8)$

37. $63\div7+2$

38. $3+6\cdot4$

39. $5-(2-3)$

40. $4+2(3-6)$

41. $(8+3)\cdot2-6$

42. $-(3-5)-6\cdot^-3$

43. $(7+3)\div10\cdot3$

44. $(6+36-2)\div8+2$

45. $1-\dfrac{32}{2\cdot4+8}+(5-9)$

46. $-4-3+2-3(5-9)$

47. $13+21\div(-7)\times3$

48. A marching band is rehearsing in rectangular formation. $\dfrac{1}{3}$ of the marching band members play percussion instruments. $\dfrac{1}{4}$ of the percussionists play the snare drum. What fraction of all the band members play the snare drum?

49. There are 400 students in a school. 40% of the students are girls. How many girls are in this school?

CHAPTER 2

Introduction to Exponents

Mathematics is a powerful language that often defines useful devices for writing things in a condensed and more easily understood way. One particular example relates to our place-value number system and the powers of ten. In this chapter we start with this place-value example and then show how to extend the notation even further.

2.1 POWERS OF TEN AND POSITIVE EXPONENTS

When writing the place-values of our number system, we are often required to write larger and larger numbers. For very large place-values the numbers can become quite large and take up quite a bit of space. There is, however, a way to condense how we write these large numbers by introducing a new device called an exponent.

Consider the place value chart below.

...	One Hundred Thousands	Ten Thousands	Thousands	Hundreds	Tens	Ones	Decimal Point
...	100,000	10,000	1,000	100	10	1	.

As we move further to the left in the place-value chart we see that the place-value gets larger and larger, and it becomes increasingly more difficult to represent these larger place-values. It is also difficult to easily distinguish between these place-values. For example consider the two large place-value numbers:

$$1000000000000000 \text{ and } 100000000000000$$

It is hard to see at first glance which number is larger and by how much. Now, we often use commas to help us distinguish the differences, and each number above could be written as:

$$1,000,000,000,000,000 \text{ and } 100,000,000,000,000$$

It is now easier to see that the first number has an extra zero, so it is ten times larger than the second. However, it still takes a large amount of effort to write down each place-value. Our Hindu-Arabic number system has place values based upon factors of ten, which is why it is called a base-10 number system. In other words if we move one place to the left in our system, it is equivalent to multiplying by a factor of ten, as shown from right to left below:

$$10 \times 10 \times 10 = 1,000, \qquad 10 \times 10 = 100, \qquad 10 \times 1 = 10, \qquad 1$$

To capture this behavior in a clear and concise manner, the concept of an exponent was created. An exponent is a super-scripted number that is placed to the right of a number, called the base, and it indicates how many factors of the base the number has in it. Thus,

$10^1 = 10$ One factor of the base 10 number
$10^2 = 10 \times 10 = 100$ Two factors of the base 10 number
$10^3 = 10 \times 10 \times 10 = 1{,}000$ Three factors of the base 10 number
$10^4 = 10 \times 10 \times 10 \times 10 = 10{,}000$ Four factors of the base 10 number
$10^5 = 10 \times 10 \times 10 \times 10 \times 10 = 100{,}000$ Five factors of the base 10 number

EXAMPLES: Rewrite the following using exponent notation.

1. 1,000,000,000,000,000

   10^{15} since there are 15 zeros which equate to 15 factors of 10

2. 10,000,000

   10^7 since there are 7 zeros which equate to 7 factors of 10

3. 100,000,000

   10^8 since there are 8 zeros which equate to 8 factors of 10

EXAMPLES: Rewrite the following by removing the exponent notation.

1. 10^3

   $10 \times 10 \times 10 = 1{,}000$ Notice the 3 zeros for the three factors of 10

2. 10^9

   $10 \times 10 \times 10 \times 10 \times 10 \times 10 \times 10 \times 10 \times 10 = 1{,}000{,}000{,}000$

                             Notice the 9 zeros for the nine factors of 10

3. 10^4

   $10 \times 10 \times 10 \times 10 = 10{,}000$ Notice the 4 zeros for the four factors of 10

Using this new exponent notation, we can rewrite the place-values as follows:

$$10 = 10^1, \qquad 100 = 10^2, \qquad 1{,}000 = 10^3, \quad 10{,}000 = 10^4, \quad 100{,}000 = 10^5, \quad \text{etc.}$$

This means we can rewrite the place-value chart as follows:

...	One Hundred Thousands	Ten Thousands	Thousands	Hundreds	Tens	Ones	Decimal Point
...	10^5	10^4	10^3	10^2	10^1	1	.

Note: An alternative form of writing the exponent is using the carat symbol on the computer or calculator, $\wedge$, instead of as a superscript. This is typically the case when using a calculator. The notation looks like this.

$$10^1 = 10\wedge1, \quad 10^2 = 10\wedge2, \quad 10^3 = 10\wedge3, \quad 10^4 = 10\wedge4, \quad 10^5 = 10\wedge5, \qquad \text{etc.}$$

Using exponential notation, we can now write our numbers in a more condensed expanded form as shown in the following examples.

EXAMPLES: Rewrite the following numbers in expanded form using exponent notation.

1. 34,365

$$34,365 = 30,000 + 4,000 + 300 + 60 + 5$$

$$= 3 \times 10,000 + 4 \times 1,000 + 3 \times 100 + 6 \times 10 + 5 \times 1$$

$$= 3 \times 10^4 + 4 \times 10^3 + 3 \times 10^2 + 6 \times 10^1 + 5 \times 1$$

2. 298,304

$$298,304 = 200,000 + 90,000 + 8,000 + 300 + 4$$

$$= 2 \times 100,000 + 9 \times 10,000 + 8 \times 1,000 + 3 \times 100 + 0 \times 10 + 4 \times 1$$

$$= 2 \times 10^5 + 9 \times 10^4 + 8 \times 10^3 + 3 \times 10^2 + 0 \times 10^1 + 4 \times 1$$

3. 6,930,496

$$6,930,496 = 6,000,000 + 900,000 + 30,000 + 400 + 90 + 6$$

$$= 6 \times 1,000,000 + 9 \times 100,000 + 3 \times 10,000 + 0 \times 1,000 + 4 \times 100$$
$$+ 9 \times 10 + 6 \times 1$$

$$= 6 \times 10^6 + 9 \times 10^5 + 3 \times 10^4 + 0 \times 10^3 + 4 \times 10^2 + 9 \times 10^1 + 6 \times 1$$

EXERCISES 2.1

Exponents and Powers of Ten

Rewrite the expressions using exponents

1. $10 \cdot 10 \cdot 10 \cdot 10 \cdot 10 \cdot 10 \cdot 10$

2. $(10)(10)(10)$

3. $10 \times 10 \times 10 \times 10 \times 10 \times 10$

4. $(10)(10)(10)(10)(10)(10)(10)$

5. $10 \cdot 10 \cdot 10 \cdot 10 \cdot 10 \cdot 10 \cdot 10 \cdot 10 \cdot 10 \cdot 10 \cdot 10$

6. $10 \times 10 \times 10 \times 10$

Rewrite without exponents, and multiply out

7. 10^3
8. 10^5
9. $^-10\text{^}4$
10. 10^9
11. $10\text{^}5$
12. $^-10^6$

Rewrite in expanded exponent notation form

13. 73
14. 56
15. 147
16. 289
17. 5,024
18. 8,290
19. 13,390
20. 130,450
21. 5,689,003
22. 21,000
23. 200,395
24. 70,402
25. 975,389
26. 8,381
27. 2,450,000,000
28. 150,003,001,202

In the next section we extend this idea to other bases and introduce some rules for simplifying more complicated terms with exponents.

2.2 POSITIVE EXPONENTS IN GENERAL

In addition to multiplying powers of 10 a number of times, we will often have to multiply other numbers by themselves several times, such as 3 multiplied by itself 4 times, or 5 factors of 3. Instead of writing $3 \cdot 3 \cdot 3 \cdot 3 \cdot 3$ all the time, we can again use the exponent or power operator.

Thus, instead of $3 \cdot 3 \cdot 3 \cdot 3 \cdot 3$, we would write 3^5 in a mathematical expression. As we have already shown, the exponent operator is simply a shorthand way of writing a number multiplied by itself a fixed number of times. The number we are multiplying is called the base of the exponent operator and the number of times we are multiplying the same number is called the exponent.

EXAMPLES:

a) $2^4 = 2 \cdot 2 \cdot 2 \cdot 2 = 16$, the base is 2 and the exponent is 4, read as "two to the fourth" or "two to the fourth power"

b) $3^6 = 3 \cdot 3 \cdot 3 \cdot 3 \cdot 3 \cdot 3 = 729$, the base is 3 and the exponent is 6, read as "three to the sixth" or "three to the sixth power"

c) $5^3 = 5 \cdot 5 \cdot 5 = 125$, the base is 5 and the exponent is 3, read as "five to the third" or "five cubed"

d) $8^2 = 8 \cdot 8 = 64$, the base is 8 and the exponent is 2, read as "eight to the second" or "eight squared"

e) $5 \cdot 5 \cdot 5 \cdot 5 \cdot 5 \cdot 5 \cdot 5 \cdot 5 \cdot 5 = 5^9$

f) $^-6 \cdot {}^-6 \cdot {}^-6 \cdot {}^-6 \cdot {}^-6 \cdot {}^-6 \cdot {}^-6 = {}^-6^7$ or $(^-6)^7$

*Note: When the superscript notation is used, you do not have to write parentheses around the negative number. However, in future courses, if the superscript notation is not used, the answers may not be the same. It all depends upon whether or not the exponent is an even or an odd number. This can cause a lot of confusion that we are avoiding for now, but it may come up again in a later course. Again, the symbol "–" is subtraction or "take the negative of" (an action or verb) whereas the symbol " $^-$ " means we are talking about a negative number (a name or noun). We should also make you aware that the symbol/button "(–)" on the calculator does not mean you are making the quantity into a negative number, instead this button simply performs the operation of "take the negative of what follows." It is just another action, and not a noun.

g) $-5^2 = -5 \cdot 5 = -25$, Notice how this is really "take the negative of 5 squared" and not "negative 5 squared." This is what was discussed in the note of part f above.

h) $\left(\dfrac{2}{3}\right)^4 = \left(\dfrac{2}{3}\right) \cdot \left(\dfrac{2}{3}\right) \cdot \left(\dfrac{2}{3}\right) \cdot \left(\dfrac{2}{3}\right) = \dfrac{2 \cdot 2 \cdot 2 \cdot 2}{3 \cdot 3 \cdot 3 \cdot 3} = \dfrac{16}{81}$

EXERCISES 2.2

Positive Exponents

Rewrite the expressions using exponents

1. $3 \cdot 3 \cdot 3 \cdot 3 \cdot 3 \cdot 3 \cdot 3$

2. $7 \cdot 7 \cdot 7$

3. $^-2 \cdot {}^-2 \cdot {}^-2 \cdot {}^-2 \cdot {}^-2 \cdot {}^-2 \cdot {}^-2$

4. $^-3 \cdot {}^-3 \cdot {}^-3 \cdot {}^-3 \cdot {}^-3 \cdot {}^-3$

5. $\left(\dfrac{3}{5}\right) \cdot \left(\dfrac{3}{5}\right) \cdot \left(\dfrac{3}{5}\right)$

6. $\left(\dfrac{1}{2}\right) \cdot \left(\dfrac{1}{2}\right) \cdot \left(\dfrac{1}{2}\right) \cdot \left(\dfrac{1}{2}\right) \cdot \left(\dfrac{1}{2}\right)$

7. $\left(\dfrac{2}{3}\right) \cdot \left(\dfrac{2}{3}\right) \cdot \left(\dfrac{2}{3}\right) \cdot \left(\dfrac{2}{3}\right) \cdot \left(\dfrac{2}{3}\right) \cdot \left(\dfrac{2}{3}\right) \cdot \left(\dfrac{2}{3}\right)$

Rewrite without exponents, and multiply out

8. 4^3

9. 2^5

10. $(^-3)^3$

11. $(^-2)^4$

12. -3^4

13. -5^2

14. $^-6^2$

15. $^-4^2$

16. $\left(\dfrac{2}{3}\right)^4$

17. $\left(\dfrac{1}{3}\right)^2$

18. $\left(\dfrac{3}{4}\right)^3$

19. -4^3

20. $^-3^4$

21. $\left(\dfrac{1}{4}\right)^3$

22. 1^8

23. -1^{12}

24. $\left(\dfrac{5}{2}\right)^4$

25. $\left(\dfrac{-2}{7}\right)^2$

2.3 BASIC RULES OF EXPONENTS

Having introduced the exponent notation, we now study this operator in more detail, starting with its definition.

Definition: When multiplying a real number by itself a number of times we define an exponential operator such that

$$a^n \ = \ a \cdot a \cdot a \cdot a \ldots a \cdot a \text{ , n factors of a}$$

Where "a" is called the base, and "n" is called the exponent,
furthermore, a and n can be any real numbers

Using this definition, some basic rules of arithmetic related to exponents necessarily follow, and we explore them below.

Consider what happens when we multiply two exponent terms with the same base,

EXAMPLE 1:

a) $3^3 \cdot 3^4 = (3 \cdot 3 \cdot 3) \cdot (3 \cdot 3 \cdot 3 \cdot 3) = 3^{3+4} = 3^7$

b) $2^5 \cdot 2^4 = (2 \cdot 2 \cdot 2 \cdot 2 \cdot 2) \cdot (2 \cdot 2 \cdot 2 \cdot 2) = 2^{5+4} = 2^9$

c) $5^3 \cdot 5^6 = (5 \cdot 5 \cdot 5) \cdot (5 \cdot 5 \cdot 5 \cdot 5 \cdot 5 \cdot 5) = 5^{3+6} = 5^9$

Looking closely, we can see a pattern. For any base, **a**, and exponents **n** and **m** we simply add the exponents:

> **Rule 1:** When multiplying two exponent terms with the same base, **a**, and exponents **n** and **m** we add the exponents.
>
> $$a^n \cdot a^m = a^{(n+m)}$$

Note: the base in each term MUST be the same for this rule to apply! We cannot combine the exponents of terms with different bases, e.g. $2^4 \cdot 3^3 = (2 \cdot 2 \cdot 2 \cdot 2) \cdot (3 \cdot 3 \cdot 3)$. We cannot rewrite this in terms of a single base raised to a single exponent.

Consider what happens when we divide two exponent terms, again with the same base,

EXAMPLE 2:

a) $\dfrac{3^5}{3^3} = \dfrac{(3 \cdot 3 \cdot 3 \cdot 3 \cdot 3)}{(3 \cdot 3 \cdot 3)} = 3^{(5-3)} = 3^2 = 9$

b) $\dfrac{2^8}{2^2} = \dfrac{(2 \cdot 2 \cdot 2 \cdot 2 \cdot 2 \cdot 2 \cdot 2 \cdot 2)}{(2 \cdot 2)} = 2^{(8-2)} = 2^6 = 64$

c) $\dfrac{5^7}{5^6} = \dfrac{(5 \cdot 5 \cdot 5 \cdot 5 \cdot 5 \cdot 5 \cdot 5)}{(5 \cdot 5 \cdot 5 \cdot 5 \cdot 5 \cdot 5)} = 5^{(7-6)} = 5^1 = 5$

We have drawn a line through the factors in the numerator and denominator that cancel out. Another pattern is evident. We subtract the exponents.

> **Rule 2:** When dividing two exponent terms with the same base, **a**, not equal to zero, and exponents **n** and **m** we subtract the exponents.
>
> $$\frac{a^n}{a^m} = a^{(n-m)}$$

At this time we will assume that **n** is always larger than **m**, so that the difference is always positive. This, however, is not always the case. In section 2.4 below, we show this for the specific base a = 10.

What happens when we raise an exponent term to a power?

EXAMPLE 3:

a) $\left(3^2\right)^3 = \left(3^2\right) \cdot \left(3^2\right) \cdot \left(3^2\right) = (3 \cdot 3) \cdot (3 \cdot 3) \cdot (3 \cdot 3) = 3^6$

b) $\left(2^4\right)^2 = \left(2^4\right) \cdot \left(2^4\right) = (2 \cdot 2 \cdot 2 \cdot 2) \cdot (2 \cdot 2 \cdot 2 \cdot 2) = 2^8$

c) $\left(2^3\right)^5 = \left(2^3\right) \cdot \left(2^3\right) \cdot \left(2^3\right) \cdot \left(2^3\right) \cdot \left(2^3\right) = (2 \cdot 2 \cdot 2) \cdot (2 \cdot 2 \cdot 2) \cdot (2 \cdot 2 \cdot 2) \cdot (2 \cdot 2 \cdot 2) \cdot (2 \cdot 2 \cdot 2) = 2^{3 \cdot 5} = 2^{15}$

The general rule is that for any base, **a**, and exponents; **n** and **m**, we multiply the exponents:

> **Rule 3:** When raising an exponent term to a power, for any base, **a,** and exponents, **n** and **m** we multiply the exponents.
> $$\left(a^n\right)^m = a^{n \cdot m}$$

What happens when we divide an exponent term by the same exponent term and apply Rule 2 above?

EXAMPLE 4:

a) $\dfrac{3^4}{3^4} = \dfrac{(3 \cdot 3 \cdot 3 \cdot 3)}{(3 \cdot 3 \cdot 3 \cdot 3)} = 3^{(4-4)} = 3^0 = 1$

b) $\dfrac{2^6}{2^6} = \dfrac{(2 \cdot 2 \cdot 2 \cdot 2 \cdot 2 \cdot 2)}{(2 \cdot 2 \cdot 2 \cdot 2 \cdot 2 \cdot 2)} = 2^{(6-6)} = 2^0 = 1$

c) $\dfrac{5^4}{5^4} = \dfrac{(5 \cdot 5 \cdot 5 \cdot 5)}{(5 \cdot 5 \cdot 5 \cdot 5)} = 5^{(4-4)} = 5^0 = 1$

The pattern that emerges here, is for any base **a** (not equal to zero), "a" raised to the zero power is equal to one.

> **Rule 4:** Any nonzero quantity raised to the zero power is always one.
> $$a^0 = 1 \quad , \quad a \neq 0$$

Suppose we want to raise a product of two numbers to a power.

EXAMPLE 5:

a) $(3 \cdot 5)^2 = 15^2 = 225 = 3^2 \cdot 5^2 = 9 \cdot 25 = 225$

b) $(2 \cdot 3)^3 = 6^3 = 216 = 2^3 \cdot 3^3 = 8 \cdot 27 = 216$

c) $(7 \cdot 3)^2 = 21^2 = 441 = 7^2 \cdot 3^2 = 49 \cdot 9 = 441$

The pattern that emerges here is that for the product of any numbers **a** and **b** raised to the power **n**, we have:

> **Rule 5:** When raising the product of two bases to an exponent **n**, we get the product of the bases raised to the exponent.
> $$(ab)^n = a^n \cdot b^n$$

Now, let's consider what happens when we raise a fractional term to a power:

EXAMPLE 6:

a) $\left(\dfrac{3}{4}\right)^3 = \left(\dfrac{3}{4}\right)\cdot\left(\dfrac{3}{4}\right)\cdot\left(\dfrac{3}{4}\right) = \dfrac{(3\cdot3\cdot3)}{(4\cdot4\cdot4)} = \dfrac{3^3}{4^3} = \dfrac{27}{64}$

b) $\left(\dfrac{5}{7}\right)^2 = \left(\dfrac{5}{7}\right)\cdot\left(\dfrac{5}{7}\right) = \dfrac{(5\cdot5)}{(7\cdot7)} = \dfrac{5^2}{7^2} = \dfrac{25}{49}$

c) $\left(\dfrac{6}{11}\right)^5 = \left(\dfrac{6}{11}\right)\cdot\left(\dfrac{6}{11}\right)\cdot\left(\dfrac{6}{11}\right)\cdot\left(\dfrac{6}{11}\right)\cdot\left(\dfrac{6}{11}\right) = \dfrac{(6\cdot6\cdot6\cdot6\cdot6)}{(11\cdot11\cdot11\cdot11\cdot11)} = \dfrac{6^5}{11^5} = \dfrac{7,776}{161,051}$

The pattern that emerges is:

Rule 6: For any fractional base $\dfrac{a}{b}$, and exponent **n**:

$$\left(\dfrac{a}{b}\right)^n = \dfrac{a^n}{b^n}, \quad b\neq0$$

Below is a summary of all the rules of exponents. You should commit these to memory.

Summary of the Rules of Exponents:

Let "**a**" and "**b**", be any real numbers, and "**n**" and "**m**" be any integers. Then

1. $a^n\cdot a^m = a^{(n+m)}$

2. $\dfrac{a^n}{a^m} = a^{(n-m)}$, provided $a\neq0$

3. $\left(a^n\right)^m = a^{n\cdot m}$

4. $a^0 = 1$, provided $a\neq0$

5. $(ab)^n = a^n\cdot b^n$

6. $\left(\dfrac{a}{b}\right)^n = \dfrac{a^n}{b^n}$, provided $b\neq0$

EXAMPLE 7: Evaluate using the rules of exponents $5^3 5^4$

$$5^3 5^4 = 5^{(3+4)} = 5^7 = 78,125$$

EXAMPLE 8: Evaluate using the rules of exponents $\dfrac{8^5}{8^3}$

$$\frac{8^5}{8^3} = 8^{(5-3)} = 8^2 = 64$$

EXAMPLE 9: Evaluate using the rules of exponents $\left(4^2 5^3\right)^0$

$$\left(4^2 5^3\right)^0 = 1$$

EXAMPLE 10: Evaluate using the rules of exponents $\left(\frac{4}{7}\right)^3 \left(\frac{4}{7}\right)^7$

$$\left(\frac{4}{7}\right)^3 \left(\frac{4}{7}\right)^7 = \left(\frac{4}{7}\right)^{10}$$

In this case, we will not multiply this out, since the numbers get quite large and we can get lost in the numbers and lose sight of the result.

EXAMPLE 11: Evaluate using the rules of exponents $\left(4^2\right)^3$

$$\left(4^2\right)^3 = 4^6 = 4,096$$

EXAMPLE 12: Evaluate using the rules of exponents $\left[\left(\frac{4}{5}\right)^3\right]^2$

$$\left[\left(\frac{4}{5}\right)^3\right]^2 = \left(\frac{4}{5}\right)^{3\cdot2} = \left(\frac{4}{5}\right)^6 = \frac{4^6}{5^6} = \frac{4,096}{15,625}$$

The above rules also apply to exponential expressions whose exponents are rational and irrational numbers, although we will not show it here.

The examples above all involve positive numbers. How do we apply the rules when the number is negative? The answer is, the same as with positive numbers, but we have to be careful! For example, $(-3)^2 = (-3)(-3) = 9$, which has the same meaning as $^-3^2 = {}^-3\cdot{}^-3 = 9$, but -3^2 really means $-1\cdot3^2 = -1\cdot3\cdot3 = -9$. This is just a subtlety of how the language of mathematics is interpreted. The parentheses tell us that the sign is grouped with the number, or the super script – symbol (⁻) tells us the number we are raising to a power is negative, so the first two examples are "a negative three squared," while the second example is "negative one times a positive three squared," or "the negative of a positive three squared." Here are some additional examples of the Rules of Exponents with negative numbers.

EXAMPLE 13: Evaluate using the rules of exponents $(-4)^3$

$$(-4)^3 = (^-4)(^-4)(^-4) = (^-64)$$

EXAMPLE 14: Evaluate using the rules of exponents -4^3

$$-4^3 = {}^-1\cdot4\cdot4\cdot4 = {}^-64$$

64

Notice how Examples 13 and 14 are really different problems, but they give the same final answer. This occurs because the exponent is an odd number. This, however, is not true if the exponent is an even number.

EXAMPLE 15: Evaluate using the rules of exponents $(^-2)^6$

$$(^-2)^6 = (^-2)(^-2)(^-2)(^-2)(^-2)(^-2) = 64$$

EXAMPLE 16: Evaluate using the rules of exponents -2^6

$$-2^6 = ^-1 \cdot 2 \cdot 2 \cdot 2 \cdot 2 \cdot 2 \cdot 2 = ^-1 \cdot 64 = ^-64$$

In these last two examples we see that the answers are different. One is positive and the other is negative.

EXERCISES 2.3

Rules of Exponents

Evaluate using the rules of exponents. Write the final answer with exponents. If the exponent is 3 or less also include the multiplied out number.

Basic Problems

1. $2^2 2^3$

2. $\left(6 \cdot 5^2\right)^0$

3. $\left(3^2\right)^3$

4. $\dfrac{15^7}{15^6}$

5. $\left(2^2\right)^4$

6. $(-2)^4$

7. $\left(5^2\right)(5)^1$

8. $(-2)^3$

9. $(-5)^0$

10. $3^2 3^4$

11. $\dfrac{4^5}{4^3}$

12. $\dfrac{6^7}{6^5}$

Intermediate to Advanced Problems

13. -3^4

14. $^-7^2$

15. $^-2^3$

16. -5^3

17. $\dfrac{3^5}{3 \cdot 3^4}$

18. $(-4)^2 (-4)^3$

19. $\dfrac{^-5^7}{^-5^3}$

20. $\left[\left(\dfrac{2}{3}\right)^2\right]^3$

21. $\left(\dfrac{4}{5}\right)^2$

22. $\left(\dfrac{3}{2}\right)^3$

23. $\dfrac{5^2}{5^2}$

24. $\dfrac{3^5}{3^4}$

25. $\left(-\dfrac{4}{7}\right)^2$

26. $\left(\dfrac{15}{19}\right)^1$

27. $(-3)^3 (-3)^3$

28. $\dfrac{(-5)^7}{(-5)^4}$

29. $\left(\dfrac{1}{7}\right)^2$

30. $\left(\dfrac{-3}{2}\right)^3$

31. $\left(\dfrac{21}{100}\right)^0$

32. $\left(\dfrac{1}{6}\right)^2$

33. $\dfrac{4^7}{4^5 4^2}$

34. $\dfrac{5^9}{5^4 5^3}$

2.4 POWERS OF TEN AND NEGATIVE EXPONENTS

Up until now we have only used positive integer exponents. After all, how could a negative exponent have any real meaning? How can you multiply a number by itself a negative number of times? However, if we look more closely at the second rule for working with exponents, we can see that a negative exponent is certainly possible if the exponent of the term in the denominator is larger than the exponent of the term in the numerator. To see this let's study Rule 2 for exponents a little further.

Now, let's look at what happens when we divide an exponent term by another exponent term with the same base, in this case 10, but raised to a higher power, and apply Rule 2 from above e.g.

EXAMPLE 1:

a) $\dfrac{10^2}{10^4}=10^{(2-4)}=10^{-2}$,but $\dfrac{10^2}{10^4} = \dfrac{(\cancel{10}\cdot\cancel{10})}{(10\cdot10\cdot\cancel{10}\cdot\cancel{10})} = \dfrac{1}{(10\cdot10)} = \dfrac{1}{10^2}$,

so $10^{-2}=\dfrac{1}{10^2}$

b) $\dfrac{10^2}{10^6}=10^{(2-6)}=10^{-4}$, but $\dfrac{10^2}{10^6} = \dfrac{(\cancel{10}\cdot\cancel{10})}{(10\cdot10\cdot10\cdot10\cdot\cancel{10}\cdot\cancel{10})} = \dfrac{1}{(10\cdot10\cdot10\cdot10)} = \dfrac{1}{10^4}$,

so $10^{-4}=\dfrac{1}{10^4}$

c) $\dfrac{10}{10^6}=10^{(1-6)}=10^{-5}$, but $\dfrac{10}{10^6} = \dfrac{(\cancel{10})}{(10\cdot10\cdot10\cdot10\cdot10\cdot\cancel{10})} = \dfrac{1}{(10\cdot10\cdot10\cdot10\cdot10)} = \dfrac{1}{10^5}$,

so $10^{-5}=\dfrac{1}{10^5}$

Now we notice that in this case, the final exponent is negative and that the following must be true:

Rule 7: Any base 10 raised to a negative exponent can be written as:

$10^{-n}=\dfrac{1}{10^n}$, or it can also be shown $10^n = \dfrac{1}{10^{-n}}$ *

---

* We introduce this extended rule here, but we will not show any examples on how to explicitly use it. Instead we leave this for the exercises.

The exponent operator has now been extended to include all integers as exponents.

Summary of the Rules of Exponents:

Let "**a**"and "**b**" be any real numbers, and "**n**" and "**m**" be any integers. Then

1. $a^n \cdot a^m = a^{(n+m)}$

2. $\dfrac{a^n}{a^m} = a^{(n-m)}$, provided $a \neq 0$

3. $\left(a^n\right)^m = a^{n \cdot m}$

4. $a^0 = 1$, provided $a \neq 0$

5. $(ab)^n = a^n \cdot b^n$

6. $\left(\dfrac{a}{b}\right)^n = \dfrac{a^n}{b^n}$, provided $b \neq 0$

7. $10^{-n} = \dfrac{1}{10^n}$ or $10^n = \dfrac{1}{10^{-n}}$

EXAMPLE 2: Evaluate using the rules of exponents $\dfrac{10^3}{10^6}$

$$\frac{10^3}{10^6} = 10^{-3} = \frac{1}{10^3} = \frac{1}{1,000}$$

EXAMPLE 3: Evaluate using the rules of exponents $\dfrac{10^5}{10^9}$

$$\frac{10^5}{10^9} = 10^{(5-9)} = 10^{-4} = \frac{1}{10^4} = \frac{1}{10,000}$$

EXAMPLE 4: Evaluate using the rules of exponents $\dfrac{10^3}{10^{-6}}$

$$\frac{10^3}{10^{-6}} = 10^{(3-{}^-6)} = 10^{3+6} = 10^9 = 1,000,000,000$$

EXAMPLE 5: Evaluate using the rules of exponents 10^{-3}

$$10^{-3} = \frac{1}{10^3} = \frac{1}{1,000}$$

A special case of Rule 7 is when n = 1. In this case we have 10^{-1}. We see that this is just another way to write the reciprocal of 10, or $\frac{1}{10}$. We will see this frequently throughout mathematics and the sciences, so we explore one of its more common forms below.

Negative exponents can also be used to extend our place–value exponent notation to include place-values to the right of the decimal point. Consider the place-value chart below.

...	Hundreds	Tens	Ones	.	Tenths	Hundredths	Thousandths	...
...	100	10	1	.	1/10	1/100	1/1000	...

In Section 2.1 we saw that we could write $10 = 10^1$, $100 = 10^2$, $1,000 = 10^3$, etc. Using the idea of a zero or a negative exponent we can now extend this notation further. The first thing we notice is that 1 can be written as 10^0 using Rule 4 for the exponents. We also just saw that 10^{-1} is equivalent to 1/10. In a similar way $1/100 = 10^{-2}$, and $1/1,000 = 10^{-3}$, etc. Thus we can now write our place-value chart using this new notation as follows.

...	Hundreds	Tens	Ones	.	Tenths	Hundredths	Thousandths	...
...	100	10	1	.	1/10	1/100	1/1000	...
...	10^2	10^1	10^0	.	10^{-1}	10^{-2}	10^{-3}	...

EXAMPLE 6:

Write the following decimal numbers in expanded exponent notation form.

a)   239.357

$$239.357 = 200 + 30 + 9 + 3/10 + 5/100 + 7/1,000$$
$$= 2 \times 100 + 3 \times 10 + 9 \times 1 + 3 \times 1/10 + 5 \times 1/100 + 7 \times 1/1,000$$
$$= 2 \times 10^2 + 3 \times 10^1 + 9 \times 10^0 + 3 \times 10^{-1} + +5 \times 10^{-2} + 7 \times 10^{-3}$$

b)   5,480.21

$$5,480.21 = 5000 + 400 + 80 + 2/10 + 1/100$$
$$= 5 \times 1000 + 4 \times 100 + 8 \times 10 + 0 \times 1 + 2 \times 1/10 + 1 \times 1/100$$
$$= 5 \times 10^3 + 4 \times 10^2 + 8 \times 10^1 + 0 \times 10^0 + +2 \times 10^{-1} + 1 \times 10^{-2}$$

EXERCISES 2.4

Simplify each expression by removing the negative exponent. Leave the final answer in exponent form.

1. 10^{-1}

2. 10^{-3}

3. 10^{-2}

4. $\dfrac{1}{10^{-2}}$

5. $\dfrac{1}{10^{-5}}$

6. $\dfrac{1}{10^{-1}}$

7. $\dfrac{1}{10^{-8}}$

8. $\dfrac{10^5}{10^8}$

9. $\dfrac{10^2}{10^6}$

10. $\dfrac{10^2}{10^{-2}}$

11. $\dfrac{10}{10^{-3}}$

12. $\dfrac{10^{-4}}{10^{-4}}$

13. $\dfrac{10^{-3}}{10^3}$

14. $\dfrac{10}{10^{-1}}$

15. $10^5 \times 10^{-4}$

16. $10^{-6} \times 10^{-7}$

17. $10^{-1} \times 10^4$

18. $10^{-2} \times 10^{-6}$

19. $10^5 \times 10^{-8}$

20. $10^{-3}\,10^{-1}$

21. $10^{-6}\,10^8$

22. $10^2\,10^{-7}$

23. $\left(10^{-6}\right)^{-2}$

24. $\left(10^{-4}\right)^{-3}$

25. $\left(10^5\right)^{-3}$

26. $\left(10^{-3}\right)^6$

27. $\left(10^8\right)^{-2}$

Write the following decimal numbers in expanded exponent notation form.

28. 4.123

29. 0.017

30. 21.005

31. 9.987

32. 0.7843

33. 5.3827

34. 0.00305

35. 30.00103

2.5 RADICALS OF REAL NUMBERS

Just as addition and multiplication had inverse operations, the same is true for the exponent (power) operator. The inverse operator of an exponent is called a radical, or root operator.

For example, consider the exponent operator defined as the square of a number, i.e.

$$3^2 = 9$$

We started with the base 3 and squared it (exponent of 2) to get the number 9.

The inverse operation starts with the number 9 and gets us back to the base of 3. The inverse operator of the squaring operator is the square root operator. It undoes the squaring operation.

More specifically, the square root operator asks the question: what number multiplied by itself gives you 9? The answer is 3.

Mathematically, we write this as

$$\sqrt{9} = 3$$

The symbol is called a **radical symbol** and the number inside the radical symbol is called the **radicand**.

Associated with every exponent, there is a root (radical) that will undo it. In this book we only focus the exponent 2, and its inverse operation square root. There are also roots to undo a^3, a^4, etc, but we will save that for your next algebra course.

EXAMPLE 1: Evaluate $\sqrt{25}$ **Solution:** $\sqrt{25} = 5$ since $5 \cdot 5 = 25$

EXAMPLE 2: Evaluate $-\sqrt{81}$ **Solution:** $-\sqrt{81} = -9 = {}^-9$ since $9\cdot9=81$, and the minus sign in front tells us to take the negative of the result.

NOTE: Here is an example where the two symbols, the minus sign, –, used for subtraction or the action of taking the negative of a number, and the negative number symbol, ⁻, used to identify a negative number, can be used interchangeably. They still mean different things, but they give the same result in the end. Thus – 9 gives the same numerical value as the number ⁻9.

EXAMPLE 3: Evaluate $\sqrt{0.09}$ **Solution:** $\sqrt{0.09} = 0.3$ since $0.3\cdot0.3=0.09$

EXAMPLE 4: Evaluate $\sqrt{7}$. Round your answer to the nearest hundredths.
Solution: $\sqrt{7} \approx 2.65$

NOTE: 7 is not a perfect square, so we must find an approximate solution using a calculator. Since $2.65\cdot2.65=7.0225\neq7$, 2.65 is only an approximate solution. We represent the occurrence of an approximate solution using the symbol, $\approx$. We need to be skilled at finding the exact solution when possible, but also recognize when and how to find an approximate solution.

EXAMPLE 5: Evaluate $^-\sqrt{19}$. Round your answer to the nearest tenths.
Solution: $^-\sqrt{19} \approx {}^-4.4$

EXAMPLE 6: Evaluate $\sqrt{\dfrac{16}{49}}$ **Solution:** $\sqrt{\dfrac{16}{49}} = \dfrac{\sqrt{16}}{\sqrt{49}} = \dfrac{4}{7}$ since $4\cdot4=16$ and $7\cdot7=49$

We should point out that sometimes the inverse operation is not true for all real numbers. There are exceptions we have to be careful about. In particular, the $\sqrt{^-4}$, is undefined. This is because, by our definition, there is no real number that when multiplied by itself will give us a ⁻4. In fact, no real number, when multiplied by itself can ever equal a negative number. Thus, the square root of any negative number is undefined for real numbers.

NOTE: In this discussion on roots and radicals we have not mentioned an aspect of their usage that will be very important in later mathematics courses. For example, when we said a square root asks the question; what number multiplied by itself equals the radicand?, we did not mention the other solution, i.e. $\sqrt{4}$ = ⁻2, since ⁻2 × ⁻2 = 4. The negative root is never mentioned, nor is it ever given on the calculator. Why? Typically the root operation had its origins in geometry where you would never have a negative solution, since length, area, and volume are never negative quantities. Also, when the root was first used in algebra, negative numbers were not even accepted as numbers themselves. Finally, most calculations today with roots yield positive results (again we are are often calculating sizes of things), since negative root interpretations are a rare occurrence. This is why to avoid any confusion, the calculator only gives the positive root. This is also called the **Principal Root**. In this course we will only consider the Principal or positive root, but in later mathematics courses we will revisit this topic further when needed.

EXERCISES 2.5

Square Roots of Real Numbers

Simplify:

1. $\sqrt{36}$

2. $\sqrt{121}$

3. $\sqrt{144}$

4. $\sqrt{100}$

5. $^-\sqrt{49}$

6. $\sqrt{169}$

7. $^-\sqrt{196}$

8. $\sqrt{^-25}$

9. $^-\sqrt{^-81}$

10. $\sqrt{^-16}$

11. $-\sqrt{16}$

12. $\sqrt{\dfrac{4}{9}}$

13. $\sqrt{\dfrac{441}{25}}$

14. $\sqrt{\dfrac{225}{289}}$

If possible, use your calculator to find the square root of the following numbers. Round your answer to the nearest hundredths.

15. $\sqrt{20}$

16. $\sqrt{73}$

17. $^-\sqrt{45}$

18. $\sqrt{^-6}$

19. $\sqrt{\dfrac{4}{5}}$

20. $^-\sqrt{\dfrac{9+5}{3+4}}$

21. $\sqrt{0.04}$

22. $\sqrt{0.006}$

23. $\sqrt{2.39}$

24. $\sqrt{13.35}$

2.6 CALCULATOR CALCULATIONS AND ORDER OF OPERATIONS WITH EXPONENTS

In addition to finding the exact values of numerical expressions, you often have to find approximate solutions. This will require the use of a calculator. Oftentimes, a calculator will be used by a student, even though the result can be found quite easily without a calculator. The problem with this is that students frequently don't know how to properly enter expressions into a calculator and, as a consequence, get the wrong answer.

In this section, we consider some common calculator mistakes students make and discuss ways to avoid them. At the same time we wish to make you more proficient using a calculator, but using it in a more knowledgeable way.

Examples of Common Calculator Mistakes.

1. Consider the problem $\dfrac{3+\sqrt{4}}{2}$.

On a calculator, some students would incorrectly enter,

$$3 + \text{sqrt}(4)/2$$

and, as a result, get the wrong answer of 4! The problem with this calculation is that you cannot divide by 2 before you add the 3 to the sqrt(4). What you have actually calculated is:

$$3+\dfrac{\sqrt{4}}{2}$$

which is not the same as the original problem. The calculator is programmed to follow PEMDAS, so you have to be careful to correctly transcribe the problem into the calculator, so that does not change the original value of the expression.

Now if we follow PEMDAS, we would simplify this as follows.

$\frac{3+\sqrt{4}}{2} = \frac{3+2}{2} = \frac{5}{2}$, or as could be written on a calculator with parentheses:

(3 + sqrt(4))/2 In this case, we use parentheses to force the calculator to compute 3 + sqrt(4) first, before we divide by 2

2. Consider the problem $\frac{2}{7-5}$

On a calculator, some students would incorrectly enter,

$$2/7 - 5$$

and, as a result, get the wrong answer of ⁻4.71428 ...! The problem with this calculation is that you cannot divide by 7 before you subtract the 5 from it. What you have actually calculated is

$$\frac{2}{7} - 5$$

which is not the same as the original problem. The calculator is programmed to follow PEMDAS, so you have to be careful to correctly transcribe the problem into the calculator, so that it does not change the original value of the expression.

Now if we follow PEMDAS we would simplify this as follows.

$$\frac{2}{7-5} = \frac{2}{2} = 1 ,$$

or as could be written on a calculator with parentheses 2/(7–5).We again can use parentheses to force the calculator to evaluate 7–5 first.

3. $5 - 2 \times 7$

In this problem, some may incorrectly work the problem on the calculator in steps working from left to right. Thus, they may start by putting 5 – 2 into the calculator, getting 3, and then multiply by 7. This is incorrect because the order of operations says we must first multiply the 2 and the 7 and then subtract that result from 5, because multiplication must be done before subtraction, or
$5 - 14 = ⁻9$

4. -2^2

In this problem, confusion arises as to whether this is a negative two squared, or the negative of two squared. This is why we introduced the notation that distinguishes the difference, -2^2 versus $^-2^2$. This is actually the negative of two squared which is $^-4$, and not $+4$.

When entering this into a calculator it is important to use the negative symbol button, usually identified by $(-)$, and not the subtraction symbol, usually written without the parentheses as, $-$. Oftentimes if you use the wrong key, you will get an error message. This is the reason why. Subtraction requires a number in front of it, while the negative symbol does not. As a result, this expression cannot be entered as it is written on a calculator. You must first square 2 and then take its negative. Again, a very common mistake when students use their calculators.

Many of the common calculator mistakes have to do with applying the correct order of operations and having a good understanding of which operation to do, and what order to do it in, especially when several operations are given together in an expression. In Chapter 1, we worked order of operations problems using the basic arithmetic operations. We now revisit order of operations with the exponent operator also included.

Order of Operations With Exponents

In the following examples, we consider order of operation problems that involve exponents.

EXAMPLES:

1. $3 - 4^2$

In words this is actually "three, subtracting, four squared," or $3 - 16 = {}^-13$

2. $-4^2 + 36 - 2\sqrt{25}$

In words, this expression is "take the negative of 4 squared, add 36, and subtract 2 times the square root of 25"

We now follow PEMDAS, and start with the square and the square root (exponents).
$$-16 + 36 - 2 \times 5$$
Next we do the multiplication,
$$-16 + 36 - 10$$
and finally we add/subtract left to right – whichever operation comes first.
$$20 - 10$$
$$= 10$$

3. $1 - (4 - 5)^3 + 15 \div 5 \times 2$

In words this expression is "one and subtract the quantity four subtracting five raised to the third power and add fifteen divided by five and then multiplied by 2."

We now follow PEMDAS, and start with the parentheses.
$$1 - ({}^-1)^3 + 15 \div 5 \times 2$$

Then the exponent.

$$1-{}^-1+15\div5\times2$$

Then do the division because it comes before multiplication as we move from left to right

$$1-{}^-1+3\times2$$

Next we multiply

$$1-{}^-1+6$$

Finally we add and subtract from left to right

$$1-{}^-1+6$$
$$=\ 1+1+6$$
$$=\ 2+6$$
$$=\ 8$$

EXERCISES 2.6

Simplify the following expressions

1. $5-22$

2. $-3+({}^-3)^2$

3. $(6-3)^3-2^2$

4. $-(3-5)^2-2+3(2+1)$

5. $16-13(2-3)$

6. $(\sqrt{25}-7)^2-3+2$

7. $12-2(6-8)^3$

8. $5+\sqrt{49}-7(2-1)^2$

9. $2\sqrt{36}-24+3(7-5)^2$

10. $-2^4+13-(3-5)$

11. $^-3^2-5+\sqrt{16}$

12. $-\sqrt{4}+3\sqrt{9}-3^3$

13. $(13-15)^2-3+5\sqrt{25}$

14. $2+(3-5)^3-21\div7\times2$

15. $^-2-(7-5)^2+2\times5\div10$

16. $-(3-5)^2-2+3(2+1)$

17. $16-13(2-3)$

18. $(7-\sqrt{25})-3+2$

19. $5+\sqrt{49}-7(1-2)$

20. $2\sqrt{36}-24+3(5-7)$

21. $-\sqrt{4}+3\sqrt{9}-3$

22. $(13-15)-3+5\sqrt{25}$

Simplify the following expressions and verify you obtain the same answer using your calculator:

23. $\dfrac{120-100}{4}$

24. $2+1.96\left(\dfrac{12}{\sqrt{9}}\right)$

25. $\dfrac{9^2+3^2+5^2+2^2}{3}$

26. $\dfrac{10-6}{2/\sqrt{25}}$

27. $(1)(0.7)+(2)(0.2)+(3)(0.1)$

28. $\dfrac{2^2-3^2+4^2-5^2}{4}$

Chapter 2 Practice Test

Rewrite using exponent notation
1. $10 \times 10 \times 10 \times 10 \times 10$

2. $(10)(10)(10)(10)(10)(10)(10)$

Simplify by writing without exponents
3. 10^8

4. -10^2

5. $10^\wedge 4$

Write in expanded form using exponent notation.
6. 56,098

7. 873,152

Rewrite the expressions using exponents
8. $5 \cdot 5 \cdot 5 \cdot 5$

9. $^-7 \cdot {}^-7 \cdot {}^-7 \cdot {}^-7 \cdot {}^-7$

10. $\left(\dfrac{3}{5}\right) \cdot \left(\dfrac{3}{5}\right) \cdot \left(\dfrac{3}{5}\right) \cdot \left(\dfrac{3}{5}\right)$

Rewrite without exponents, and multiply out
11. 5^3

12. $(^-3)^4$

13. -2^4

14. $^-4^4$

15. $\left(\dfrac{3}{5}\right)^3$

Evaluate using the rules of exponents.
16. $3^2 3^4$

17. $\left(4^5\right)^6$

18. $\dfrac{7^9}{7^7}$

19. $\left(\dfrac{-3}{5}\right)^2$

20. -3^0

Simplify each expression by removing the negative exponent.
21. 10^{-1}

22. $\dfrac{1}{10^{-3}}$

23. $\dfrac{10^{-2}}{10^{-3}}$

Write the following decimal numbers in expanded exponent notation form.
24. 12.17

25. 45.375

Simplify:
26. $\sqrt{64}$

27. $^-\sqrt{36}$

28. $\sqrt{144}$

29. $\sqrt{^-16}$

Use your calculator to find the square root of the following numbers. Round your answer to the nearest hundredths.
30. $\sqrt{33}$

31. $\sqrt{40}$

Simplify the following expressions

32. $7^2 - 13$

34. $2 + (3-5)^3 - 21 \div 7 \times 2$

35. $^-5 + (6-9)^3 - 4 \times 5 \div 6 \times 2$

33. $^-1^2 + 3 - \sqrt{49}$

Simplify the following expressions and verify you obtain the same answer using your calculator:

36. $\dfrac{75-30}{9}$

37. $\dfrac{(20-15)^2}{15} + \dfrac{(10-7)^2}{7}$

38. $\dfrac{5.3-6}{1.8/\sqrt{(9)}}$

CHAPTER 3

Variables, Terms, Expressions and Polynomials

This is where arithmetic ends and algebra begins. It is with the concept of the variable that the true power of arithmetic becomes unleashed through the algebra. In this chapter, we introduce the variable concept, along with related concepts. We show how all the rules of arithmetic apply to these new representations of numbers. We then introduce a particular form of variable expressions called polynomials, and show how to work with them. We should also point out that when arithmetic ends and algebra begins, it becomes more conducive (easier) to use the alternative notation of a negative number. Thus, in some of the problems we will use the subtraction symbol, –, in front of a number, or a variable, and not always use the super scripted symbol, ⁻, to represent a negative quantity. We will, however, not do this right away, but will do this at times and try to point out on those occasions when we do this. In algebra, however, you should get used to this dual use of the minus symbol, because it is essentially the only notation that is used in algebra. The super scripted symbol is virtually never used in algebra. It was initially helpful to distinguish the operation of subtraction, from that of a negative number, but when it comes to algebra, it becomes very difficult to continue like this.

3.1 VARIABLES, TERMS, AND EXPRESSIONS

The word "vary", obviously describes something that changes. In algebra, this **something** is a number. Now, instead of the fixed numbers of arithmetic, we want to talk about numbers more abstractly. We don't want to consider the exact number two, three, one-half, etc. Instead we wish to talk about numbers more generally.

For example, if you wanted to discuss the dimensions of a room, you would say the room has a certain length and width, and that together, the length and width provide the dimensions of the room. You could then compare rooms of different sizes, depending upon what their respective lengths and widths were. In talking about the various lengths and widths, you are actually using the fundamental concept of algebra called a variable. Since, we may not know the precise length or width of the room ahead of time, we give these unknown values names, such as length and width. These values will vary depending upon the room we are talking about. Now, rather than using the whole name to designate length and width, in mathematics we typically use the abbreviations L for length, and W for width to represent these varying numerical quantities.

Think about other numerical quantities that you work with, without even knowing their values – such as someone's height or weight, the temperature during the day, the price of a shirt at various stores, the cost of a particular item of food at different supermarkets, or the value of a particular stock on any given day. You are constantly using the concept of a variable without even thinking about it. Any time you think about something that can be quantified numerically, you are taking into consideration that the value may vary. This is the concept of a variable. Any numerical quantity that can change can be thought of as a variable.

In this chapter, we wish to build upon this intuitive concept and extend it so that it becomes much more useful as a predictive tool. Wouldn't it be nice to see a trend in how a price was changing, and be able to react or do something about it?

Doctors know that it's helpful to measure and compare various chemical levels in peoples' bodies to ascertain what levels of which chemicals are good, and which are bad. Without the concept of variable, our health would deteriorate! We wouldn't be able to keep track of changes in our bodies.

What makes algebra very useful, but at the same time more confusing, is when we change the focus from what the variable actually represents – such as time, distance, money, or temperature, and instead focus on its properties as a variable. Thus, we strip away the **physical** meaning and are then able to focus on how this new abstract concept of a variable can be manipulated to obtain new and interesting results, which can then be re-applied and re-interpreted in terms of its original **physical** meaning at a later date, should we so desire. This is both a curse and a blessing.

The convention has been to choose letters near the end of the alphabet, such as x, y, z, to represent variables that do not have a concrete meaning assigned to them. Variables that do have a specific meaning are sometimes assigned the value of the first letter of the word used to describe them – such as L for length, W for width, t for time, A for amount or area, V for volume, C for cost or circumference, or P for profit or pressure.

In this chapter, we will investigate the arithmetic properties of these unassigned/abstract variables. To simplify the explanation, we will only consider problems involving a single variable. In the later part of the book, we will begin to look at problems with two or more variables, but for now it is best to develop this new language using only a single variable.

Definition: A **Variable** is a letter used to represent an unspecified number.

EXAMPLES: x, y, z, a, b, c, N, W, etc.

As with any number, we can add, subtract, multiply and divide variables. Thus, if we have a variable, x, then we can multiply x by a number, say three, and obtain a new number that can be written as either $3 \cdot x$, or $3 \times x$ (which is rarely used, since the multiplication symbol looks too much like the variable "x."), or $3 * x$, or as we will prefer to write it, 3x. You should make the distinction that this is really 3 times an unknown quantity x, and not simply a new unknown 3x quantity. The 3 and the x are separable.

Furthermore, addition, subtraction, and division can be represented as: $x + 2$ or $2 + x$, $x - 2$ or $2 - x$, $x \div 2$ or x/2, or $\frac{x}{2}$, respectively.

To help our discussion move forward, we need to begin to start attaching names to these new quantities we begin to develop. By combining variables with arithmetic, we create what we call terms and expressions.

Definition: A **term** is either a single number, or a product of numbers and variables.

EXAMPLES: $^-13$, $\dfrac{7}{12}$, $2x$, $\dfrac{3}{5}x$, or $3x^2 = 3 \cdot x \cdot x$

Definition: An **expression** is simply a sum of terms.

EXAMPLES:
$2 + 3x$,
$4 - 3x$ which can be rewritten as $4 + (-3x)$, or $4 + {}^-3x$,
$2x - 5 + x$, which can be rewritten as $2x + {}^-5 + x$,

$\dfrac{1}{3}x + 4 - 3x$ which can be rewritten as, $\dfrac{1}{3}x + 4 + ({}^-3x)$

$\dfrac{x}{2} + 7 - 3 + 4x$ which can be rewritten as $\dfrac{x}{2} + 7 + ({}^-3) + 4x$

Definition: The **numerical coefficient** (also called the coefficient) of a term is the fixed-number part of the term.

EXAMPLES:
The term $^-11$ has a numerical coefficient of $^-11$,
The term $6x$, has a numerical coefficient of 6,
The term 25, has a numerical coefficient of 25,
The term $\dfrac{3}{5}x$, has a numerical coefficient of $\dfrac{3}{5}$, or
The term $\dfrac{x}{3}$, has a numerical coefficient of $\dfrac{1}{3}$, since $\dfrac{x}{3}$ can be written as $\dfrac{1}{3}x$

We can combine terms and expressions using the operations of arithmetic, just as we do with specific numbers. The only difference is that we now have to define a new concept called **like terms**.

Definition: Like terms (also called similar terms) are two terms with the same exact variable part, but they may have a different numerical coefficient.

EXAMPLES:
^-5x and $\dfrac{x}{7}$ are like terms with coefficients of $^-5$ and $\dfrac{1}{7}$ respectively
$3x^2$ and $^-x^2$ are like terms with coefficients 3 and $^-1$ respectively

78

$\frac{2}{3}x^2$ and $8x^2$ are like terms with coefficients $\frac{2}{3}$ and 8 respectively

To add or subtract terms and expressions, we simply add or subtract the numerical coefficients of the like terms.

EXAMPLES: Simplify each expression by combining like terms

a) $3b + 7b$ 　　　　　　　　Solution: $3b + 7b = (3+7)b = 10b$

b) $-z + 4z$ 　　　　　　　　Solution: $-z+4z = -1z+4z = {}^-1z+4z = ({}^-1+4)z = 3z$

c) $-6n - 3n$ 　　　　　　　Solution: $-6n-3n = {}^-6n + {}^-3n = ({}^-6+{}^-3)n = {}^-9n$

d) $5x - 13 + 3x - 4$ 　　　Solution: $5x-13+3x-4 = (5+3)x+({}^-13+{}^-4)$
$$= 8x+{}^-17 = 8x-17$$

Note: In the last part we are writing the sum of the ${}^-17$ as a difference with $+17$.

e) $2x^2 - 4x + 6 - 3x^2 - 5x + 1$ 　　Solution: $2x^2-4x+6-3x^2-5x+1$
$$= (2+{}^-3)x^2+({}^-4+{}^-5)x+(6+1)$$
$$= {}^-1x^2+{}^-9x+7 = -x^2-9x+7$$

Note: In the last part we are writing the sum of the ${}^-9x$ as a difference with $+9x$.

f) $x^3 + x + 5x^3 - x^2 + 7x - 16$ 　Solution: $(1+5)x^3 - x^2 + (7+1)x - 16 = 6x^3 - x^2 + 8x - 16$

Note: The negative superscript for ${}^-16$ was omitted in the solution here, due to lack of like terms for ${}^-x^2$ and ${}^-16$.

EXERCISES 3.1

Combining Like Terms

Simplify each expression by combining like terms

1. $5x - 3x$

2. $13n + 12n$

3. $-5x + 3x$

4. $5a + 3a$

5. $x + 4x$

6. $13x - x$

7. $6v - 3v + 3$

8. $6r + 3 + 2r - 7$

9. $-12x - 3x$

10. $-16n - 9n$

11. $-2x + 13 + 4x$

12. $6t - 5 + 3t - 7$

13. $-32 + 16b + 32 - 16b$

14. $9c - 13 + 13c$

15. $-3x + 2 - 3x + 2$

16. $5x^2+9x-12+13x^2+7x+9$ 20. $12x-4x^2-13+8x^2-11x$ 24. $9x^3-6x^2-1+7x^3+x^2-3x$

17. $-2x^3+5x^2+7x-3x^3-x^2$ 21. $x^3+x^2-7+6x^3-3x^2$

18. $6x^2-9x-4x^3-2x^2-x+11$ 22. $3x^2-6x+2+x^2-9x-7$ 25. $\frac{1}{2}x^2-\frac{2}{3}x-3+\frac{2}{3}x^2+\frac{5}{3}x+9$

19. $7x^2-2x-9-7x^2+5x-7$ 23. $-4x^2-9+x^2-7x+4$

26. $\frac{1}{2}x^3+\frac{1}{3}x^2-7x-3+\frac{3}{2}x^3-\frac{2}{3}x^2+5x+4$ 29. $-\frac{3}{5}x^2+\frac{7}{9}x-13+\frac{3}{2}x^3-\frac{2}{3}x^2-\frac{5}{9}x+24$

27. $\frac{4}{9}x^2-\frac{2}{7}x-\frac{1}{6}+\frac{2}{3}x^3-\frac{2}{3}x^2+\frac{2}{7}x+\frac{4}{9}$ 30. $\frac{1}{2}+\frac{5}{6}x^2-7x-x^3-\frac{3}{2}x^3-\frac{2}{3}x^2+2x+4$

28. $\frac{1}{5}x^3+\frac{1}{4}x^2-2x-\frac{1}{3}+\frac{3}{10}x^3-\frac{2}{3}x^2+9x+\frac{4}{3}$

3.2 THE DISTRIBUTIVE PROPERTY

The next arithmetic operation involving variables we introduce, is multiplying expressions by numbers. This is called the distributive property, and will be used often when working problems involving variables. We already introduced the distributive property for fixed numbers. In this section we generalize the concept and apply it to variable expressions.

To multiply an expression by a fixed number, you must "distribute" the constant to all terms in the expression, and then multiply it with each of the separate terms in the expression.

Examples of the distributive property:

Multiplying a two term expression, containing positive coefficients, by a positive number:
$$2(x+3)=2\cdot x+2\cdot 3=2x+6$$

Multiplying a two term expression, containing positive coefficients, by a negative number:
$$^-2(x+1)=^-2\cdot x+^-2\cdot 1=-2x+^-2=-2x-2$$

Multiplying a two term expression, containing positive and negative coefficients, by a negative number:
$$^-4(4x-3)=^-4(4x+^-3)=^-4\cdot(4x)+^-4\cdot^-3=^-16x+12=-16x+12$$

Multiplying a two term expression, containing positive coefficients, by a negative sign:

$$-(x+1)=^-1(x+1)=^-1\cdot(x)+^-1\cdot(1)=-x+^-1=-x-1$$

In General,

$$a(bx+c)=abx+ac$$

or

$$(bx + c)a = bxa + ca$$

$$= abx + ac, \quad \text{by the Commutative Property of Multiplication}$$

EXAMPLES: Simplify each expression.

a) $3(x+3)$ Solution: $3(x+3) = 3\cdot x + 3\cdot 3 = 3x + 9$

b) $^-2(2a-4)$ Solution: $^-2(2a-4) = {}^-2(2a+{}^-4) = {}^-2\cdot 2a + {}^-2\cdot{}^-4 = {}^-4a + 8$

c) $(3n-7)4$ Solution:

$$(3n-7)4 = (3n+{}^-7)4 = 3n\cdot 4 + {}^-7\cdot 4 = 12n + {}^-28 = 12n - 28$$

Note: In the last part we are writing the sum of the $^-28$
as a difference with +28.

d) $3x + 2(x-3)$ Solution: $3x + 2(x-3) = 3x + 2(x+{}^-3) = 3x + 2\cdot x + 2\cdot{}^-3$

$$= 3x + 2x - 6$$

Note: In the last part we are writing the sum of
the $^-6$ as a difference with +6.

$$= 5x - 6 \qquad \text{Combine like terms}$$

e) $x + 5(x-6) + 3$ Solution: $x + 5(x-6) + 3 = x + 5(x+{}^-6) + 3$

$$= x + 5x + 5\cdot{}^-6 + 3$$
$$= x + 5x + {}^-30 + 3$$
$$= 6x - 27 \quad \text{Combine like terms}$$

Note: In the last part we are writing the sum of
the $^-27$ as a difference with +27.

f) $-4x - 3(x-2) + 5$ Solution: $-4x - 3(x-2) + 5 = -4x + {}^-3(x+{}^-2) + 5$

$$= {}^-4x + {}^-3x + {}^-3\cdot{}^-2 + 5$$
$$= {}^-7x + 6 + 5$$
$$= -7x + 11 \quad \text{Combine like terms}$$

g) $x - (3-x) - 4$ Solution: $x - (3-x) - 4 = x + {}^-1(3+{}^-1x) + {}^-4$

$$= x + {}^-1\cdot 3 + {}^-1\cdot{}^-1x + {}^-4$$
$$= x + {}^-3 + x + {}^-4$$
$$= 2x - 7 \quad \text{Combine like terms}$$

h) $3(x^2+6x-8)$ Solution: $3(x^2+6x-8) = 3x^2+3\cdot6x+3\cdot{}^-8$

$$= 3x^2+18x+{}^-24$$
$$= 3x^2+18x-24$$

EXERCISES 3.2

Using the Distributive Property
Simplify each expression
Basic Problems

1. $-5(3x+2)$

2. $4(5+4x)$

3. ${}^-4\cdot(2x+9)$

4. $(5a+2)6$

5. $7(-b+5)$

6. $-5(2a-3)$

7. $(3x+2)\cdot9$

8. $(-6n-4)\cdot({}^-2)$

9. $(2-x)3$

10. $3(3r+5)$

11. $-7(3x+7)$

12. $-8(1-4v)$

13. $-5\cdot(1-c)$

14. $(1+x)8$

15. $(4-r)13$

16. $\dfrac{2}{3}(6x+12)$

17. $-\dfrac{1}{5}(-x+10)$

18. $-4(\dfrac{3}{8}x+\dfrac{1}{12})$

19. $\dfrac{2}{7}(-42x+\dfrac{21}{4})$

Intermediate and Advanced Problems

20. $4x+3(2x-6)$

21. $-x+5(x+3)$

22. $11x+8(2x+1)$

23. $8x-(x-3)$

24. $6x-4(3x-2)$

25. $-7x+2(3-x)$

26. $13-(2-4x)+3x$

27. $-8+5(x-5)-5x$

28. $-16+6(x+3)-5x$

29. $-2x+4(3-2x)-17$

30. $-(x+3)+2x-13$

31. $3(2-x)+6-4x$

32. $4(1-x)-9+4x$

33. $-2(x-5)+8-13x$

34. $4x-3(2-3x)-1-7x$

35. $-\dfrac{4}{5}(10x-15)+2x$

36. $\dfrac{2}{3}(-12x-18)-19$

37. $-3x-\dfrac{1}{6}(-18x-24)$

38. $-\dfrac{5}{9}(-45x-\dfrac{18}{5})+\dfrac{1}{2}x$

39. $4(x^2-3x+5)$

40. $6(2x^2+4x-2)$

41. ${}^-2(3x^2-4x-5)$

42. ${}^-5(x^2-x-6)$

43. $\dfrac{2}{3}(6x^2-12x+9)$

3.3 POLYNOMIALS

In the last section, we performed some basic operations on expressions with variables, and we saw that some fairly complicated looking expressions are possible. In this chapter we begin to classify these expressions into groups that have similar properties. The first group of expressions is called polynomials. We will define what polynomials are, and how to manipulate them under the rules of arithmetic in this chapter. The goal is to get comfortable working with them, since they are some of the most important expressions in algebra. Polynomials are very useful when we construct mathematical models to approximate important phenomena in the world, so we need to be well versed in their language.

Definition: A **polynomial** is a sum of terms whose variables have exponents that are whole numbers, and are not located in the denominator.

> In general, we can write an arbitrary polynomial with unspecified coefficients as:
>
> $$a_n x^n + a_{n-1} x^{n-1} + a_{n-2} x^{n-2} + ... + a_1 x + a_0$$
>
> This looks like a fairly complicated representation of a polynomial, and at this level of algebra it is. If you don't understand it, or if you find it very confusing don't worry, you don't have to follow it. For the more daring among us, the n represents a whole number, and the a_n, a_{n-1}, a_{n-2}, a_1, and a_0 are just numerical coefficients that are associated with an exponent of the variable that is the same as the subscript. The subscript is just part of the name of the coefficient, and has no other more complicated meaning.

Examples of Polynomial Expressions:

$3x - 1$, $5x^2 - 7$, $3x^2 - 2x + 17$, $-x^3 + 4x^2 - 9x + 6$, $\dfrac{23}{35}$

$23x^8 + 14x^5 - 3$, $\dfrac{2}{3}x^2 - \dfrac{7}{5}x + \dfrac{1}{2}$, which can also be written as $\dfrac{2x^2}{3} - \dfrac{7x}{5} + \dfrac{1}{2}$

Non-Polynomial Expressions:

$\dfrac{1}{x}$, $5x - 3 + \dfrac{7}{x^3}$, $\dfrac{2x - 3}{x + 1}$, $7x^4 - \dfrac{3}{2x^3} + \dfrac{4}{x}$, $\sqrt{x - 3}$, $2x^{-2} + x^{-1} + 7$

<u>Some Terminology</u>

Polynomials are distinguished between each other by two methods. The first is by the number of terms that a polynomial has. The second, as we shall show below, is related to the exponent of the variable terms in the polynomial.

The Number of Terms in a Polynomial

Let's start simple. If a polynomial has only one term it is called a **monomial** ("mono" for "one"), if it has two terms it is called a **binomial** ("bi" for "two"), if it has three terms it is called a **trinomial** ("tri" for "three"), and if it has four or more terms it is referred to as a **polynomial** ("poly" for "many"). We should also note that we can also call a monomial, binomial, or a trinomial, a polynomial too. They are ALL polynomials!

Examples of each type:

Monomials: $2x$, $\quad$ 32, $\quad$ $\dfrac{24x^7}{5}$, $\quad$ $\dfrac{4}{3}\pi x^2$

Binomials: $2x - 1$, $\quad$ $x^3 + 5x$, $\quad$ $1 - x^{12}$, $\quad$ $\sqrt{2}x^2 - 3x$

Trinomials: $2x^2 - 5x + 3$, $\quad$ $x^4 + x^2 + x$, $\quad$ $x - 6x^3 + \pi$, $\quad$ $x^2 - 4x + 2$

The Degree of a Polynomial

Another way to characterize a polynomial is in terms of its largest exponent. The most basic polynomial you can write is one that is constant (a simple number by itself). This is called a zeroth-degree polynomial, or polynomial of degree 0.

Examples of Polynomials of Degree 0:

16, $\quad$ -32, $\quad$ π, $\quad$ $\dfrac{5}{9}$, $\quad$ $\sqrt{2}$

A polynomial with an x term with no exponent (really an exponent of one, x^1, but we don't write the one), is called a first degree polynomial. This type of polynomial is called a linear polynomial (**linear** for short) and can be written as:

$$ax + b,$$

where a and b are numerical coefficients, and can be any real number, except that $a \neq 0$. This polynomial is also said to be a polynomial of degree 1.

Examples of Linear Polynomials:

$3x - 1$, $\quad$ $\dfrac{2}{5}x$, $\quad$ $\sqrt{3}x + 7$, $\quad$ $-6x + \sqrt{2}$, $\quad$ $\pi x - 4$

Higher Degree Polynomials

If the largest exponent is a two, the polynomial is called a quadratic, or 2nd degree polynomial.
If the largest exponent is a three, the polynomial is called a cubic, or 3rd degree polynomial.
If the largest exponent is a four, the polynomial is called a quartic, or 4th degree polynomial.
A 5th degree polynomial is a quintic, etc.

EXAMPLES: Identify the polynomials by their degree and number of terms.

1. $3x - 2$ $\qquad\qquad\qquad\qquad$ A first degree binomial

2. $4z^5$ $\qquad\qquad\qquad\qquad$ A fifth degree monomial

3. $16w^2 - 4w + 3$ $\qquad\qquad\qquad$ A second degree (quadratic) trinomial

4. $2a^7 - 3a^2 + 4a$ $\qquad\qquad\qquad$ A seventh degree trinomial

5. $4y^5 - 3y^3 - 5y^2 + 2y - 1$ $\qquad\quad$ A fifth degree (quintic) polynomial

6. $3z^9 - 2z$ A ninth degree binomial

7. -6 A zeroth degree monomial

8. $14b^4 - 3b + 1$ A fourth degree (quartic) trinomial

9. $8x^3 - 27$ A third degree (cubic) binomial

10. $3x^6 - 2x^5 + 4x^4 + 3x^2 - 10$ A sixth degree polynomial

The Leading Coefficient of a Polynomial

The numerical coefficient of the term with the highest degree in a polynomial is called the **leading coefficient**.

EXAMPLES: Identify the leading coefficient of the polynomials.

1. $3x - 2$ The leading coefficient is 3

2. $4z^5$ The leading coefficient is 4

3. $16w^2 - 4w + 3$ The leading coefficient is 16

4. $2a^7 - 3a^2 + 4a$ The leading coefficient is 2

5. $4y^5 - 3y^3 - 5y^2 + 2y - 1$ The leading coefficient is 4

6. $3z^9 - 2z$ The leading coefficient is 3

7. $6 - 12x^2$ The leading coefficient is ⁻12

8. $2a - a^5 + 13a^2 - 8$ The leading coefficient is ⁻1

Descending Powers of a Polynomial

Polynomials are often written with the term with the highest degree first, and then the term with the next highest degree, etc. A polynomial written this way, is said to be written in **descending powers**.

EXAMPLES: Rewrite the polynomial in descending powers

1. $5x - 13 + 2x^2 - x^3$ In descending powers: $-x^3 + 2x^2 + 5x - 13$

2. $z - 3z^3 + 4z^5$ In descending powers: $4z^5 - 3z^3 + z$

3. $2 + 16w^2 - 4w$ In descending powers: $16w^2 - 4w + 2$

4. $-3a^2 + 4a + 2a^7$ In descending powers: $2a^7 - 3a^2 + 4a$

5. $4y^5 - 3y^3 - 5y^2 + 2y - 1$ In descending powers: $4y^5 - 3y^3 - 5y^2 + 2y - 1$

EXERCISES 3.3

Terminology

Identify each polynomial by the degree and the number of terms. Also find the leading coefficient, and if not already written this way, write in descending powers.

1. $2x^2 - x$

2. -6

3. $-3a + 8$

4. $12 - 2w$

5. $-3x^6 - 4x + 4x^9 + 1$

6. $16m^3 - 8m^9 - 3m + 12$

7. $^-2$

8. $5x^2 + 2x - 3$

9. $-7 + 2p + 3p^4 - 5p^3$

10. $5x^3 - 13x^7$

11. $1 - z^2$

12. $-x^2 + 7x + 3$

13. $-14x^8$

14. $3n$

15. $3a^2 - 6a$

16. $42d - 16d^2 - 56d^4$

17. $v - 4 - 2v^5 - 3v^4$

18. $13x^4 + 2x^2 - 1$

19. $-6b^4 + 6b^3$

20. $2x - 3$

21. $1 - 3x$

22. $64r^8$

23. $-5 + \dfrac{1}{2}x$

24. $2x + \dfrac{3}{4}$

25. $-3w^3 + 2w - 5 + 15w^5$

26. $15 + 4x^3 - \dfrac{2}{3}x^6$

27. $24x^2 - 13 + 22x - x^3$

28. $1 - \dfrac{w^2}{2}$

29. $13 + b + 44b^2 + \dfrac{2b^3}{3}$

3.4 ADDING AND SUBTRACTING POLYNOMIALS

Now that we know how to refer to the different types of polynomials, we can now learn how to work with them under the rules of arithmetic. We start with addition and subtraction and then move on to multiplication and division in the following sections.

To add or subtract two polynomials, we simply add or subtract the like terms of the polynomials.

EXAMPLE 1: Add the two polynomials: $(2x - 3) + (5x + 7)$

$(2x-3)+(5x+7)$

$= 2x - 3 + 5x + 7$ We can remove the parentheses

$= 2x + {}^-3 + 5x + 7$ Rewrite subtraction as adding a negative

$= 2x + 5x + {}^-3 + 7$ Reorder using the commutative rule of addition

$= (2+5)x + ({}^-3+7)$ Combine like terms

$= 7x + 4$ Add the coefficients of the terms

EXAMPLE 2: Subtract the two polynomials: $\left(3x^2 - 2x + 5\right) - \left(4x + 7\right)$

$$\left(3x^2 - 2x + 5\right) - \left(4x + 7\right)$$ Distribute the subtraction to each term

$$= 3x^2 - 2x + 5 - 4x - 7$$ 4x becomes -4x, and 7 becomes -7

$$= 3x^2 - 2x - 4x + 5 - 7$$ Reorder the terms

$$= 3x^2 + {}^-2x + {}^-4x + 5 + {}^-7$$ Rewrite subtraction as adding a negative

$$= 3x^2 + \left({}^-2 + {}^-4\right)x + \left(5 + {}^-7\right)$$

$$= 3x^2 + {}^-6x + {}^-2$$ Combine like terms

$$= 3x^2 - 6x - 2$$ Rewrite as a subtraction problem

EXAMPLE 3: Add the two polynomials: $\left(x^3 - 1\right) + \left(4x^4 - 2x^3 + 5x + 9\right)$

$$\left(x^3 - 1\right) + \left(4x^4 - 2x^3 + 5x + 9\right)$$

$$= x^3 - 1 + 4x^4 - 2x^3 + 5x + 9$$ Remove the parentheses

$$= 4x^4 + x^3 - 2x^3 + 5x - 1 + 9$$ Reorder the terms

$$= 4x^4 + x^3 + {}^-2x^3 + 5x + {}^-1 + 9$$ Rewrite subtraction as adding a negative

$$= 4x^4 + \left(1 + {}^-2\right)x^3 + 5x + \left({}^-1 + 9\right)$$ Combine like terms

$$= 4x^4 + {}^-1x^3 + 5x + 8$$

$$= 4x^4 - x^3 + 5x + 8$$ Rewrite as a subtraction problem

EXAMPLE 4: Subtract the two polynomials: $\left(2x^2 - 3x + 1\right) - \left(5x^3 - 3x^2 - 4x + 1\right)$

$$\left(2x^2 - 3x + 1\right) - \left(5x^3 - 3x^2 - 4x + 1\right)$$ Remove the parentheses

$$= 2x^2 - 3x + 1 - 5x^3 + 3x^2 + 4x - 1$$ Distribute subtraction to 2nd pair of parentheses

$$= -5x^3 + 2x^2 + 3x^2 - 3x + 4x + 1 - 1$$ Reorder the terms

$$= -5x^3 + 2x^2 + 3x^2 + {}^-3x + 4x + 1 + {}^-1$$ Rewrite subtraction as adding a negative

$$= -5x^3 + \left(2 + 3\right)x^2 + \left({}^-3 + 4\right)x + \left(1 + {}^-1\right)$$ Combine like terms

$$= -5x^3 + 5x^2 + x$$

EXERCISES 3.4

Adding and Subtracting Polynomials

Simplify each expression by adding and subtracting the polynomials as indicated.

1. $\left(2x - 1\right) + \left(3x + 2\right)$

2. $\left(1 - 5x\right) + \left(3x - 2\right)$

3. $\left(13x - 9\right) - \left(5x - 3\right)$

4. $\left(5x^2 - 2x\right) + \left(3x^2 + 5x\right)$

5. $\left(3x^3 - 7x\right) + \left(3x^2 + 6x\right)$

6. $\left(4x^2 - 7x + 9\right) - \left(5x^2 - 4x - 3\right)$

7. $\left(-a^2 + 3a - 2\right) - \left(a^2 + 6a - 2\right)$

8. $\left(2w^2 - 7w + 13\right) + \left(w^2 - 1\right)$

9. $\left(5n^4 - 13n^2\right) - \left(6n^5 - 13n^2 + 3\right)$

10. $\left(7a + 3\right) + \left(7a - 3\right)$

11. $\left(3p - 4\right) - \left(3p + 4\right)$

12. $\left(2b^2 - 1\right) + \left(4b^2 + 1\right)$

13. $\left(3x^2 - 2\right) - \left(3x^2 - 2\right)$

14. $\left(3w^5 - 2w\right) - \left(2w - 3w^5\right)$

15. $\left(4r^2 - 3r + 7\right) - \left(16 - 4r + 3r^2\right)$

16. $\left(5n - 7n^2 + 2n^4\right) - \left(2n^4 - 5n + 7n^2\right)$

17. $\left(7 - 15p + 3p^2 - 13p^3\right) + \left(10p^3 + 10p - 8\right)$

18. $\left(-x^3 + 5x^2 + 3x - 16\right) + \left(-3x - 2x^2 + x^3\right)$

19. $\left(21x^7 - 15x^4 + 32x - 9\right) - \left(4x^4 - 9 + 17x^7 + 8x\right)$

20. $\left(5x^2 - 2x + 4\right) - \left(2x^2 + 7x - 3\right) + \left(-x^2 + 6x + 8\right)$

21. $\left(x^2 + 5x + 1\right) - \left(2x^2 - 3x - 6\right) - \left(4x^2 - x - 3\right)$

22. $\left(7t^5 - 3t^4 - 1\right) - \left(6t^5 + 2t^3 - 3t + 1\right) + \left(-t^5 - 4t\right)$

23. $\left(12x^4 - 3x + 2\right) - \left(2x^2 + 1\right) - \left(8x^4 + 3x - 5\right)$

24. $\left(3x^6 - 2x^4 + 5x^2\right) - \left(2x^6 + 3x^4 - 6x^2\right) - \left(x^6 + x^4 - x^2\right)$

3.5 MULTIPLYING POLYNOMIALS

In this section we need a generalization of the law of exponents from Chapter 2 when multiplying variables. The rule we will use is as follows:

$$x^n \cdot x^m = x^{n+m}$$

For example: $x \cdot x = x^2$, and $x \cdot x^2 = x^{1+2} = x^3$, etc.

Multiply a Polynomial by a Monomial Term:

To multiply a monomial by a polynomial, we distribute the monomial term-by-term to each term in the polynomial, and then multiply the constants and variables using the rules of exponents.

EXAMPLE 1: Multiply the monomial times the binomial: $2x\left(x - 3\right)$

$$2x\left(x - 3\right)$$
$$= 2x\left[x + \left(-3\right)\right] \quad \text{Rewrite as an addition problem}$$

$$= 2x\left[x + \left(-3\right)\right] \quad \text{Distribute the 2x to the two terms}$$
$$= 2x \cdot x + 2x \cdot \left(-3\right) \quad \text{in the binomial as indicated by the arrows}$$
$$= 2x \cdot x + 2x \cdot \left(-3\right) \quad \text{Multiply the numerical coefficients and the variables}$$
$$= 2 \cdot x \cdot x + 2 \cdot \left(-3\right) \cdot x$$

88

$$= 2x^{1+1} + (-6x)$$
$$= 2x^2 - 6x \qquad \text{Rewrite as a subtraction problem}$$

EXAMPLE 2: Multiply the monomial times the binomial: $5x^2\left(3x^3 + 7x\right)$

$$5x^2\left(3x^3 + 7x\right)$$

$$= 5x^2\left(3x^3 + 7x\right) \qquad \text{Distribute the } 5x^2 \text{ to the two terms}$$
$$= 5x^2 \cdot 3x^3 + 5x^2 \cdot 7x \qquad \text{in the binomial as indicated by the arrows}$$
$$= 5 \cdot 3x^2x^3 + 5 \cdot 7x^2 \cdot x \qquad \text{Multiply the numerical coefficients and the variables}$$
$$= 15x^{2+3} + 35x^{2+1}$$
$$= 15x^5 + 35x^3$$

Multiply Two Binomial Factors:

To multiply a binomial by a binomial, we distribute each term of the first binomial to each term in the second binomial and then multiply the constants and variables using the rules of exponents.

This is equivalent to multiplying the **F**irst terms in each binomial, then multiplying the **O**uter terms in each binomial, then multiplying the **I**nner terms, and finally multiplying the **L**ast terms. Some people refer to this as the FOIL method, and use the letters to help remember what to multiply. We prefer to use the distributed arrow approach, since, as you will see below, it applies to multiplication of ALL polynomials, not just binomials. You, however, should use whatever technique you feel most comfortable with.

EXAMPLE 1: Multiply the binomials: $(2x + 3)(x + 2)$

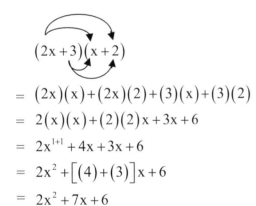

$$(2x + 3)(x + 2)$$

Distribute the terms in the leading binomial as shown by the arrows. Multiply the first term by the two following terms, then move to the second term and do the same. Or use the FOIL method.

$$= (2x)(x) + (2x)(2) + (3)(x) + (3)(2)$$
$$= 2(x)(x) + (2)(2)x + 3x + 6$$
$$= 2x^{1+1} + 4x + 3x + 6$$

Multiply the coefficients and variables

$$= 2x^2 + \left[(4) + (3)\right]x + 6$$

Combine the two like linear terms

$$= 2x^2 + 7x + 6$$

EXAMPLE 2: Multiply the binomials: $(3x - 4)(2x + 6)$

$$\left[(3x) + (-4)\right](2x + 6)$$

Rewrite the subtraction as an addition of the opposite sign, and distribute as shown by the four arrows. In this example we have chosen to use the minus sign in front to represent a negative number.

$$= (3x)(2x) + (3x)(6) + (-4)(2x) + (-4)(6)$$

$$= (3)(2)(x)(x)+(3)(6)x+(-4)(2)x+(-24)$$ Multiply the coefficients and the variables

$$= 6x^{1+1}+18x+(-8)x-24$$

$$= 6x^2+\left[(18)+(-8)\right]x-24$$ Combine the like linear terms

$$= 6x^2+10x-24$$

Multiply Polynomials in the Form (ax + b)²:

To multiply a polynomial in the form $(ax+b)^2$, you should first rewrite it in its alternative form: $(ax+b)(ax+b)$ and then use the method above.

EXAMPLE 1: Multiply: $(2x+1)^2$

$$(2x+1)^2 = (2x+1)(2x+1)$$ Rewrite

Distribute using method above

$$(2x+1)(2x+1)$$

$$= (2x)(2x)+(2x)(1)+(1)(2x)+(1)(1)$$ Multiply coefficients and variables

$$= 4x^2+2x+2x+1$$ Combine like terms

$$= 4x^2+4x+1$$

EXAMPLE 2: Multiply: $(4x-5)^2$

$$(4x-5)^2 = (4x-5)(4x-5)$$ Rewrite

$$(4x-5)(4x-5)$$ Distribute using method above

$$= (4x)(4x)+(4x)(-5)+(-5)(4x)+(-5)(-5)$$ Multiply coefficients and variables.

$$= 16x^2-20x-20x+25$$

$$= 16x^2-40x+25$$ Combine like terms

Again, in this last example we have chosen to use a minus symbol in front of a number to represent a negative number. To highlight this distinction, though, we have placed parentheses around the number.

EXERCISES 3.5

Multiply the Polynomials

Simplify each expression by multiplying the polynomials as indicated.

Basic Problems

1. $3x(2x-1)$

2. $-4x^2(5x+2)$

3. $7x^3(1-2x)$

4. $-2a(a^2+4)$

5. $-7(a+7)$

6. $(n+6)(n-3)$

7. $(3x+4)(2x+2)$

8. $(p+6)(p+3)$

9. $(x+3)(x+7)$

10. $(2c+1)(c-3)$

11. $(3x-5)(x+7)$

12. $(5w+2)(w-2)$

13. $(8t-9)(7t-7)$

14. $(12b+1)(3b-4)$

15. $(7x-3)(x-4)$

16. $(5a-9)(2a-10)$

17. $(3x-8)(x-2)$

18. $(x+1)(x-1)$

19. $(2a-3)(2a+3)$

20. $(6w+4)(6w-4)$

21. $(5r-9)(5r+9)$

22. $(2x+1)^2$

23. $(a+3)^2$

24. $(2c+6)^2$

25. $(3w+5)^2$

26. $(x-1)^2$

27. $(3x-6)^2$

28. $(2w-3)^2$

29. $(5a-7)^2$

30. $(9x-8)^2$

31. $4x(x^2-3x+20)$

32. $-2x^2(-x^2+4x-7)$

33. $3x^3(3x^2-5x-9)$

34. $6x^2(-3x^2+8x-5)$

35. $-x^3(5x^2-7x-6)$

3.6 FACTORING POLYNOMIALS

A large part of learning the language of Algebra is learning how to perform operations, and then learning how to undo those operations. In the last section we learned how to multiply two polynomial expressions. We will now learn how to undo that operation. To undo multiplication, we must perform an operation called factoring. Factoring is simply **multiplication in reverse**.

Greatest Common Factor

We will start by undoing multiplication by a single monomial term. This is called factoring out the greatest common factor (GCF). The first step is to find the GCF. The GCF is the **greatest factor** that is common to all terms in the expression.

To obtain the GCF, we must find the greatest numerical factor in common with all terms in the expression, and then take the variable term to the lowest power of all the terms in the expression. The product of these two factors is the GCF. We then "factor the GCF out of these terms," and write the final factored form as a product of the GCF with the remaining terms. To factor out the GCF means that we are performing the reverse of the distributive property, as the examples below illustrate.

EXAMPLE 1: Factor the expression: $5x + 15$

Find the GCF:
We notice that 5 is the largest common numerical factor of both 5 and 15, and since the second term does not contain a variable, there isn't a variable term in common, so the GCF = 5.

Factor out the GCF:
We can then rewrite this expression as:

$$5x + 15 = (5)x + (5)3 \quad \text{Factor each term}$$
$$= 5(x + 3) \quad \text{Factor out the GCF}$$

Thus, $5(x + 3)$ is considered the factored form of $5x + 15$.

EXAMPLE 2: Factor the expression: $9x^3 + 12x^2$

Find the GCF:
We notice that 3 is the largest common numerical factor of both 9 and 12, since 3 divides evenly (no remainder) into both 9 and 12.

Furthermore, the second term has the lowest exponent (2) associated with the variable, x, so the greatest variable factor is x^2. Thus, the GCF = $3 x^2$ = 3 x x.

Factor out the GCF:
We now **factor** this monomial out of the two terms. That is, we divide the terms by this factor and write what remains.

Dividing 9 by 3, we obtain 3, and dividing x^3 by x^2 we obtain x. That is, $9x^3$ is equivalent to $(3x^2)(3x)$.

Doing the same for the second term, we have: 12 divided by 3 is 4 and x^2 divided by x^2 is 1. That is, $12x^2$ is equivalent to $(3x^2)(4)$.

$$9x^3 + 12x^2 = (3x^2)(3x) + (3x^2)(4) \quad \text{Rewrite the terms with the GCF factors}$$
$$= 3x^2(3x + 4) \quad \text{Factor out the GCF}$$

$3x^2(3x + 4)$ is called the factored form of the expression, $9x^3 + 12x^2$.

Note: If we multiply the factored form out, using the distributive property, we always obtain the original expression. Thus, we have undone the operation of multiplication by factoring.

EXAMPLE 3: Factor the expression: $27x^4 - 15x^2 - 3x$

<u>Find the GCF:</u>

Since 3 is the largest numerical factor of 27, ⁻15, and ⁻3, and x is the largest variable factor of x^4, x^2, and x , the GCF = 3x.

<u>Factor out the GCF:</u>

So, $27x^4 - 15x^2 - 3x$, is equivalent to $(3x)(9x^3) - (3x)(5x) - (3x)(1)$

Now, factoring out the 3x, we obtain: $3x(9x^3 - 5x - 1)$

Factoring Trinomials

We learned that whenever we multiplied two linear binomial terms we almost always ended up with a trinomial expression.

$$(x+3)(x-4) = x^2 - 4x + 3x - 12 = x^2 - x - 12$$,

$$(2x+3)(x-2) = 2x^2 - 4x + 3x - 6 = 2x^2 - x - 6$$, or

$$(x+5)(x-5) = x^2 - 5x + 5x - 25 = x^2 - 25$$

(Note: The last result is not a trinomial, but we will address this later)

Thus, since we are trying to factor a trinomial it makes sense to assume that it was obtained as a product of two binomial terms. This is how we will develop this type of factoring technique.

We start with trinomials of the form: $x^2 + bx + c$

An example of a trinomial with a one as a leading coefficient is: $x^2 + 3x + 2$

Since the coefficient of the x^2 term is 1, we know that the binomial factors must be of the form:
$$(x+m)(x+n)$$

We now need to find a procedure to determine what "m" and "n" are equal to. Recall the procedure for multiplying binomials from above. This would yield:

$$(x+m)(x+n)$$
$$xx + mx + nx + mn$$
$$x^2 + (m+n)x + mn$$

Thus, the product mn must equal 2, and the sum $m+n$ must equal 3. We have to find two numbers whose product is 2 and whose sum is 3.

Since, $2(1) = 2$, and $2 + 1 = 3$ the numbers are 1 and 2.

$$\text{So, } x^2 + 3x + 2 = (x+1)(x+2)$$

Whenever there is a one for the leading coefficient, the factoring approach will be the same. We need to find two numbers whose product equals the last numerical term, and whose sum is the middle coefficient.

EXAMPLE 1: $x^2 + 5x + 4$

Since, $4 = 4(1)$ and $4 + 1 = 5$, the numbers we are looking for are 1 and 4. Therefore,
$$x^2 + 5x + 4 = (x+1)(x+4)$$

We should also note that since multiplication is commutative, we could also write the solution as: $(x+4)(x+1)$.

EXAMPLE 2: $x^2 + 4x + 4$

Since, $4 = 2(2)$ and $2 + 2 = 4$, the numbers we are looking for are 2 and 2. Therefore,
$$x^2 + 4x + 4 = (x+2)(x+2)$$

We should also note that using the properties of exponents, we can also write this as $(x + 2)^2$

In the last two examples, all the coefficients were positive so the numbers we were looking for were also positive. Now we'll consider other possibilities.

Two Special Cases:
Case 1: The middle term is negative and the last term is positive. In this case both of the numbers must be negative.

EXAMPLE 3: $x^2 - 6x + 5$

We are now looking for two numbers whose product is 5 and whose sum is ⁻6. The numbers are ⁻5 and ⁻1, since ,⁻5 × ⁻1 = 5, and ⁻5 + ⁻1 = ⁻6. The factored form is:
$$(x-5)(x-1)$$

Case 2: The last term is negative. Then the numbers must have opposite signs (one positive and one negative.)

EXAMPLE 4: $x^2 + 4x - 12$

We are now looking for two numbers whose product is (–12) and whose sum is (4). The numbers are (6) and (–2), since $(6)(-2) = -12$, and $[6 + (-2)] = 4$. The factored form is:
$$(x+6)(x-2)$$

In this last example, we have chosen to use the alternative notation for a negative number, i.e. (–12) instead of ⁻12.

EXAMPLE 5: $x^2 - x - 6$

We are now looking for two numbers whose product is (-6) and whose sum is (-1). The numbers are (-3) and (2), since $(-3)(2) = -6$, and $(-3) + (2) = -1$. The factored form is:

$$(x - 3)(x + 2)$$

In this last example, we have chosen to use the alternative notation for a negative number, i.e. (-6) instead of $^-6$.

EXAMPLE 6: $x^2 + 7x + 9$

We are looking for two numbers whose product is 9, and whose sum is 7. Since, there are no integers for which this is true, we say that this trinomial does not factor over the integers, or it is prime over the integers. Its only factors are 1, and itself. The resulting answer is just the original trinomial, and so it is considered "prime."

$$x^2 + 7x + 9$$

Difference of Two Squares

You may recall that in the previous section we sometimes multiplied two binomials and the result was not a trinomial, because the middle terms canceled out. For example, consider the product:

$$(2x + 3)(2x - 3)$$

If we multiply this out we obtain:

$$4x^2 - 9$$

We note that the x terms –6x and +6x canceled out.

To undo this type of multiplication we use what is called the Difference of Squares formula. This formula states that if we have a difference of two **perfect square** terms the factored form is given as follows:

$$A^2 - B^2 = (A + B)(A - B)$$

The A and B terms can be any expression we want.

First we define what a perfect square number is. A perfect square number is a number that factors into a product of two of the same numbers,

For example,

4 is a perfect square,	since $2 \cdot 2 = 4$
9 is a perfect square,	since $3 \cdot 3 = 9$
16 is a perfect square,	since $4 \cdot 4 = 16$
25 is a perfect square,	since $5 \cdot 5 = 25$
36 is a perfect square,	since $6 \cdot 6 = 36$
49 is a perfect square,	since $7 \cdot 7 = 49$
etc.	

If a variable is present, then any variable raised to an even power is also a perfect square variable term.

For example,

x^2 is a perfect square,	since $x \cdot x = x^2$
x^4 is a perfect square,	since $x^2 \cdot x^2 = x^4$
x^6 is a perfect square,	since $x^3 \cdot x^3 = x^6$
x^8 is a perfect square,	since $x^4 \cdot x^4 = x^8$
x^{10} is a perfect square,	since $x^5 \cdot x^5 = x^{10}$
x^{12} is a perfect square,	since $x^6 \cdot x^6 = x^{12}$
etc.	

We now illustrate how to use the formula with a few examples.

EXAMPLE 1: Factor the difference of squares: $x^2 - 16$

We recognize that both x^2 and 16 are perfect squares, so we can rewrite the expression as:

$$(x)^2 - (4)^2$$

Comparing this to the formula above, we see that $A = x$, and $B = 4$. If we now substitute these values for A and B into the right-hand-side of the formula above we have:

$$(x)^2 - (4)^2 = (x + 4)(x - 4)$$

This is the factored form.

EXAMPLE 2: Factor the difference of squares: $9w^2 - 49$

We recognize that both $9w^2$ and 49 are perfect squares, so we can rewrite the expression as:

$$(3w)^2 - (7)^2$$

Comparing this to the formula above, we see that $A = 3w$, and $B = 7$. If we now substitute these values for A and B into the right-hand-side of the formula above we have:

$$(3w)^2 - (7)^2 = (3w + 7)(3w - 7)$$

This is the factored form.

EXAMPLE 3: Factor the difference of squares: $x^4 - 25$

We recognize that both x^4 and 25 are perfect squares, so we can rewrite the expression as:

$$(x^2)^2 - (5)^2$$

Comparing this to the formula above, we see that, $A = x^2$ and $B = 5$. If we now substitute these values for A and B into the right-hand-side of the formula above we have:

$$(x^2)^2 - (5)^2 = (x^2 - 5)(x^2 + 5)$$

This is the factored form.

We should note that the **sum of two perfect squares** does NOT factor. So expressions such as:

$$x^2 + 9, \qquad 4b^2 + 25, \qquad \text{and } w^2 + 100,$$

cannot be factored further and are considered **prime**.

EXERCISES 3.6

Greatest Common Factor

Factor the following expressions by factoring out the greatest common factor.

Basic Problems

1. $9x + 12$
2. $21x + 7$
3. $15c + 60$
4. $8x + 24$
5. $3w - 9$
6. $32x - 16$
7. $18y - 6$
8. $14n - 28$

Intermediate and Advanced Problems

9. $3x^2 + 6x + 9$
10. $14b^2 + 7b + 21$
11. $5x^2 + 25x + 15$
12. $36a^2 + 18a + 6$
13. $6x^3 - 3x^2$
14. $15x^4 - 10x^2$
15. $24x^5 + 8x^2$
16. $9a^3 - 27a^2$
17. $13p^3 - 26p$
18. $7k^4 + 49k^8$
19. $5x^3 + 20x^2 - 10x$
20. $16t^4 - 32t^3 + 16t^2$
21. $5y^3 + 25y^6 + 15y^9$

Trinomials

Factor each completely

Basic Problems

22. $x^2 + 3x + 2$
23. $a^2 + 6a + 5$
24. $y^2 + 4y + 3$
25. $z^2 - z - 6$
26. $m^2 + 8m + 15$
27. $x^2 - 2x - 15$
28. $n^2 - 7n - 18$
29. $c^2 - 3c - 10$
30. $x^2 - 6x + 9$
31. $v^2 - 7v + 6$
32. $w^2 - 9w + 20$
33. $z^2 - 13z + 36$
34. $k^2 + 2k - 24$
35. $x^2 + 4x - 12$
36. $b^2 - 5b + 6$
37. $z^2 - 10z + 16$

Difference of Squares

Factor each completely.

Basic Problems

38. $b^2 - 1$
39. $x^2 - 9$
40. $w^2 - 25$
41. $a^2 - 36$
42. $y^2 - 49$
43. $4z^2 - 9$
44. $25x^2 - 36$
45. $16b^2 - 9$
46. $9x^2 - 64$
47. $4r^2 - 1$
48. $9c^2 - 4$
49. $25x^2 - 16$

Intermediate and Advanced Problems

50. $x^4 - 49$
51. $4x^4 - 9$
52. $x^2 + 25$
53. $9x^4 + 16$
54. $x^6 - 36$
55. $25x^6 - 4$
56. $4x^2 + 9$
57. $x^6 - 64$
58. $9x^4 - 25$
59. $x^6 - 1$
60. $16x^4 - 49$
61. $25x^2 + 1$

3.7 EVALUATING EXPRESSIONS

Another important skill needed when working with polynomials or other types of expressions, is how to evaluate these expressions for a specific numerical value of the variable in the expression. We often need to find the fixed numerical value of an expression, given a fixed value for the variable in the expression. To do so, we simply replace the variable with the specified number and perform the arithmetic calculations as we did in Chapter 1.

EXAMPLE 1: Evaluate, $2x - 1$ for $x = 3, -2, 7$

Solution: $x = 3$, $\qquad 2(3) - 1 = 6 - 1 = 5$

Solution: $x = -2$, $\qquad 2(-2) - 1 = -4 - 1 = -5$

Solution: $x = 7$, $\qquad 2(7) - 1 = 14 - 1 = 13$

EXAMPLE 2: Evaluate, $5 - 3x$ for $x = 1, -4, 0$

Solution: $x = 1$, $\qquad 5 - 3(1) = 5 - 3 = 2$

Solution: $x = -4$, $\qquad 5 - 3(-4) = 5 - (-12) = 5 + 12 = 17$

Solution: $x = 0$, $\qquad 5 - 3(0) = 5 - 0 = 5$

EXAMPLE 3: Evaluate, $2x - 3$ for $x = 1, \dfrac{1}{2}, -\dfrac{3}{2}$

Solution: $x = 1$, $\qquad 2(1) - 3 = 2 - 3 = -1$

Solution: $x = \dfrac{1}{2}$, $\qquad 2\left(\dfrac{1}{2}\right) - 3 = \dfrac{2}{2} - 3 = 1 - 3 = -2$

Solution: $x = -\dfrac{3}{2}$, $\qquad 2\left(-\dfrac{3}{2}\right) - 3 = \dfrac{-6}{2} - 3 = -3 - 3 = -6$

EXAMPLE 4: Evaluate, $2x^2 + 4x + 7$ for $x = 4, -1, -3$

Solution: $x = 4$, $\qquad 2(4)^2 + 4(4) + 7 = 2(16) + 16 + 7 = 32 + 16 + 7 = 55$

Solution: $x = -1$, $\qquad 2(-1)^2 + 4(-1) + 7 = 2(1) - 4 + 7 = 2 - 4 + 7 = 5$

Solution: x = – 3, $\qquad 2(-3)^2 + 4(-3) + 7 = 2(9) + (-12) + 7 = 18 - 12 + 7 = 13$

EXAMPLE 5: Evaluate, $- x^2 - 3x + 12$, for x = –1, 0, 5

Solution: x = – 1, $\qquad -(-1)^2 - 3(-1) + 12 = -(1) - (-3) + 12 = -1 + 3 + 12 = 14$

Solution: x = 0, $\qquad -(0)^2 - 3(0) + 12 = -(0) - (0) + 12 = 0 - 0 + 12 = 12$

Solution: x = 5, $\qquad -(5)^2 - 3(5) + 12 = -(25) - (15) + 12 = -25 - 15 + 12 = -28$

EXAMPLE 6: Evaluate, $x^3 - 1$ for x = –1, 0, 2

Solution: x = – 1, $\qquad (-1)^3 - 1 = -1 - 1 = -2$

Solution: x = 0, $\qquad (0)^3 - 1 = 0 - 1 = -1$

Solution: x = 2, $\qquad (2)^3 - 1 = 8 - 1 = 7$

EXERCISES 3.7

Evaluating Expressions

Evaluate each using the value given.

Basic Problems

1. $2x - 3$; use x = –1

2. $1 - 4a$; use a = 7

3. $-3 - 12n$; use n = –3

4. $5x - 3 + 2x - 4$; use x = 4

5. $(4a + 9)2 - 7$; use a = 2

6. $-2(2x - 3)$; use x = –7

7. $4 - (1 - 2b)$; use b = –5

8. $(4 - 7x)$; use x = 0

9. $5 - 6a$; use a = – 6

10. $4w^2 - 9$; use w = 3

11. $r^2 - 16$; use r = –4

12. $d^2 - 25$; use d = 5

13. $5 - 7x$; use x = –2

14. $2y + 5 - 6y - 4$; use y = 3

15. $3(1 - x) + 4$; use x = 4

16. $-(3 - x) - 4$; use x = –3

17. $2y - 6 + 3y - 7$; use y = 5

18. $w^2 - 3w$; use w = –2

19. $4 - 2x^2$; use x = –1

20. $4v - 9 - v$; use v = 7

21. $2 + (3 - 2z)$; use z = 5

22. $2x - 3(x - 2)$; use x = 5

23. $2r^2 - 9$; use r = 2

24. $8 - 3y + 2(y - 4)$; use y = –4

Intermediate and Advanced Problems

25. $3x - 7$; use $x = -\dfrac{4}{3}$

26. $2 - 5r$; use $r = \dfrac{3}{5}$

27. $4+8p$; use $p=-\dfrac{1}{4}$

28. $6y+12$; use $y=\dfrac{1}{2}$

29. $10w-8$; use $w=-\dfrac{1}{5}$

30. $-2-(4+3n)-7$; use $n=8$

31. $4(4x-6)-16x+24$; use $x=3$

32. $4n-6+4(7n-3)$; use $n=4$

33. $-(p+3)+(12-3p)-32$; use $p=-6$

34. $(3x-2)-(2-5x)$; use $x=0$

35. $16+c-3c+(1-2c)$; use $c=-1$

36. $(b-3)3-2b$; use $b=-5$

37. x^2-3x+1 ; use $x=2$

38. $2m^2+m-7$; use $m=-2$

39. $-3b^2+5b-9$; use $b=-1$

40. $3x^3+2$; use $x=-6$

41. $-2n^3+5$; use $n=-1$

42. $4b^3-6b^2-24$; use $b=3$

43. $-a^3-3a+2$; use $a=-1$

44. $4s^3-7s^2$; use $s=2$

Chapter 3 Practice Test

Simplify each expression by combining like terms

1. $5x-3x+7-6$

2. $13n-6+12n+9$

3. $-5x+3x-6-x-3$

4. $-2-5a-4+3a$

Use the Distributive Property to expand each expression and simplify

5. $-5(3x+2)$

6. $2(4-2x)$

7. $3-4(5+4x)$

8. $3x-4\cdot(2x+9)+2$

9. $(5a+2)6+2a-3$

10. $-2+c-(3-c)$

Identify each polynomial by the degree and the number of terms. Also find the leading coefficient, and place in descending powers, if not in that form already.

11. $-x+2x^2$

12. -6

13. $-3a+8$

14. $-3x^6-4x+1+4x^9$

15. $-3+5x^2+2x$

16. $3p^4-5p^3+2p-7$

Simplify each expression by adding and subtracting the polynomials as indicated.

17. $(2x-1)+(3x+2)$

18. $(1-5x)+(3x-2)$

19. $(13x-9)-(5x-3)$

20. $(5x^2-2x)+(3x^2+5x)$

21. $(2b^2-1)+(4b^2+1)$

22. $(3x^2-2)-(3x^2-2)$

23. $(4r^2-3r+7)-(16-4r+3r^2)$

24. $(5n-7n^2+2n^4)-(2n^4-5n+7n^2)$

100

Simplify each expression by multiplying the polynomials as indicated.

25. $3x(2x-1)$

29. $-7(a+7)$

32. $(7x-3)(x-4)$

26. $-4x^2(5x+2)$

30. $(n+6)(n-3)$

33. $(5a-9)(2a-10)$

27. $7x^3(1-2x)$

31. $(9x-8)^2$

34. $(3x-8)(x-2)$

28. $-2a(a^2+4)$

35. $(5w-3)(5w+3)$

Factor the following expressions completely.

36. $9x+12$

40. b^2+6b+9

43. $3w-9$

47. $4b^2+4$

37. $21b^2+7b$

41. $15c+60$

44. $32z^4-16z^3$

48. $36t^2-9$

38. $16w^2-25$

42. y^2+4y+3

45. $x^2-8x+15$

49. $-8x-24$

39. z^2-5z+6

46. a^2-36

Evaluate each using the value given.

50. $2x-3$; use $x=-1$

54. $-2y-9$; use $y=-\dfrac{1}{2}$

57. $-b^2+5b-6$; use $b=-1$

51. $1-4a$; use $a=7$

58. w^2-36; use $w=-6$

52. $13+14z$; use $z=-3$

55. $3a+15$; use $a=\dfrac{2}{3}$

53. $4x-3+2x-4$; use $x=\dfrac{1}{2}$

56. $3x^2-2x+7$; use $x=-2$

CHAPTER 4

Units of Measure– Just the Basics

In the Introduction of this text, we discussed the important process of measuring. We introduced measuring early on, since it is a common theme, and we intend to use it throughout this book. Measuring is a critical skill that serves to unify the concepts of numbers, algebra, and geometry. What we did not do in our Introduction, however, was to talk in depth about the two measuring scales used to do our measurements. We wanted to first develop your skills with numbers and basic algebra before we approached the sometimes difficult concept of units and unit conversions. In fact, understanding number concepts and algebra is a key skill needed to do conversions. In this chapter we now try to rectify that omission and present the idea of what these measuring scales are, where they come from, and how to change (convert) within a measuring scale, and from one measuring scale to another.

As we stated in the Introduction, a measurement involves two parts; a number and a unit. In the previous chapters, we focused on gaining a better understanding of the number concept, both fixed numbers and variables. We now turn our attention to the other aspect of measurement as shown in the chart below, that of units or attributes.

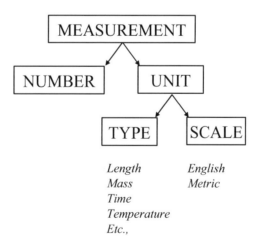

4.1 UNITS/ATTRIBUTES

An attribute is simply what we are measuring. Another word for attribute is the unit of measure, or simply, unit for short.

Some common attributes or units are:

Size: length, area, and volume	Amount (mole)
Weight	Density
Distance	Forces (including pressure)
Temperature	Energy
Luminosity (Brightness)	Velocity
Electric Potential	Time
pH level	Mass

Each of the above quantities is a particular type of unit. Each of the above units falls into one of two different categories. It is either a fundamental unit or a derived unit.

Fundamental Units

A fundamental unit (also called a base unit) is a unit that is **dimensionally-independent** from the other units. In other words, they are fundamentally different and cannot be derived from another unit. There are seven fundamental or base units, and they are:

> Length
> Mass
> Time
> Temperature
> Amount (mole)
> Electric Charge
> Luminosity

We'll only focus our attention on the first five fundamental attributes, as these are more common in our day-to-day living, and in the courses we'll see in college.

Length
Length is the fundamental geometric measure, which is associated with the linear size or location of an object – How long is it? How tall is it? How far away is it?

Mass
Mass measures the amount of matter in an object. Sometimes it is mistakenly referred to as weight. Weight, on the other hand, is the gravitational attraction between the earth and the mass of the object. The object's mass is attracted to the earth, and the earth is attracted to the object, and the mutual strength of this attraction is called the weight of the object. A strange fact is that an object can have mass, yet still be weightless. For example, if we go way out into space, we float, as we are weightless, but we still have a "bulkiness" associated with us. This bulkiness is our mass.

Since most of the measurements we come in contact with are done on earth, we won't worry about the distinction. Here on earth, if two things have the same mass, they will also have the same weight. Even though we might treat mass and weight the same, technically we should distinguish between the two. If you took a bowling ball or a car to the moon, its weight would be less, but its mass would remain the same.

Time

Time measures the passing of events. It tells us **how long** something takes to happen.

It is interesting to note that time, like length and mass, is an independent quantity. However, Einstein has so strangely shown us, these quantities are intimately connected. Time can change based upon an object changing its location (length), or if it is in the presence of a very large mass. These are very strange results indeed, and even scientists still do not fully understand them!

Temperature

Temperature is a quantity that measures the degree of **hotness** or **coldness** of an environment or a body. The hotness is a measure of how much **vibrational-energy** (kinetic energy) is contained within the molecules of the material in question. The more the atoms vibrate, the higher the temperature.

Molecules that do not move, do not have any measurable vibrational-energy and are considered to have a temperature that is called absolute zero. In practice, this never occurs. There really is no absolute zero!

Mole

A mole (mol) is the amount of a chemical substance that contains a number of either, atoms, molecules, ions, electrons, or photons, equivalent to the number of atoms in 12 grams of carbon-12. This number is given as Avogadro's constant, which has a value of 602,214,085,700,000,000,000,000 per mol (In Chapter 9 we'll see how to write this number more compactly). Note, that on May 20, 2019 Avogadro's constant is being redefined to 602,214,076,000,000,000,000,000 per mol. The mole is used throughout chemistry as a way to express amounts of reactants and products of chemical reactions. As we just stated, the number of molecules per mole is known as Avogadro's constant, and is defined such that the mass of one mole of a substance, expressed in grams, is equal to the mean relative molecular mass of the substance.

Derived Units

Everything else we measure we obtain from the above fundamental quantities, and these units are called derived units. The fundamental units provide the basis from which the units for all other quantities are derived. The letters in the alphabet give rise to all of the words in the English language, and similarly, the base units create all the other types of units we need and use in our calculations. We obtain these derived units by multiplying and dividing the fundamental units by each other.

EXAMPLES:

Velocity is distance (length) divided by time (length/time), e.g. miles per hour (mph) or meters per second (mps).

Volume is the product of three length measurements (length^3), e.g. cubic inches (in^3) or cubic meters (m^3). Note how we use exponents to simplify the representation.

Density is mass divided by volume, where volume is another derived unit. Thus, we can express density in terms of fundamental units as mass divided by length cubed (mass/length^3), e.g. grams per centimeter cubed (g/cm^3) or kilograms per meter cubed (kg/m^3).

Examples of other derived quantities:

Area is the product of length times length (width).
Acceleration is length divided by seconds squared.
Force is mass times acceleration (another derived unit).
etc.

Any time we measure an attribute, there are two parts to consider. The first is the number or numerical part, and the second is the unit part. The unit part is really two things. The first is the type or attribute we are measuring, and the second is the scale to which we are referencing it. For now we will not spend a lot of time on the scale part of the unit concept, but will instead focus on the previous exposure you have had to the different scales for length, such as; inches, feet, mile, meter, centimeter, millimeter, and kilometer, as well as for mass, such as; ounce, pound, gram, and kilogram. In the next sections, we spend more time introducing the two major scales; the metric and the English, and show in detail how they are connected and how we can translate (convert) between and within both systems. However, our primary focus in this book will be on the metric system, since that is the system needed in our scientific study. It is a shame that here in the US we have to learn two different scales. This places another unnecessary burden on our children in a subject that is already quite complicated and we wonder why our students seem to be behind other nations when it comes to mathematics. We handicap our children and then wonder why they can't compete!

What you should take away from this part of the book, is just to have a very vague understanding of units. You need to know that a unit is an attribute, and there are roughly 5 important units/attributes we need to be aware of; length, mass (weight), time, amount (mole), and temperature. You should know specifically that some of the units for length are; inches, feet, yards, miles, meters, centimeters, millimeters, and kilometers. These are just a few of the important ones. There are others, but these you should know are all used to measure a length. You should also know some of the units used to measure mass such as; pound, ton, gram, kilogram, and milligram. These are all units to measure how much "stuff" something has. Again, there are others, but these are the important ones, for now. The units for time are fairly self-explanatory such as: seconds, minutes, hours, days, months, and years. The units for amount, typically used in chemistry, are moles. The units for temperature are either degrees Fahrenheit (^{o}F), or degrees Celsius (^{o}C).

By multiplying the unit of length we get two very important derived units; area and volume. We want you to fully understand these two very important derived units; Some common units for area are; square inches (in^2), square feet (ft^2), square meters (m^2), and square centimeters (cm^2). Common units for volume are; cubic inches (in^3), cubic feet (ft^3), cubic meters (m^3), and cubic centimeters (cm^3).

In the sections that follow, we intend to more thoroughly introduce the idea of units, along with the two scales used to better quantify and translate within and between these two scales.

EXERCISES 4.1

---

1. Come up with a list of at least 10 attributes of an object or an event that you can measure.

2. For the quantities in the table below, list all the units that can be used to measure the given attributes. Choose these units from the list beneath this table.

Different unit names for the same attribute	
Attribute	**Unit names**
length	
area	
volume	
mass (weight)	
temperature	
time	
density	
velocity	

board-feet	feet per second	light-years	milliseconds
centigrade	fluid ounces	liters	minutes
centigrams	gallons	meters	nanometers
centimeters	grams per	meters per second	nanoseconds
centuries	cubic centimeter	metric tons	ounces
cubic centimeters	grams per liter	micrograms	pints
cubic yards	grams per milliliter	microns	pounds per cubic foot
cups	grams	mils	quarts
days	hours	miles	square inches
decades	inches	miles per hour	square centimeters
°Celsius	kilograms	milligrams	tons
°Fahrenheit	kilometers	millennia	yards
feet	kilometers per hour	milliliters	

3. Given the units from problem 2 above, identify which are fundamental and which are derived.

4.2 MEASUREMENT SCALES (SYSTEMS OF UNITS)

All measurements are relative and are made up of two distinct parts. The first part answers the how-much? question. It is the numerical quantification of the measurement such as 1, 2, 3, 6.5, or 10.75. The second part answers the "of what?" question. It identifies what we are measuring — it is the **units** part of our measurement. The units part, however, is also made up of two parts. The first is the particular attribute (unit) we are measuring such as length, mass, time, and temperature; and the second part is the scale we use to measure this attribute. In this case we have two distinct scales; English (also called Imperial) and metric.

Early on, we gave a very informal introduction of the different scales, simply by using them and relying on the fact that you have heard of them throughout your courses in mathematics and science in high school and earlier. We would now like to take a more complete look at the units we use and, in particular, the scales we use to measure them with.

Once we know "what" we are measuring, the next question is, "what scale" are we using to measure it? To be useful, a measurement must be comparable to some standard, i.e. some non-changing reference that we all agree upon and use. Over the years, trade and commerce between nations has driven us towards two common systems of measurements. The two systems are the English (or British) and the metric system. The metric system is by far the most dominant, due to its simplicity and the logical interrelationships between its units. The English System (no longer used in England) is what is primarily used in the US, and in just about NO other nation in the world. In fact, as of the writing of this book, there are only 3 nations that use the English system. They are the United States, Liberia and Myanmar. The rest of the ENTIRE world uses the metric system. Even in the US, the US Congress passed the Metric Conversion Act that then President Gerald R. Ford signed into law on December 23, 1975. It declared the metric system to be "the preferred system of weights and measures for United States trade and commerce", but permitted the use of United States customary units in non-business activities, yet we still follow the English system here.

Our primary focus in this book will be on the metric system, but for completeness we introduce the English system as well, mainly to see the two systems side-by-side, and to perhaps highlight why the rest of the world is using the more logical metric system approach. Additionally there is the practical reasons that we in the US still use this system, so we need to know how to calculate with it and understand how to use it.

The English System: An ad-hoc, confused, and sometimes humorous, system of measurements.

Though no longer used in England, many of the units in this system have been used for centuries and were originally based on common objects or human body parts, such as the yard and foot. We have all been trained extensively in the English system of measurement. Learning terms such as: inch, foot, yard and mile, or dry ounce, pound, and ton, or fluid ounce, pint, quart, and gallon. Additionally, there are all the crazy conversion factors needed to go from one to the other – twelve inches to a foot, three feet to a yard, 5,280 feet in a mile, sixteen dry ounces to a pound, 2,000 pounds to a ton, sixteen fluid ounces to a pint, two pints to a quart, four quarts to a gallon, 21, 24 or 44 gallons to a barrel. The list goes on and on.

Though practical when they were first created, the standards for the English units were not fixed and the standard could vary from object to object or from person to person. Eventually, the system was standardized, but the numerical relationship between common units in the system, as we'll see, was quite complicated.

In the English system the first four of the seven fundamental units are:

Quantity	English Base Unit
Length	foot (ft)
Mass	pound (lb)
Time	second (s)
Temperature	Fahrenheit (°F)

From these units many different types of units are derived. As you can see from all the different multiplication factors discussed above, the English system of measurement is quite complicated. To illustrate this somewhat, but not entirely, English conversions are provided in the tables below:

Length
12 inches = 1 foot
3 feet = 1 yard
220 yards = 1 furlong
8 furlongs = 1 mile
5280 feet = 1 mile
1760 yards = 1 mile

Area
144 sq. inches = 1 square foot
9 sq. feet = 1 square yard
4840 sq. yards = 1 acre
640 acres = 1 square mile
1 sq. mile = 1 section
36 sections = 1 township

Volume
1728 cu. inches = 1 cubic foot
27 cu. feet = 1 cubic yard

Capacity (Dry)
2 pints = 1 quart
8 quarts = 1 peck
4 pecks = 1 bushel

Capacity (Liquid)
16 fluid ounces = 1 pint
4 gills = 1 pint
2 pints = 1 quart
4 quarts = 1 gallon (8 pints)

Mass
437.5 grains = 1 ounce
16 ounces = 1 pound (7000 grains)
14 pounds = 1 stone
100 pounds =1 hundredweight (cwt)
20 cwt = 1 ton (2000 pounds)

Troy Weights
24 grains = 1 pennyweight
20 pennyweights = 1 ounce (480 grains)
12 ounces = 1 pound (5760 grains)

The purpose of introducing these conversions is to illustrate just how complicated the English system is. We intend to use only a few of the more basic conversions from above, so you do not need to know the vast majority of these. They are shown primarily for illustration purposes. In addition to the English system, there is the alternative metric system.

The Metric System: A system based upon logic and reason

Since our real number system is all based upon powers of ten, why not have our measurement system conform to the same pattern? This way we won't have all these crazy numbers to memorize to try and change from one measure to another. This is quite logical, and makes sense. That is the basis of the metric system. It is just like our base–10 (place–values with powers of 10) number system.

Most people don't want to change over to the metric systems, because they haven't been trained to **think** in metric. Over the years we have learned to get a sense for how big an inch, foot, and mile. are, or how cold 28^0F is, and we are comfortable with these quantities. However, if you ask the average person to perform calculations between the various units, mistakes are inevitable, because the conversions, as shown earlier, are quite complicated and appear to be without a rational reason.

We need to be trained to think in metric units, so that we can then easily compute using the metric system, and move away from the English system. Here are some ways to think-metric:
- A parking meter, the height of a six year old boy, four basketballs placed side by side, a baseball bat, 4 pieces of dry spaghetti laid end to end, the distance from an adult man's nose to his finger tips when he puts his arms out sideways, seven one dollar bills placed end to end, a stack of $100 in dimes, are all about a meter in length.

- The width of a small fingernail, or the width of a black key on the piano, is about the size of a centimeter (one one-hundredths of a meter).

- The width of the ball on a ballpoint pen, thickness of a dime, paperclip, or credit card, are about the size of a millimeter (one one-thousandths of a meter).

- The weight of a small paperclip, or a kidney bean is about a gram.

Once we establish a comfortable point of reference, it is so much easier to do calculations using the metric system.

The first four of the seven fundamental units for the metric system are:

Quantity	Metric Base Unit
Length	meter (m)
Mass	kilogram (kg)
Time	second (s)
Temperature	Kelvin (K), or Celsius (^{o}C)

The larger and smaller versions of each unit in the metric system do not require a series of complicated conversion tables to represent, as we needed for the English system. Instead, the related measurements are created either by multiplying, or dividing, the base unit by factors of ten. This feature provides a great convenience to users of the system, as it eliminates the need for such cumbersome calculations as dividing by 16 (to convert ounces to pounds) or by 12 (to convert inches to feet). Similar calculations in the metric system are performed simply by shifting the decimal point in the numerical part of the measurement. The metric system is a base-10 or decimal system. How these units are subdivided or multiplied, is through a series of prefixes that specify the size relative to the standard measure for each of the fundamental units: meter, gram and liter.

Size of Units and Prefixes

If you want to measure the distance between cities, the meter might be considered too small. You wouldn't measure the distance from New York to Los Angeles in meters any more than you would measure it in yards or feet. You really need a larger unit of measure. Similarly, you wouldn't use a meter stick to measure distances at the cellular or atomic level. You really need a much smaller unit of measure. To accommodate very large and very small measures, a series of word prefixes are used to signify multiples or fractions of the fundamental meter unit. To simplify matters, all the related larger units are multiplied by factors of 10 and all the smaller units are divided by multiples of 10. Some commonly used prefixes with their meanings and numerical equivalences are:

Prefix	Meaning	Abbreviation	Number	Exponent Form
Tera-	trillion	T	1,000,000,000,000	10^{12}
Giga-	billion	G	1,000,000,000	10^{9}
Mega-	million	M	1,000,000	10^{6}
kilo-	thousand	k	1,000	10^{3}
centi-	hundredths of	c	0.01	10^{-2}
milli-	thousandths of	m	0.001	10^{-3}
micro-	millionths of	μ	0.000001	10^{-6}
nano-	billionths of	n	0.000000001	10^{-9}
pico-	trillionths of	p	0.000000000001	10^{-12}

Prefixes within the same units are related to one another. There are 1000 millimeters in a meter, 100 centimeters in a meter, and 1000 meters in a kilometer, i.e.

$$100 \text{ centimeters} = 1 \text{ meter}$$
$$1000 \text{ meters} = 1 \text{ kilometer}$$

A little less obvious, but something that can be shown is:
$$10 \text{ millimeters} = 1 \text{ centimeter}$$

Whenever we make a measurement, we need to pick the appropriate prefix depending upon the size of the unit we are measuring. For example, the correct metric unit for distances between cities is a kilometer (km). That prefix, kilo-, means that we're talking about a unit that is 1,000 times larger than the basic unit of a meter. Thus a kilometer is 1,000 times the length of one meter. If you know how many meters it is between New York and Portland, it's very easy to figure out the distance in kilometers (divide by 1,000). That is much easier than with English units. In the metric system, the conversions only involve factors of ten.

The scientific community has adopted the metric system and has agreed upon a set of standards called the International Standard or SI system of units. Since just about all the countries of the world have adopted this standard, we need to be familiar with the two different measurement systems – the English and the metric systems. However, in the world of science the metric system is the system of choice, and the system that every scientist should be intimately familiar with.

> **Careful, we need to be consistent! Even scientists have made BIG mistakes!** In December 1999 the Mars Polar Lander was to land on Mars to collect samples of material and take photographs. Scientists waited for the first transmission from the NASA probe, but they never came. It was later discovered that some US scientist had mistakenly used English units when programming a portion of the navigation system, but the rest of the system had been programmed in metric units. They believe that this mix-up in units caused the crash of the Lander, and the loss of nearly $120 Million!

4.3 CONVERTING UNITS

In this section we learn how to convert units, both within a single system, as well as converting units between the English and the metric systems. We introduce the approach by using something called **dimensional analysis**. Dimensional analysis is a process where we convert a problem by using only conversion factors to obtain the final unit type. We start by understanding what the beginning unit is and then we need to determine the sequence of multiplication factors to obtain the desired unit after all the unit (dimension) cancellations. We illustrate this approach through examples.

<u>Converting Units Within a Single System</u>

The process of converting in either the English or the metric is the same. The only difference is that for the English system the conversion factors are varied, while for the metric they are all powers of 10. We'll start with the metric system, as this is easier than the English system.

Conversions Within the Metric System

To convert similar type units within the metric system we must first identify the size of the unit we are converting from, and the size of the unit we wish to convert to. This will tell us what conversion factors to use. Next, we choose the conversion factor that will convert our unit, e.g. meter, gram or liter. We then determine the conversion factor that will convert the meter, gram or liter unit into our desired unit. We then multiply by the appropriate conversion factors and write the new unit designation.

A conversion factor is a ratio of two equivalent physical quantities expressed in different units. For example, we know that 1,000 mm = 1 m. If we divide both sides by 1,000 mm we have:

$$\frac{1,000\text{ mm}}{1,000\text{ mm}} = \frac{1\text{ m}}{1,000\text{ mm}} \quad \text{or} \quad 1 = \frac{1\text{ m}}{1,000\text{ mm}}$$

The conversion factor is $\dfrac{1\text{ m}}{1,000\text{ mm}}$ and, as we have shown, this equals 1. We can continue in the same way and derive conversion factors between the units and derive the following conversion tables below:

Converting to meter (also gram, or liter): "From" the unit in the denominator

From Unit	To Unit	Unit Name	Conversion Factor
pm pg pL	m g L	picometer picogram picoliter	$\dfrac{1\,(\text{m},\text{g},\text{L})}{1{,}000{,}000{,}000{,}000\ \text{p}\,(\text{m},\text{g or L})}$
nm ng nL	m g L	nanometer nanogram nanoliter	$\dfrac{1\,(\text{m},\text{g},\text{L})}{1{,}000{,}000{,}000\ \text{n}\,(\text{m},\text{g or L})}$
μm μg μL	m g L	micrometer microgram microliter	$\dfrac{1\,(\text{m},\text{g},\text{L})}{1{,}000{,}000\ \mu\,(\text{m},\text{g or L})}$
mm mg mL	m g L	millimeter milligram milliliter	$\dfrac{1\,(\text{m},\text{g},\text{L})}{1{,}000\ \text{m}\,(\text{m},\text{g or L})}$
cm cg cL	m g L	centimeter centigram centiliter	$\dfrac{1\,(\text{m},\text{g},\text{L})}{100\ \text{c}\,(\text{m},\text{g or L})}$
km kg kL	m g L	kilometer kilogram kiloliter	$\dfrac{1{,}000\,(\text{m},\text{g},\text{L})}{1\ \text{k}\,(\text{m},\text{g or L})}$

Converting from meter (also gram, or liter): "To" the unit in the numerator

From Unit	To Unit	Unit Name	Conversion Factor
m g L	pm pg pL	picometer picogram picoliter	$\dfrac{1{,}000{,}000{,}000{,}000\ \text{p}\,(\text{m},\text{g or L})}{1\,(\text{m},\text{g},\text{L})}$
m g L	nm ng nL	nanometer nanogram nanoliter	$\dfrac{1{,}000{,}000{,}000\ \text{n}\,(\text{m},\text{g or L})}{1\,(\text{m},\text{g},\text{L})}$

From Unit	To Unit	Unit Name	Conversion Factor
m g L	μm μg μL	micrometer microgram microliter	$\dfrac{1,000,000\,\mu\,(m,g\,\text{or}\,L)}{1\,(m,g,L)}$
m g L	mm mg mL	millimeter milligram milliliter	$\dfrac{1,000\,m\,(m,g\,\text{or}\,L)}{1\,(m,g,L)}$
m g L	cm cg cL	centimeter centigram centiliter	$\dfrac{100\,c\,(m,g\,\text{or}\,L)}{1\,(m,g,L)}$
m g L	km kg kL	kilometer kilogram kiloliter	$\dfrac{1\,k\,(m,g\,\text{or}\,L)}{1,000\,(m,g,L)}$

We illustrate the process with several examples:

EXAMPLE 1: Convert 15 mm into m.

We are converting from mm to m. We identify the appropriate conversion factor from the table above as $\dfrac{1\ m}{1{,}000mm}$, since we are converting from mm to m. **The "from" unit is always in the denominator.** Now multiply and cancel the units as follows:

$$15\ \text{mm} = \frac{15\ \cancel{\text{mm}}}{1}\cdot\frac{m}{1{,}000\ \cancel{\text{mm}}} = 0.015\ m$$

Notice that the mm units cancel to leave the desired m units. This is what dimensional analysis is. We choose the conversion factor so that the wanted dimension does not cancel, but the other units do. This, as you can see, is identical to how we multiplied fractions in Chapter 1. If a quantity in is both the numerator and denominator (top and bottom of the expression) it cancels out.

EXAMPLE 2: Convert 32.5 ng into mg.

We are converting from ng to mg. In using the table, we will need to use two conversion formulas, since we do not have a single conversion that takes us from ng to mg. Thus, using dimensional analysis, we first go from ng to g and then from g to mg.

We identify the appropriate conversion factors from the tables above as $\dfrac{1g}{1{,}000{,}000{,}000ng}$ (we obtain this simply by replacing the m for meter in the above table with g for gram everywhere), since we are converting from ng to g, and then $\dfrac{1{,}000mg}{1g}$, since we are converting from g to mg. In both cases the "from" unit is always in the denominator.

Now multiply and cancel the units as follows:

$$32.5\,\text{ng} = \frac{32.5\,\cancel{\text{ng}}}{1} \cdot \frac{1\,\text{g}}{1,000,000,000\,\cancel{\text{ng}}} \cdot \frac{1,000\,\text{mg}}{1\,\cancel{\text{g}}}$$

$$= \frac{(32.5)\,1,000\,\text{mg}}{1,000,000,000}$$

$$= 0.0000325\,\text{mg}$$

Notice that the ng and g units cancel to leave the mg units.

EXAMPLE 3: Convert 0.75 L into mL.

We are converting from L to mL. We identify the appropriate conversion factor from the table above as, $\dfrac{1,000\ \text{mL}}{1\text{L}}$ since we are converting from L to mL. The "from" unit is always in the denominator. Now multiply and cancel the units as follows:

$$0.75\ \text{L}\ =\ \frac{0.75\ \cancel{\text{L}}}{1} \cdot \frac{1,000\ \text{mL}}{1\ \cancel{\text{L}}}\ =\ 750\ \text{mL}$$

Notice that the L units cancel to leave the mL units.

EXAMPLE 4: Convert 47.35 cm into pm.

We are converting from cm to pm. We identify the appropriate conversion factors from the tables above as $\dfrac{1\ \text{m}}{100\ \text{cm}}$, since we are converting from cm to m, and then $\dfrac{1,000,000,000,000\ \text{pm}}{1\ \text{m}}$, since we are converting from m to pm. In both cases the "from" unit is always in the denominator. Now multiply and cancel the units as follows:

$$47.35\ \text{cm} = \frac{47.35\ \cancel{\text{cm}}}{1} \cdot \frac{1\ \cancel{\text{m}}}{100\ \cancel{\text{cm}}} \cdot \frac{1,000,000,000,000\ \text{pm}}{1\ \cancel{\text{m}}}$$

$$= \frac{(47.35)\,1,000,000,000,000\,\text{pm}}{100}$$

$$=\ 473,500,000,000\,\text{pm}$$

Notice that the cm and m units cancel to leave the pm units.

Conversions Within the English System

We now show how to convert within the English system. We derive the conversion factors in the same way we have already shown for the metric conversions. This leads to the following conversion tables:

114

Convert larger to smaller units:

Convert "from" "to"	Conversion Factor
feet to inches	$\dfrac{12 \text{ inches}}{1 \text{ foot}}$
yard to feet	$\dfrac{3 \text{ feet}}{1 \text{ yard}}$
miles to feet	$\dfrac{5280 \text{ feet}}{1 \text{ mile}}$
pint to fl. ounces	$\dfrac{16 \text{ fl. ounces}}{1 \text{ pint}}$
quart to pints	$\dfrac{2 \text{ pints}}{1 \text{ quart}}$
gallon to quarts	$\dfrac{4 \text{ quarts}}{1 \text{ gallon}}$
pounds to ounces	$\dfrac{16 \text{ ounces}}{1 \text{ pound}}$
tons to pounds	$\dfrac{2000 \text{ pounds}}{1 \text{ ton}}$

Convert smaller to larger units:

Convert "from" "to"	Conversion Factor
inches to feet	$\dfrac{1 \text{ foot}}{12 \text{ inches}}$
feet to yards	$\dfrac{1 \text{ yard}}{3 \text{ feet}}$
feet to miles	$\dfrac{1 \text{ mile}}{5280 \text{ feet}}$
fl. ounces to pints	$\dfrac{1 \text{ pint}}{16 \text{ fl. ounce}}$
pints to quarts	$\dfrac{1 \text{ quart}}{2 \text{ pints}}$

Convert "from" "to"	Conversion Factor
quarts to gallons	$\dfrac{1 \text{ gallon}}{4 \text{ quarts}}$
ounces to pounds	$\dfrac{1 \text{ pound}}{16 \text{ ounces}}$
pounds to tons	$\dfrac{1 \text{ ton}}{2000 \text{ pounds}}$

NOTE: In the English System we use the same word ounce for either a dry or a liquid quantity. To distinguish between the two types of measurements we will use the term fl. ounce to distinguish a fluid ounce measurement from a dry ounce measurement. Whenever you see fl. ounce, this implies a liquid measure.

EXAMPLE 1: Convert 72 feet to yards.

We are converting from feet to yards. We identify the appropriate conversion factor from the table above as $\dfrac{1 \text{ yard}}{3 \text{ feet}}$. Now multiply and cancel the units as follows:

$$72 \text{ feet} = \frac{72 \text{ feet}}{1} \cdot \frac{1 \text{ yard}}{3 \text{ feet}}$$
$$= \frac{72 \text{ yards}}{3}$$
$$= 24 \text{ yards}$$

Notice that the feet units canceled to leave the yard units.

EXAMPLE 2: Convert 2.5 miles to feet.

We are converting from miles to feet. We identify the appropriate conversion factor from the table above as $\dfrac{5280 \text{ feet}}{1 \text{ mile}}$, Now multiply and cancel the units as follows:

$$2.5 \text{ miles} = \frac{2.5 \text{ miles}}{1} \cdot \frac{5280 \text{ feet}}{1 \text{ mile}}$$
$$= \frac{2.5 (5280 \text{ feet})}{1}$$
$$= 13,200 \text{ feet}$$

Notice that the mile units canceled to leave the feet units.

EXAMPLE 3: Convert 468 fl. ounces to gallons.

We are converting from ounces to gallons. Using dimensional analysis, we must first convert ounces to pints, then pints to quarts. and finally quarts to gallons.

We identify the appropriate conversion factors from the tables above as $\dfrac{1 \text{ pint}}{16 \text{ fl. ounce}}$, $\dfrac{1 \text{ quart}}{2 \text{ pints}}$, and

$\dfrac{1 \text{ gallon}}{4 \text{ quarts}}$.

Now multiply and cancel the units as follows:

$$468 \quad \text{fl. ounces} = \frac{468 \text{ fl. ounces}}{1} \cdot \frac{1 \quad \text{pint}}{16 \text{ fl. ounces}} \cdot \frac{1 \quad \cancel{\text{quart}}}{2 \quad \text{pints}} \cdot \frac{1 \quad \text{gallon}}{4 \quad \cancel{\text{quarts}}}$$

$$= \frac{468 \quad \text{gallons}}{(16)\,(2)\,(4)}$$

$$= \frac{468 \quad \text{gallons}}{128}$$

$$= 3.65625 \quad \text{gallons}$$

Notice that the ounces, pints, and quarts units canceled to leave the gallons units.

EXAMPLE 4: Convert 1.25 tons to ounces.

We are converting from tons to ounces. We must first convert tons to pounds, and then pounds to ounces. We identify the appropriate conversion factors from the tables above as

$$\frac{2000 \text{ pounds}}{1 \text{ ton}}, \text{ and } \frac{16 \text{ ounces}}{1 \text{ pound}}$$

Now multiply and cancel the units as follows:

$$1.25 \text{ tons} = \frac{1.25 \text{ tons}}{1} \cdot \frac{2000 \text{ pounds}}{1 \text{ ton}} \cdot \frac{16 \text{ ounces}}{1 \text{ pound}}$$

$$= \frac{(1.25)(2000)(16) \text{ ounces}}{1}$$

$$= 40{,}000 \text{ ounces}$$

Notice that the tons and pounds units cancel to leave the ounces units.

Converting Between the English and Metric Systems

This tends to be the most complicated, since it involves many different types of units. For this we will derive two conversion tables. The first is metric to English, and the second is English to metric. The conversion factors will be obtained as was done previously. Since 1 meter = 3.28 feet we divide both sides by 1 meter to obtain the conversion factor from meters to feet as:

$$\frac{1 \text{ meter}}{1 \text{ meter}} = \frac{3.28 \text{ feet}}{1 \text{ meter}} \quad \text{or} \quad \frac{1 \cancel{\text{ meter}}}{1 \cancel{\text{ meter}}} = 1 = \frac{3.28 \text{ feet}}{1 \text{ meter}}$$

We can continue in this way to derive the conversion tables:

English to Metric

Convert "from" "to"	Conversion Factor
inch to cm	$\dfrac{2.54 \text{ cm}}{1 \text{ inch}}$
feet to meter	$\dfrac{1 \text{ m}}{3.28 \text{ feet}}$
yard to meter	$\dfrac{0.9144 \text{ m}}{1 \text{ yd}}$
fl. ounce to mL	$\dfrac{29.574 \text{ mL}}{1 \text{ fl.ounce}}$
ounce to g	$\dfrac{28.35 \text{ g}}{1 \text{ ounce}}$
lb to kg	$\dfrac{1 \text{ kg}}{2.2046 \text{ lb}}$
mile to km	$\dfrac{1.6093 \text{ km}}{1 \text{ mile}}$
mile to m	$\dfrac{1,609.3 \text{ m}}{1 \text{ mile}}$
quart to L	$\dfrac{1 \text{ L}}{1.06 \text{ quarts}}$
gallon to L	$\dfrac{1 \text{ L}}{0.264 \text{ gallon}}$
cubic inches to L	$\dfrac{1 \text{ L}}{61 \text{ in}^3}$
mL to cubic cm	$\dfrac{1 \text{ cm}^3}{1 \text{ mL}}$
L to cubic cm	$\dfrac{1,000 \text{ cm}^3}{1 \text{ L}}$
cubic inch to cubic cm	$\dfrac{16.387 \text{ cm}^3}{1 \text{ in}^3}$

Metric to English

Convert "from" "to"	Conversion Factor
cm to inch	$\dfrac{1 \text{ in}}{2.54 \text{ cm}}$
meter to feet	$\dfrac{3.28 \text{ feet}}{1 \text{ meter}}$
meter to yard	$\dfrac{1 \text{ yd}}{0.9144 \text{ m}}$
mL to fl. ounce	$\dfrac{1 \text{ fl. ounce}}{29.574 \text{ mL}}$
g to ounce	$\dfrac{1 \text{ ounce}}{28.35 \text{ g}}$
kg to lb	$\dfrac{2.2046 \text{ lb}}{1 \text{ kg}}$
km to mile	$\dfrac{1 \text{ mile}}{1.6093 \text{ km}}$
m to mile	$\dfrac{1 \text{ mi}}{1{,}609.3 \text{ m}}$
L to quart	$\dfrac{1.06 \text{ quarts}}{1 \text{ L}}$
L to gallon	$\dfrac{0.264 \text{ gallon}}{1 \text{ L}}$
L to cubic inches	$\dfrac{61 \text{ in}^3}{1 \text{ L}}$
cubic cm to ml	$\dfrac{1 \text{ mL}}{1 \text{ cm}^3}$
cubic cm to L	$\dfrac{1 \text{ L}}{1{,}000 \text{ cm}^3}$
cubic cm to cubic inch	$\dfrac{1 \text{ in}^3}{16.387 \text{ cm}^3}$

The process will be to choose a sequence of conversion factors where each factor cancels out one of the units and this continues until you end up with the desired unit. We shall illustrate this through several examples.

EXAMPLE 1: Convert 2 quarts into milliliters.

The process requires us to change quarts to L and then L to mL. This requires two conversion factors of

$$\frac{1 \ L}{1.06 \ \text{quarts}} \quad \text{and} \quad \frac{1,000 \ \text{mL}}{1L}$$

Now multiply by the conversion factors and cancel the units to obtain:

$$2 \ \text{quarts} = \frac{2 \ \cancel{\text{quarts}}}{1} \cdot \frac{1 \ \cancel{L}}{1.06 \ \cancel{\text{quarts}}} \cdot \frac{1,000 \ \text{mL}}{1 \ \cancel{L}}$$

$$= \frac{(2)(1)(1,000) \ \text{mL}}{1.06}$$

$$= 1886.79 \ \text{mL}$$

EXAMPLE 2: Convert 3.5 miles into cm.

The process requires us to change miles to m, and them m to cm. This requires two conversion factors of

$$\frac{1,609.3 \ \text{m}}{1 \ \text{mile}} \quad \text{and} \quad \frac{100 \ \text{cm}}{1 \ \text{m}}$$

Now multiply by the conversion factors and cancel the units to obtain:

$$3.5 \ \text{miles} = \frac{3.5 \ \cancel{\text{miles}}}{1} \cdot \frac{1,609.3 \ \cancel{\text{m}}}{1 \ \cancel{\text{mile}}} \cdot \frac{100 \ \text{cm}}{1 \ \cancel{\text{m}}}$$

$$= \frac{(3.5)(1,609.3)(100) \ \text{cm}}{1}$$

$$= 563,255 \ \text{cm}$$

EXAMPLE 3: Convert 49,650 g to tons.

The process requires us to change grams to ounces, then ounces to pounds and then pounds to tons. This requires three conversion factors of $\frac{1 \ \text{ounce}}{28.35 \ \text{g}}$, $\frac{1 \ \text{lb}}{16 \ \text{ounces}}$ and $\frac{1 \ \text{ton}}{2,000 \ \text{lb}}$.

Now multiply by the conversion factors and cancel the units to obtain:

$$49,650 \text{ g} = \frac{49,650 \text{ g}}{1} \cdot \frac{1 \text{ \cancel{ounce}}}{28.35 \text{ g}} \cdot \frac{1 \text{ \cancel{lb}}}{16 \text{ \cancel{ounce}}} \cdot \frac{1 \text{ ton}}{2,000 \text{ \cancel{lb}}}$$

$$= \frac{(49,650)(1)(1) \text{ ton}}{(1)(28.35)(16)(2,000)}$$

$$= 0.0547 \text{ ton}$$

EXAMPLE 4: Convert 450 cubic centimeters to mL.

The process requires us to change cc's (cubic centimeters) to cubic inches, then cubic inches to L, then L to mL.

This requires three conversion factors, $\dfrac{1 \text{ in}^3}{16.387 \text{ cc}}$, $\dfrac{1 \text{ L}}{61 \text{ in}^3}$, and $\dfrac{1,000 \text{ mL}}{1 \text{ L}}$.

Now multiply by the conversion factors and cancel the units to obtain:

$$450 \text{ cc} = \frac{450 \text{ \cancel{cc}}}{1} \cdot \frac{1 \text{ \cancel{in}^3}}{16.387 \text{ \cancel{cc}}} \cdot \frac{1 \text{ \cancel{L}}}{61 \text{ \cancel{in}^3}} \cdot \frac{1,000 \text{ mL}}{1 \text{ \cancel{L}}}$$

$$= \frac{(450)(1)(1)(1,000) \text{ mL}}{(1)(16.387)(61)(1)}$$

$$= 450 \text{ mL}$$

We have shown that 1 cc = 1 mL!

Temperature Measurements and Conversions

Temperature measurements are made to determine the amount of heat flow in an environment. To measure temperature, it is necessary to establish relative scales of comparison. Three temperature scales are in common use today. The general temperature measurements we use on a day-to-day basis in the United States are based on the Fahrenheit scale. In science, the Celsius scale and the Kelvin scale are used. The figure below shows a comparison of the three scales.

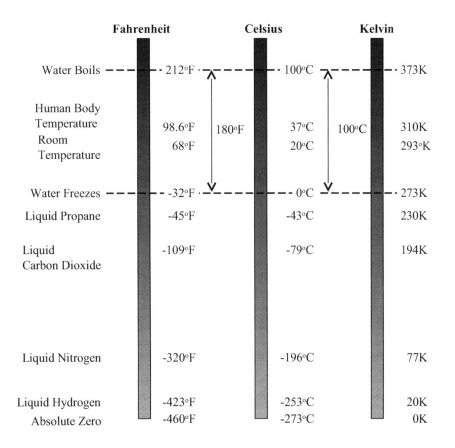

The Fahrenheit scale, named for its developer, was devised in the early 1700's. This scale was originally based on the temperatures of human blood and salt-water, and later on the freezing and boiling points of water. Today, the Fahrenheit scale is a secondary scale defined with reference to the other two scientific scales. The symbol °F is used to represent a degree on the Fahrenheit scale. About thirty years after the Fahrenheit scale, Anders Celsius, a Swedish astronomer suggested that it would be simpler to use a temperature scale divided into one hundred degrees between the freezing and boiling points of water. For many years, his scale was called the centigrade scale. In 1948 an international conference of scientists named it the Celsius in honor of its inventor. The Celsius degree, °C, was defined as 1/100 of the temperature difference between the freezing point and boiling point of water.

In the 19th century, an English scientist, Lord Kelvin, established a more fundamental temperature scale that used the lowest possible temperature as a reference point for the beginning of the scale. The lowest possible temperature, sometimes called absolute zero, was established as 0 K (zero Kelvin). This temperature is 273.15°C below zero, or ⁻273.15°C. Accordingly, the Kelvin degree, K, was chosen to be the same as a Celsius degree so that there would be a simple relationship between the two scales.

Note that the degree sign (°) is not used when stating a temperature on the Kelvin scale. Temperature is stated simply as Kelvins (K). The Kelvin was adopted by the 10th Conference of Weights and Measures in 1954 and is the SI unit of thermodynamic temperature. Note that the degree Celsius (°C) is the SI unit for expressing Celsius temperature and temperature intervals. The temperature interval of one degree Celsius equals one Kelvin exactly. Thus, 0°C = 273.15 K by definition.

To convert from one unit system to another, the following formulas are used:

Conversion	Formulas
°F to °C	$°C = \left(\dfrac{5}{9}\right)(°F - 32)$ or $°C = \dfrac{(°F - 32)}{1.8}$
°C to °F	$°F = 1.8(°C) + 32$ or $°F = \left(\dfrac{9}{5}\right)(°C) + 32$
°C to K	$K = °C + 273.15$
K to °C	$°C = K - 273.15$

EXAMPLE 1: Convert 65° Fahrenheit to Celsius.

$$°C = \frac{(65°F - 32)}{1.8}$$
$$°C = \frac{33}{1.8}$$
$$°C = 18.3°$$

EXAMPLE 2: Convert 39° Celsius to Fahrenheit.

$$°F = 1.8(39°C) + 32$$
$$°F = 70.2 + 32$$
$$°F = 102.2°$$

EXAMPLE 3: Convert 28° Celsius to Kelvin.

$$K = 28°C + 273.15$$
$$K = 301.15$$

EXERCISES 4.3

Find the following. If an exact answer is not possible, then round all answers to the hundredths place, unless stated otherwise.

Conversions Within the Metric System: Convert the following measurements.

1. 22 m to mm
2. 275 g to kg
3. 15 cm to m
4. 0.15 m to cm

5. 0.034 m to mm
6. 0.049 m to μm
7. 3,200 km to m
8. 68 m to km

9. 22 L to mL
10. 26 nm to μm
11. 9 mm to cm
12. 0.2 cm to mm

Conversions Within the English System: Convert the following measurements.

13. 2 miles to feet
14. 4.5 yards to feet
15. 15 quarts to gallons

16. 1.25 miles to feet
17. 6 yards to inches
18. 1.5 miles to yards

19. 74 fl. ounces to quarts
20. 2.5 yards to inches

Conversions Between the English and Metric Systems: Convert the following measurements.

21. 22 inches to cm
22. 2 miles to km
23. 38 km to miles
24. 4952 meters to miles
25. 237 feet to meters
26. 47 mL to fl. ounces
27. 5 gallons to Liters

28. 68.9 cm to inches
29. 597 meters to feet
30. 75 yards to meters
31. 69 fluid ounces to milliliters
32. 190 pounds to kilograms
33. 67 mm to feet
34. 1843 ounces to kg

35. 2350 μm to inches
36. 197 mm to inches
37. 16 kilometers to feet
38. 0.5 liters to ounces
39. 45,000 grams to pounds
40. 0.007 years to minutes

Temperature Conversions: Convert the following measurements.

41. -25° C to °F
42. 38° C to °F
43. 78°F to °C

44. -10°F to °C
45. 307.5 K to °C and to °F
46. 205° F to K

47. 41°F to °C and to K
48. 77°F to °C and to K

Applications:

49. One of the largest hailstones ever measured weighed 0.77 kg. How many pounds does that represent?
50. The geostationary weather satellites orbit above the equator at an altitude of 22,375 miles. What is it in km?
51. The National Weather Service often uses a radar frequency whose wavelength is 10 cm to detect precipitation. What is the wavelength in inches?

52. The air temperature in the upper regions of a thunderstorm may reach 210K. What is the temperature in $^\circ$C and $^\circ$F?

53. A man is 5 feet 11 inches tall. How many centimeters is this man's height? How many meters is the man's height?

54. A patient has a fever of 108° F. What is this in $^\circ$ C ?

55. An adult human being has about 5,200,000 red blood cells (RBCs) per μL of blood. If Joe has 5.05 L of blood, how many RBCs does he have?

56. If the field diameter of a microscope lens is known to be 4.65 mm, and you can fit 9 copies of an object across the diameter if they are placed end-to-end, how long is the object in μm?

57. You are viewing a cell that is 300 μm in diameter. How many millimeters is this?

4.4 THE ALGEBRA OF UNITS

Measurements are algebraic quantities and may be mathematically manipulated, subject to the rules of algebra. When working problems with measurements, it should be noted that the units follow the same algebraic rules as the numerical values, i.e.

$$(cm) \times (cm) = cm^2$$

$$\frac{ft^3}{ft} = ft^2$$

$$\frac{1}{yr} = yr^{-1}$$

Arithmetic of Units

Adding and Subtracting Measurements with Units
To add or subtract measurements with units, the units must be the same (type and scale.) Then to add you simply combine the numerical part of the measurement and keep the units as they are.

EXAMPLES:
1. 2 kg + 3 kg = 5 kg   Same type units (mass), and same scale.

2. 412 m – 12 m = 400 m   Same type units (length), and same scale.

If the units are not the same scale, you have to convert them first before doing the calculation.

EXAMPLES:
1. 4.021 kg + 324 g   Same type units (mass), but different sub-scales.
    They are the same overall scale, however, within the English and the metric scales there are sub-scales for the same type of units. Thus, in this case, we must convert either g to kg or kg to g.

$$4.021 \text{ kg} + 324 \text{ g} = 4.021 \text{ kg} + 0.324 \text{ kg} \quad \text{Convert the g term to kg}$$
$$= 4.345 \text{ kg} \qquad \text{Now combine like terms}$$

2. $8.455 \text{ Gm} - 6 \text{ Mm}$ Same type units (length), but different sub-scales.

    They are the same type of unit, but you have to change to the same sub-scale before you can add their coefficients. We must convert either Gm to Mm or Mm to Gm.

$$8.455 \text{ Gm} - 6 \text{ Mm} = 8.455 \text{ Gm} - 0.006 \text{ Gm} \quad \text{Convert the Mm term to Gm}$$
$$= 8.449 \text{ Gm} \qquad \text{Now combine like terms}$$

Note: It does NOT make any sense to add or subtract different types of units. For example you could not add: 3m + 4g, or 7L + 5cm, since it doesn't make any sense to add a length to a mass or a volume to a length, the units must be the same type and scale if you are asked to add or subtract them.

Multiplying and Dividing Measurements with Units

To multiply or divide measurements with units, you multiply or divide the numerical parts and the units part separately. The units follow the exponent laws of algebra under multiplication and division. Unlike addition and subtraction, the units are not required to be the same type, but if they are the same type, they must also be the same scale, if the units are to be canceled. Sometimes the units cancel without having to have all the same scale, as we'll show in an example below. However, this is more of the exception than the rule. We demonstrate this concept through some examples.

EXAMPLES:

1. $2 \text{ m} \times 5 \text{ m}$ Same type, and same scale units

$$2 \text{ m} \times 5 \text{ m} = (2 \times 5)(\text{m} \times \text{m})$$
$$= 10 \text{ m}^2$$

Here we multiplied two lengths to get an area.

2. $15 \text{ kg} \times 9.8 \text{ m/s}^2$ All are different types of units (same scale not required)

$$15 \text{ kg} \times 9.8 \text{ m/s}^2 = (15 \times 9.8)\left(\frac{\text{kg} \times \text{m}}{\text{s}^2}\right)$$
$$= 147 \frac{\text{kg m}}{\text{s}^2}$$

The units here are the newly derived units of force.

<u>Units may also cancel</u>

3. $15 \text{ m} / 5 \text{ m}$ Same type and same scale units

$$15\ m / 5\ m = \frac{15\ \cancel{m}}{5\ \cancel{m}}$$

$$= 3 \qquad \text{No units!}$$

In this problem we have no units. What we have actually found is the relative sizes of these two lengths to each other. 15 m, is 3 times longer than 5 m.

4. 4.00 m / s × 24.52 in / m   Some units are the same type, and the same scale.

$$4.00\ \ m/s \times 24.52\ \ in/m = (4.00 \times 24.52) \left(\frac{m}{s} \times \frac{in}{m} \right)$$

$$= 98.08\ in/s$$

Measurements are subject to algebraic laws and can be added, subtracted, multiplied, and divided. However, when the type of measurement is the same, we need to be sure that the scale is also the same.

5. 10kg / 250g   Same type units, but different sub-scales (kg and g).

We must convert either the kg to g or the g to kg before we can carry out the division. Let's convert 250g to kg.

$$250\ g \times \frac{1\ kg}{1,000\ g}\ =\ \frac{250}{1,000}\ kg\ =\ 0.25\ kg$$

$$\frac{10\ kg}{250\ g}\ =\ \frac{10\ \cancel{kg}}{0.25\ \cancel{kg}}\ =\ 40$$

The units cancel, because this is simply a relative comparison. It means that the 10kg object is 40 times heavier than the 250 g object.

EXERCISES 4.4

Perform the indicated operations. If an exact answer is not possible, then round all answers to the hundredths place, unless stated otherwise.

1. 36 cm − 15 cm

2. 6.9 kg + 1.5 kg

3. 24 mg − 15.3 mg

4. 57.6 km + 153 km

5. 1.5 nm − 0.3 nm

6. 2 cm + 4 m

7. 3.7 kg − 400 g

8. 378 mL + 4 L

9. 2 L + 46 mL

10. 2 mm + 40 μm

11. 2.5 cm − 4.3 mm

12. 13.9 m − 64.3 cm

13. 3 m × 12 m

14. 3 kg × 5.2 kg

15. 5 mm × 13 mm

16. 22 cm × 3 cm × 1 cm

17. 1.5 mm × 2.5 mm × 2.0 mm

18. 2 cm × 3 mm × 4 cm

19. 4.4 m × 75 cm × 2 m

20. 3.0 kg × 9.8 m/s^2

21. 4.0 kg × 9.8 m/s^2

22. 35 kg / 7 kg

23. 4.2 miles / 6 miles

24. 3.6 mm / 3 mm

25. 76 mL / 2 mL

26. 22 kg / 11 m^3

27. 16 lb / 8 in^3

28. 44 mL / 8 cm^3

29. 14.2 mm^2 / 2 mm

30. 5 cm^3 / 10 mm

31. 2 L^2 / 500 mL

32. 15 cm^2 / 5 m

33. 12.8 kg^3 / 40 g

34. 6 m/s × 12 s/m

35. 14 kg/m × 3 m

36. 42.5 g/cm × 2 cm

37. 15 g ÷ 3 cm^3

38. 33.3 kg ÷ 11.1 m^3

39. 24 mm^2 ÷ 8 cm

40. 1.5 μm^2 ÷ 3 nm

Chapter 4 Practice Test

Convert the following measurements. If an exact answer is not possible, then round all answers to the thousandths place, unless stated otherwise.

1. 67 mm to feet

2. 1843 ounces to kg

3. 1,600 mm to m

4. 2350 micrometers (μm) to inches.

5. 205° F to ° C

6. ⁻25° C to °F

7. 22 inches to cm

8. 2 miles to km

9. 38 km to miles

10. 4952 meters to miles

Perform the indicated operations. If an exact answer is not possible, then round all answers to the thousandths place, unless stated otherwise.

11. $36\,\text{cm} - 0.15\,\text{m}$

12. $6.9\,\text{kg} + 1.5\,\text{kg}$

13. $24\,\text{mg} - 15.3\,\text{mg}$

14. $57{,}600\,\text{m} + 153\,\text{km}$

15. $1.5\,\text{nm} - 0.3\,\text{nm}$

16. $42.5\,\text{g}/\text{cm} \times 2\,\text{cm}$

17. $15\,\text{g} \div 3\,\text{cm}^3$

18. $33.3\,\text{kg} \div 11.1\,\text{m}^3$

19. $24\,\text{mm}^2 \div 8\,\text{cm}$

20. $1.5\,\mu\text{m}^2 \div 3\,\text{nm}$

Solve the following problems. If an exact answer is not possible, then round all answers to the thousandths place, unless stated otherwise.

21. Weather satellites typically give visible light and infrared images of the earth. Infrared images are often taken at a wavelength of 10 micrometer (μm). What is the wavelength in inches?

22. If the field diameter of a microscope lens is known to be 1.86 mm, and you can fit 3.5 copies of an object across the diameter if they are placed end-to-end, how long is the object in μm?

23. An adult human being has about 5,200,000 red blood cells (RBCs) per μL of blood. If Sal has 4.75 L of blood, how many RBCs does she have?

24. You are viewing a cell that is 272.5 μm in diameter. How many millimeters is this?

CHAPTER 5

Basic Geometry

Geometry is essentially about one unit of measure, that of length, and its two products: area and volume. Another aspect associated with geometry is that of angle. Now an angle is not technically a unit or an attribute. Instead an angle is a measure of the relative orientation of two objects (lines), with respect to each other. In this chapter we focus on gaining a better understanding of this subject we call geometry, since it is critical for any college level math or science class.

5.1 WHAT IS GEOMETRY?

Geometry probably had its first start in humans, for practical reasons. Just as many animals know that the shortest path between two objects is a straight line, so too would humans see this. It is quicker to cut across a field than to walk around its perimeter to get to the other side. Also, questions of – How can we divide up a food item, that may have a complicated shape, so that we can share it with others fairly? – were bound to come up and needed to be addressed to everyone's satisfaction.

It is in ancient Egypt and Sumeria (Mesopotamia) that the beginnings of our modern geometric concepts probably arose. Perhaps the early Egyptians and Sumerians visualized lines coming together, so they formed a new concept, that of an angle. It was a way to describe how two straight edges meet, or how to compare how rapidly different pyramids rose to the heavens. It could also be very useful in constructing such pyramids, and an almost limitless numbers of other structures, too. The corners of many buildings needed to be a right (ninety degree) angle to help assure their stability. If the angles of blocks were not correct, you could not properly stack them to create walls that would not collapse.

Furthermore, in 3000 BCE in Egypt, farmland needed to be measured so that taxes could be assessed. What made matters difficult was that the Nile river flooded on an annual basis. Consequently, the land areas had to be recalculated every year, so that taxes could be assessed accurately. Furthermore, the shapes of the pieces of property could be quite complex. To measure these parcels of land; fairly and accurately, geometric formulas for area were created. Formulas for the areas of squares, rectangles, triangles, circles and other more complicated shapes were needed. It was out of this practical necessity that modern geometry was born.

It is not surprising that geometry began as an experimental tool for controlling and measuring the spatial patterns in the world. This is how we try and decipher Nature's secrets – we experiment. The early Sumerians (predecessors of the Babylonians) had simple formulas for finding lengths, areas and volumes of certain shaped objects. The Babylonians, too, are accredited with many geometric findings and ideas, including the concept of a degree, and that a circle traces out 360 degrees. All these early cultures seemed to aid and affect each other in the development of geometry.

Many formulas were created and used by both the Babylonians and Egyptians to build pyramids and construct great buildings. To them, geometry was first and foremost a tool. It was nothing more than a set of formulas and concepts, for them to be able to build and shape their environment. Geometry was born out of our need to understand and control our environment.

Eventually, the Ancient Greeks would transform this discipline, and create one of the most influential subjects in the history of human thought and reason. In fact, the word geometry comes from the Greek word **geometrein**, which means "earth measuring." Eventually Euclid in Ancient Greece, organized this subject matter into a book called the Elements, that is second only to the Bible in its number of editions and publications, and created the groundbreaking approach to mathematics and science called the axiomatic approach.

In this chapter, we only explore the practical tool aspect of geometry. We will not explore the intricate details of the axiomatic approach, nor will we study its impact on the development of scientific reasoning that came out of Euclid's Elements, called Euclidean Geometry. We will simply state that the axiomatic approach to understanding our world is to break down whatever we are studying into its fundamental parts (which begin with things we cannot define, but must simply take for granted) and then from these we define other things along with rules (axioms) on how these things relate to and interact with each other. It is a step–by–step approach to inquiry.

5.2   THE CONCEPT OF DIMENSION

As Euclid and those that came before him observed, it is possible to organize and make sense out of the spatial or geometric patterns that are around us. We begin the process by starting out systematically, and slowly. We first transport ourselves into a perfect world of "regular" geometric forms. We look around and we "see" lines, rays, angles, squares, triangles, spheres, cones, rectangles, cylinders, circles, etc. They all seem so different, yet at the same time have similar traits. The first question to arise is: Can we describe this spatial world?

We look around at our geometric world and the first things we see are "solid" objects. We can reach out and touch them. These are things we have come to call spheres, cubes, pyramids, cones, etc., but how can we begin to describe and define them in terms that others will understand? What do they all have in common?

First, all solids have boundaries, where a boundary can be thought of as a transition region between the solid and the space around it. We call these boundaries **surfaces**. This is where things now begin to get a bit strange. A surface can stretch across an area and wrap itself around a solid, but it has NO thickness. It is an imaginary boundary between the solid and the rest of the world. In the real world, a surface cannot exist without a solid, but in our abstract geometrical world we need this strange object to exist on its own to be able to describe our solid. So we must now introduce our first undefined, but necessary, object: the surface. We'll describe it as an object with a length (something we've yet to discuss, but which we have an intuitive sense of) and a width (another length, but in a different "direction," also called breadth), but with no depth (or thickness). This is really strange if you think about it. Something with zero thickness can't really exist, so it isn't really there!

Now if we consider this strange new object called a surface by itself, we observe that it too has boundaries. The "edges" of the surface we'll call lines or curves, and the **lines** or **curves** bound or contain the surface. Edges, lines or curves are objects with length, but no breadth, and with no depth. Again, in reality we can never see a line, since they have zero thickness and zero depth, but we'll let them into our geometric world as another undefined, but necessary, object. Now, if we look at the line, we see that it too has a boundary at its ends. We'll call these boundaries, **points**. What is a point? We can't really say. Is it a spot in space that has no size or shape, or it is something without a part? To us it is also undefinable, but without it we cannot even begin to fully describe our world. It is, however, our starting point, no pun intended.

A point has no length, no breadth, and no depth. This is the most fundamental object in our system. Now a point can either be an object itself (that which has no part, according to Euclid), or it can be a location in space. The space description is less confusing and causes fewer problems, but we need the concept of a point to be defined as an object as well, so that we can define other objects in the world in terms of it.

The undefinable concept we are really dancing around in the previous paragraphs is that of dimension. Dimension is the most basic and fundamental undefined quantity in geometry. What we are really saying above, is that an object of dimension zero is a point, an object of dimension one is a line (length), an object of dimension two is a surface (area), and an object of dimension three is a solid (volume) – finally something we can see! This thing that we cannot define (nor can we explain it in terms of something we've already defined) is the concept of dimension. Dimension is the true undefinable quantity, since any definition we try, would have to reference another undefined term. For example, take a common definition for dimension from dictionary.com.

> **Dimension:** A property of space; extension in a given direction: A straight line has one dimension, a parallelogram has two dimensions, and a parallelepiped (solid) has three dimensions.

In this definition, we needed the term "space", as well as the concept of "extension in a direction", to define dimension. However, to define space and direction we really need the concept of dimension. So this really isn't a definition at all – it is circular. We are trying to define a concept using a different name for the same concept.

It turns out that we could go round and round forever and never get anywhere with anything. Like it or not, to make any progress, we will have to agree that there are things that we simply cannot define. We instead must agree that they exist and move forward. We allow our intuition to guide us, so that we accept that we already know what space and direction are, even though we cannot define it in terms of different concepts we already know. If you think about it, this is true of anything. Anything we try to describe or understand, must inevitably lead back to something that we all have to take as understood, without a true definition. We have to start somewhere, and that somewhere is accepting that the concept of dimension is NOT definable, but in some sense, it is describable.

For our purposes in this book, we will view the idea of dimension more concretely, and simply. We accept that our world has 3 unique directions (right–left, forward–back, and up–down) that we can move in, or as we say it has 3–dimensions. The distance we move in each of these directions is quantified by a length. Thus, there are 3 unique length directions in our world. If we are restricted to move in only one direction, we call the length, dimension one. If we can move in two distinct directions, we say this is two dimensions, and if we can move in three distinct directions, we say this is three dimensions. If we cannot move anywhere, we call this point a zero dimension. Objects that exist in these various dimensions have uniquely defined units associated with; length, area and volume. We introduced units for these common types of objects back in Chapter 4.

Whereas in Chapter 4 we discussed four fundamental units; length, mass, time and temperature, in this chapter on geometry we only need one unit: length, and its derived units of area and volume. We need to pay close attention to the proper units for each, as indicated by their dimension.

A **length** (dimension 1) has basic units of length such as; meter (m), centimeter (cm), inches (in), foot (ft), mile (mi).

An **area** (dimension 2) has basic units of length squared, which is why it is a 2 dimensional object. These units are of the form; square meters (m^2), square centimeters (cm^2), square inches (in^2), square feet (ft^2), square miles (mi^2).

A **volume** (dimension 3) has basic units of length cubed, which is why it is called a 3 dimensional object. These units are of the form; cubic meters (m^3), cubic centimeters (cm^3), cubic inches (in^3), cubic feet (ft^3).

In the next section we introduce some common formulas associated with useful 1–, 2–, or 3– dimensional objects.

EXERCISES 5.2

Determine the dimension of each object described below, and give a possible example of units to measure it.

1. The circumference of a circle
2. A bag of cement
3. The surface of a desk
4. The length of a string
5. A bucket full of water
6. A sheet for a bed
7. The distance to the sun
8. The size of a planet
9. Cargo space of an SUV
10. Laptop screen size (as advertised)
11. Waist measurement for jeans
12. Fuel tank capacity of a school bus
13. Supermarket receipt
14. Kitchen cabinet door
15. Length of Olympic size swimming pool
16. The capacity of a washing machine
17. The four walls of a room
18. Topsoil dumped from a truck
19. A path around a lake
20. A pool cover
21. A bottle of soda
22. The interior space of an automobile

5.3 BASIC GEOMETRIC FORMULAS

We are going to learn how to calculate the lengths and areas associated with some basic planar figures, as well as volumes of some basic three-dimensional figures.

Aside from the length of a side of an object, another common length we are asked to find is the perimeter. The **perimeter** is the length along the outside or boundary of the object. You can visualize this as taking a flexible tape measure and wrapping it around the outside of the object, and then pulling the tape measure straight and reading how far around it is. See the figure below showing the idea of a perimeter of a planar object.

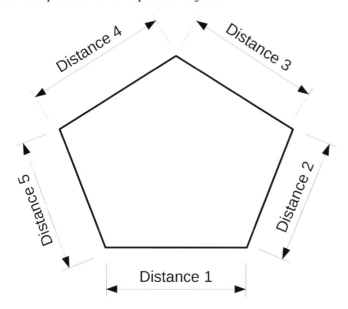

Perimeter is the sum of all the distances around the object. Its units are always a length.

Another way to think of finding or measuring the perimeter, is to imagine using unit length straight edges, and laying them end to end around the boundary of the object, and counting how many of these lengths we have. That is the measure of the perimeter. The unit length we use could be any measure we want, such as: inch, foot, mm, and cm.

The other common quantity we calculate is the **area** of an object. In area calculations, we count the number of squares, that are of a fixed unit size, that will completely cover the object. This is why units of area are units squared. Sometimes we have an exact fit, although other times we cannot fit an entire square into our object, and can only use part of it. In this case, we see decimal or fractional areas. See the figure below.

11	12	13	14	15
6	7	8	9	10
1	2	3	4	5

Area counts the number of squares it takes to completely cover an object

It takes 15 squares (square units) to cover this rectangle so its area is 15 units ^{2}

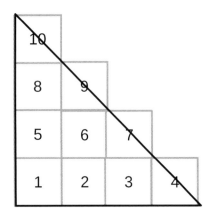

It takes less than 16 whole squares to cover this circle. In fact it takes a little over 12 and one half squares, so its area is around 12.5 units ^{2}

It takes 6 whole squares and 4 half squares for a total of 8 squares to cover this triangle so its area is 8 units ^{2}

The last quantity that we are interested in calculating is the **volume**. The volume counts how many cubes of a fixed size it takes to completely fill our solid, or three-dimensional object. See the figure below to get the general idea.

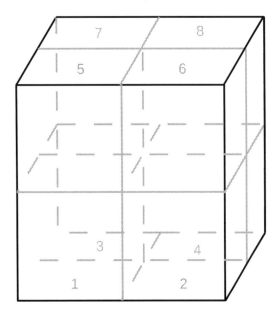

It takes 8 whole cubes to fill this larger cube, so its volume is 8 units ^{3}

We could go about "covering" quantities we are asked to measure with either lines, squares or cubes, but that would be fairly difficult and time consuming, and may not be accurate enough for some of our needs. This is the power of geometry. It provides formulas that give us the same results as above, but much more easily, and our answers can be as precise as we like. What Euclid's Elements gave us was a way to prove a whole host of perimeter, area and volume calculations in the form of formulas.

Some of these formulas are given below:

Object	Figure	Formulas
Square		Perimeter = $4 \times s$ Area = s^2
Rectangle		Perimeter = $2L + 2W$ Area = $L \times W$
Triangle		Perimeter = $a + b + c$ Area = $\frac{1}{2} \times b \times h$
Parallelogram		Perimeter = $2a + 2b$ Area = $b \times h$

Object	Figure	Formulas
Trapezoid		Perimeter = add all sides Area = $\frac{1}{2} \times (\text{base}\,1 + \text{base}\,2) \times h$
Circle		Circumference = $2\,\pi\,r$ Area = $\pi\,r^2$
Prism		Volume = $L \times W \times H$
Cylinder		Volume = $\pi\,r^2\,h$

Object	Figure	Formulas
Sphere	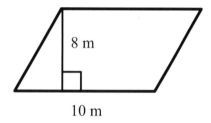	Volume = $\dfrac{4}{3}\pi r^3$

EXAMPLE 1: Find the area of the given parallelogram.

8 m

10 m

Area of the parallelogram $= b \times h$
$= 10\,\text{m} \times 8\,\text{m}$
$= 80\,\text{m}^2$

EXAMPLE 2: If the base and height of a parallelogram are 12 cm and 6 cm respectively, what is its area?

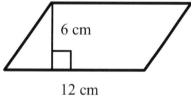

6 cm

12 cm

$$\text{Area} = 12\,\text{cm} \times 6\,\text{cm} = 72\,\text{cm}^2$$

EXAMPLE 3: Given a rectangle with length 6.3 in and width 3.5 in. find the perimeter and area of this rectangle.

6.3 in

3.5 in

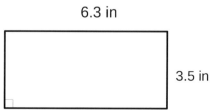

Perimeter = 2L+2W = 2(6.3in) + 2(3.5in) = 12.6in + 7in = = 19.6 in

Area = LW = (6.3in)(3.5in) = 22.05 in^2

EXAMPLE 4: Find the circumference of a circle if its radius is 5.28 cm. (Use the approximation for $\pi = 3.14$) Round your answer to the nearest hundredths.

Circumference $= 2\,\pi\,r$

$$C=2(3.14)(5.28\,cm) = 32.97\,cm$$

The radius of the circle is 32.97 cm.

EXAMPLE 5: Find the area of a trapezoid with the dimensions given on the diagram.

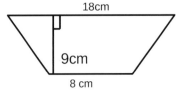

18cm

9cm

8 cm

Area of the trapezoid $= \dfrac{1}{2}\times(\text{base 1}+\text{base 2})\times h$

$$= \dfrac{1}{2}\times(18\,cm+8\,cm)\times9\,cm$$

$$= 117 \text{ cm}^2$$

EXAMPLE 6: Find the volume of a rectangular box with width 15cm, length 20cm, and height 10cm.

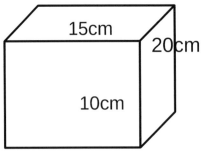

15cm

20cm

10cm

Volume $=$ L×W×H

$= 20\,cm\times15\,cm\times10\,cm$

$= 3000 \text{ cm}^3$

Since this is three dimensions, the units are cubic centimeters.

EXAMPLE 7: Find the volume of a cylinder with radius 8.4 inches and height 11 inches. Use the approximation $\pi = 3.14$.

$$
\begin{aligned}
\text{Volume} \ &= \pi\, r^2 h \\
&= 3.14(8.4 \text{ in})^2(11\text{in}) \\
&= 2437.1424 \text{ in}^3
\end{aligned}
$$

or $2{,}437.14$ in^3 rounded to two decimal places.

EXAMPLE 8: Find the area of a rectangle that measures 2 feet in length by 13 inches in width.

To determine the area, we must multiply the length by the width. The measurements are in different units, and cannot be multiplied directly. We need to convert one of the units so that both units are the same type. If we convert the first unit to inches, we have:

$$
\begin{aligned}
2 \text{ ft} \times 13 \text{ in} &= \left(2 \text{ ft} \times \frac{12 \text{ in}}{\text{ft}}\right) \times 13 \text{ in} \\
&= 24 \text{ in} \times 13 \text{ in} \\
&= 312 \text{ in}^2
\end{aligned}
$$

If we instead convert the second unit to feet, we have:

$$
\begin{aligned}
2 \text{ ft} \times 13 \text{ in} &= 2 \text{ ft} \times \left(13 \text{ in} \times \frac{\text{ft}}{12 \text{ in}}\right) \\
&= 2 \text{ ft} \times 1.08\,\overline{3} \text{ ft} \\
&= 2.1\,\overline{6} \text{ ft}^2
\end{aligned}
$$

Both of these results are equal in size – they are just expressed in different units. Which one we prefer depends upon the problem at hand. If we are buying a product that is sold as in^2 we would choose the first value. However, if the product is sold as ft^2, we would choose the second.

EXERCISES 5.3

Answer the following. If an exact answer is not possible, then round all answers to the thousandths place, unless stated otherwise.

Basic geometric formulas

1. Find the perimeter and area of each of the following figures:

a) square

b) rectangle

c) trapezoid

d) circle

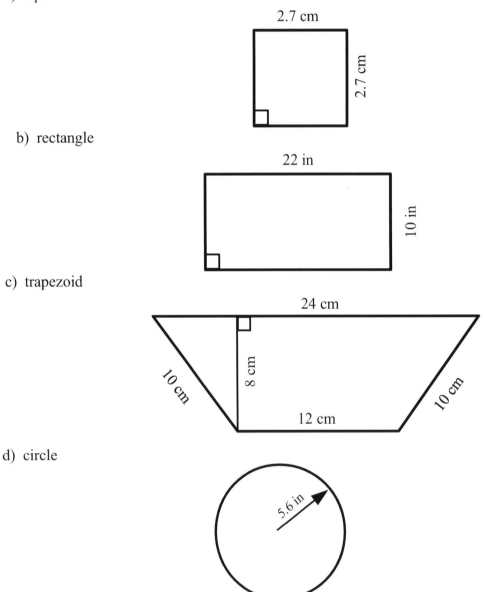

2. Find the area of a parallelogram having a base of 18 inches and height of 8 inches. Draw a parallelogram and label the dimensions.

3. The area of a triangle is 49 cm^2 and the height is 14 cm. Find the base of this triangle. Draw a diagram of the triangle.

4. Find the area of the Norman window by calculating the areas of the rectangle and semi-circle and adding the two areas. Approximate your answer using π = 3.14

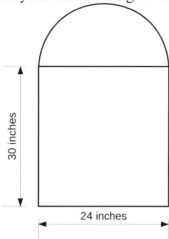

30 inches

24 inches

5. Find the volume of a cube given one side.

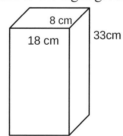

20cm

6. Find the volume of a rectangular prism to the right given the sides as shown.

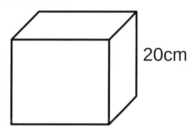

8 cm

18 cm

33cm

7. Find the volume of the cylinder below. Approximate your answer using π = 3.14.

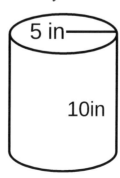

5 in

10in

8. Find the volume of a sphere with radius 27 inches. Approximate your answer using π = 3.14.

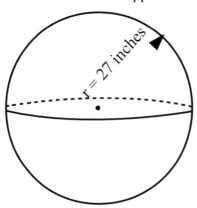

9. A flower garden is circular in shape with radius 8 feet. Tony wants to build a walkway 2 feet wide around the garden. What is the area of the walkway?

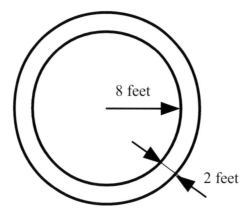

10. Find the volume of a rectangular solid that measures 3 feet in length by 18 inches in width, by 6 inches in height.

11. Find the area of a rectangle that measures 15 mm in length by 2 cm in width.

12. Find the area of a rectangle that measures 3.5 m in length by 64 cm in width.

13. Find the volume of a rectangular solid that measures 3 cm by 2 cm by 5 cm.

14. Find the area of a circle that is 30 mm in diameter.

15. Find the area of a circle with a radius of 6 cm.

16. Find the area of the following trapezoid.

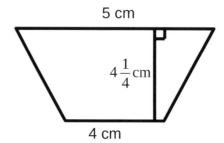

17. The bases of a trapezoid are 4.6 cm and 10.5 cm. The height of the trapezoid is 3.4 cm. Find the area of this trapezoid.

18. Find the circumference of a circle that has a radius of 3.4 cm. Use $\pi = 3.14$.

19. If the radius of a circle is 5.2 inches, what is the circumference? Use $\pi = 3.14$

20. If the radius of a circle is 2.8 cm, find the area of this circle. Use $\pi = 3.14$.

21. If the radius of a circle is 4.5 cm, find the circumference of this circle. Use $\pi = 3.14$.

22. A cylinder has a radius of 2.5 inches and its height is 4.1 inches. Find the volume of this cylinder. Use $\pi = 3.14$

23. A cylinder has a radius of 9.2 cm and its height is 14.4 cm. Find the volume of this cylinder. Use $\pi = 3.14$

24. The radius of a cylinder is 8 cm, and the height is 1.6 cm. Find the volume of the cylinder. Use $\pi = 3.14$

25. The radius of a sphere is 5.2 inches. Find the volume of the sphere. Use $\pi = 3.14$. Round your answer to the nearest tenths

26. The radius of a sphere is 7.4 cm. Find the volume of the sphere. Use $\pi = 3.14$. Round your answer to the nearest tenths

27. If a circular gear has a diameter of 8 inches, what is its radius?

28. If a cube has a side of length 28 cm, what is the volume of this cube? What is the total surface area of this cube?

29. If a cylinder has a radius of 4 cm and height of 8 cm, what is the volume of this cylinder? Use the formula for the volume of a cylinder $V = \pi r^2 h$ where r is the radius and h is the height of the cylinder. Use $\pi = 3.142$.

5.4 ANGLES IN GEOMETRY

In addition to length, area, and volume, another feature we must be aware of, is how objects are oriented with respect to each other. We call this measure of orientation an **angle**. Angles are most easily visualized by sketching two intersecting lines and identifying a measure of how their orientation differs from each other. See the figure below:

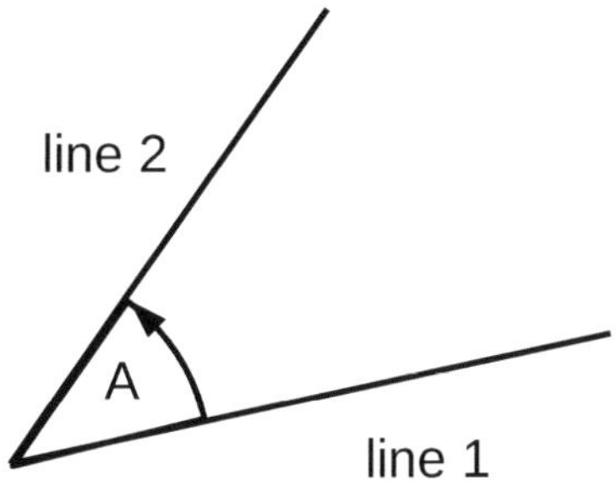

line 2

A

line 1

The value of "A" is a measure of the relative orientation of the two lines. We call this the angle between line 1 and line 2

There are many ways to create a scale for a measure of "A." The most common way is to define something called a degree. The degree symbol is a superscript $^\circ$ to the right of a number, such as 30° (30 degrees). Next we need to give the degree symbol a value. As the angle "A" increases, the arc in the drawing traces out a circle. Since circles come back and repeat over each other, there is a cyclical aspect to an angle measure. Although nobody knows for certain, it is believed that since the number of days in a year is close to 360 and since our years cycle around and repeat, the first culture to introduce the degree measure, the Babylonians, used one degree to represent $1/360^{th}$ around the circle. Or more to the point, there are 360 degrees in a complete circle, just as there are about 360 days in a year.

A device called a protractor is used to measure angles. See figure below.

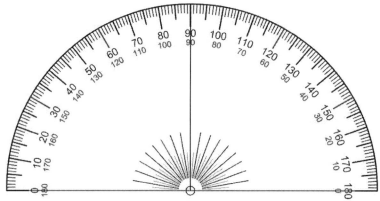

You simply place one line along the bottom of the device, and then read the angle where the other line crosses on the protractor.

Having defined what we mean by an angle and how to measure it, we introduce some terms and aspects related to this concept.

Some common angles and their measures:

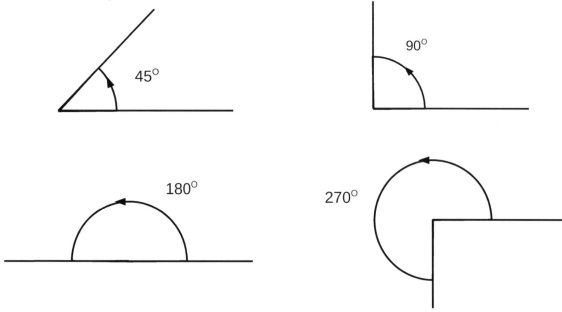

Some terminology and useful facts regarding angles:

1. A 90° angle is also called a **right angle**.

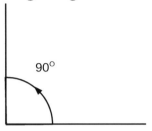

2. A small square where the angle would go indicates a 90° angle.

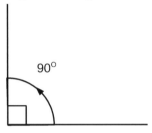

3. The interior angles of any triangle add up to 180°.

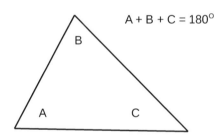

4. Angles on a line add to 180°.

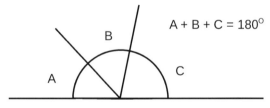

5. When two lines cross, opposite angles are equal.

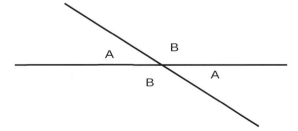

6. A triangle that has one angle equal to 90° is called a **right triangle**.

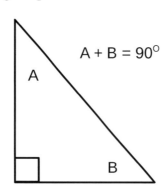

$$A + B = 90^\circ$$

7. In an **equilateral triangle** (all sides of equal length), the three interior angles are all 60°.

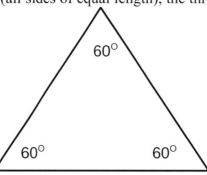

EXERCISES 5.4

Find the missing angle/angles.

1.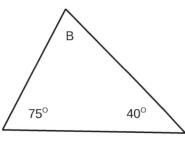

2.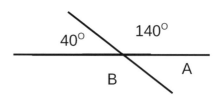

3.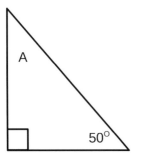

4.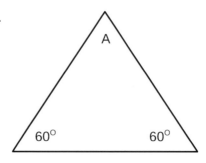

5.

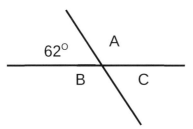

62° A

B C

6.

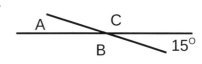

A C

B 15°

7.

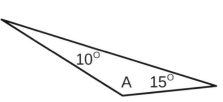

10°

A 15°

8.

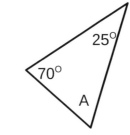

25°

70°

A

9.

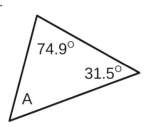

74.9°

31.5°

A

10.

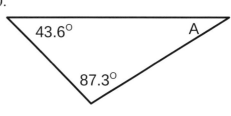

43.6° A

87.3°

11.

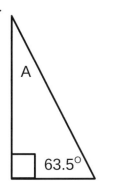

A

63.5°

12.

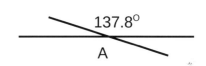

137.8°

A

13.

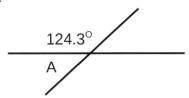

124.3°

A

14.

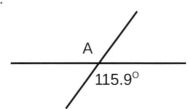

A

115.9°

15.

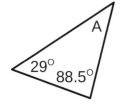

A

29° 88.5°

16.

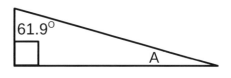

61.9°

A

17.

A

14.7°

18.

A   C

B   13.8°

19.

79.6°

19°   A

20.

A

136.9° 16°

5.5 UNITS IN GEOMETRY

As we said earlier, geometry has only three quantities that require units; length, area and volume. We already learned how to convert units of length in Chapter 4. We will now take a look at converting areas and volumes.

Area Conversions

Two common conversions of area values in geometry are the following:

1. ft^2 to in^2 We can see this visually in the figure (not drawn to scale) below:

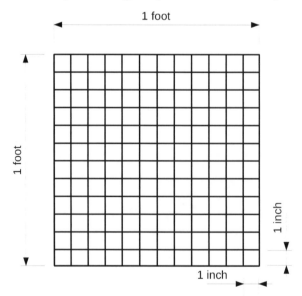

1 foot

1 foot

1 inch

1 inch

A 1 ft^2 area is equivalent to $12\,\text{in} \times 12\,\text{in} = 12^2\,\text{in}^2 = 144\,\text{in}^2$. We can see that there are 144 square inch boxes contained in the square foot above. So we have the conversion that:

$$1 \text{ ft}^2 = 144 \text{ in}^2$$

This also provides the conversion factors:

$$\frac{1\,\text{ft}^2}{144\,\text{in}^2} \quad \text{and} \quad \frac{144\,\text{in}^2}{1\,\text{ft}^2}$$

That can be used to convert from either in^2 to ft^2, as in the first factor, or from ft^2 to in^2, as in the second.

2. cm^2 to mm^2

We can see this visually in the figure (not drawn to scale) below:

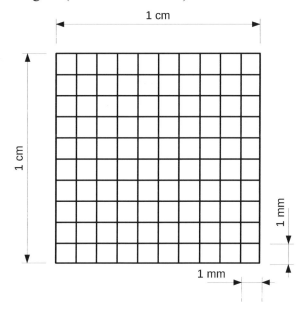

A 1 cm^2 area is equivalent to $10\,\text{mm} \times 10\,\text{mm} = 10^2\,\text{mm}^2 = 100\,\text{mm}^2$. We can see that there are 100 square mm boxes contained in the square cm above. So we have the conversion that:

$$1 \text{ cm}^2 = 100 \text{ mm}^2$$

This also provides the conversion factors:

$$\frac{1\,\text{cm}^2}{100\,\text{mm}^2} \quad \text{and} \quad \frac{100\,\text{mm}^2}{1\,\text{cm}^2}$$

That can be used to convert from either mm^2 to cm^2, as in the first factor, or from cm^2 to mm^2, as in the second.

EXAMPLE 1: Convert 8 ft² to in².

We use the conversion factor above to convert from ft² to in².

$$8\,\text{ft}^2 \;=\; 8\,\cancel{\text{ft}^2}\times\frac{144\,\text{in}^2}{1\,\cancel{\text{ft}^2}} \;=\; 8\times144\,\text{in}^2 \;=\; 1152\,\text{in}^2$$

EXAMPLE 2: Convert 32 cm² to mm².

We use the conversion factor above to convert from cm² to mm².

$$32\,\text{cm}^2 \;=\; 32\,\cancel{\text{cm}^2}\times\frac{100\,\text{mm}^2}{1\,\cancel{\text{cm}^2}} \;=\; 32\times100\,\text{mm}^2 \;=\; 3{,}200\,\text{mm}^2$$

Volume and Volume Conversions

A volume is a three dimensional object, so we expect to see it measured in cubic units. This is true when we measure solids, but not always true when we measure liquids. For example a gallon of milk is still a volume measure, but the gallon unit of measure does not explicitly show cubic units. Here we'll show both types of units as well as some common conversions.

Two common conversions of solid volume values in geometry are the following:
1. ft³ to in³

We can see this visually in the figure (not drawn to scale) below:

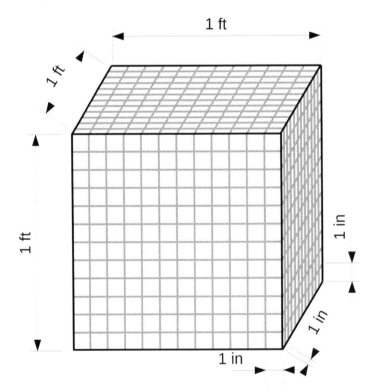

A 1 ft³ volume is equivalent to $12\,\text{in} \times 12\,\text{in} \times 12\,\text{in} = 12^3\,\text{in}^3 = 1{,}728\,\text{in}^3$. We can see that there are 1,728 cubic inch cubes contained in the cubic foot above. So we have the conversion that:

$$1\ \text{ft}^3 = 1{,}728\ \text{in}^3$$

This also provides the conversion factors:

$$\frac{1\,\text{ft}^3}{1{,}728\,\text{in}^3} \quad \text{and} \quad \frac{1{,}728\,\text{in}^3}{1\,\text{ft}^3}$$

That can be used to convert from either in³ to ft³, as in the first factor, or from ft³ to in³, as in the second.

2. cm³ to mm³

A picture of the relation between the two would be similar to the image above. The only difference is that we would have 10 instead of 12 cubes on each side. Thus, a 1 cm³ volume is equivalent to $10\,\text{mm} \times 10\,\text{mm} \times 10\,\text{mm} = 10^3\,\text{mm}^3 = 1{,}000\,\text{mm}^3$. We can visualize that there would be 1,000 mm cubes contained in the cubic cm above. So we have the conversion that:

$$1\ \text{cm}^3 = 1{,}000\ \text{mm}^3$$

This also provides the conversion factors:

$$\frac{1\,\text{cm}^3}{1{,}000\,\text{mm}^3} \quad \text{and} \quad \frac{1{,}000\,\text{mm}^3}{1\,\text{cm}^3}$$

That can be used to convert from either mm³ to cm³, as in the first factor, or from cm³ to mm³, as in the second.

3. m³ to cm³

In this problem, we don't use a visual aid, but instead use what we learned in the last chapter on units. We know that 1m = 100cm. If we now replace the meters in the cubic meters with centimeters, we have:

$$\text{m}^3 = \text{m} \times \text{m} \times \text{m} = 100\,\text{cm} \times 100\,\text{cm} \times 100\,\text{cm} = 100^3\,\text{cm}^3 = 1{,}000{,}000\,\text{cm}^3$$

This says that one cubic meters is equal to one million cubic centimeters!

$$1\ \text{m}^3 = 1{,}000{,}000\ \text{cm}^3$$

This also provides the conversion factors:

$$\frac{1{,}000{,}000\,\text{cm}^3}{1\,\text{m}^3} \quad \text{and} \quad \frac{1\,\text{m}^3}{1{,}000{,}000\,\text{cm}^3}$$

That can be used to convert from either m³ to cm³, as in the first factor, or from cm³ to m³, as in the second.

EXAMPLE 1: Convert 5 m³ to cm³.

We use the conversion factor above to convert from m³ to cm³.

$$5\,\text{m}^3 \;=\; 5\,\cancel{\text{m}^3}\times\frac{1,000,000\,\text{cm}^3}{1\,\cancel{\text{m}^3}} \;=\; 5\times1,000,000\,\text{cm}^3 \;=\; 5,000,000\,\text{cm}^3$$

EXAMPLE 2: Convert 3 ft³ to in³.

We use the conversion factor above to convert from ft³ to in³.

$$3\,\text{ft}^3 \;=\; 3\,\cancel{\text{ft}^3}\times\frac{1,728\,\text{in}^3}{1\,\cancel{\text{ft}^3}} \;=\; 3\times1,728\,\text{in}^3 \;=\; 5,184\,\text{in}^3$$

EXAMPLE 3: Convert 0.5 cm³ to mm³.

We use the conversion factor above to convert from cm³ to m³.

$$0.5\,\text{cm}^3 \;=\; 0.5\,\cancel{\text{cm}^3}\times\frac{1,000\,\text{mm}^3}{1\,\cancel{\text{cm}^3}} \;=\; 0.5\times1,000\,\text{mm}^3 \;=\; 500\,\text{mm}^3$$

Liquid Volume Measures and Conversions

Liquid volume measures do not contain cubic units. Instead, we have two fundamental liquid volume measures for the English and metric systems. In the English system, we use a gallon and in the metric system, we use a liter. We can, however, relate these liquid measures to solid measures that have the cubic units.

Liquid to solid volume conversions

In the metric system, 1 liter (L) is equivalent to 1,000 cm³. This is a fairly straightforward conversion, and was designed to be this simple. This leads to the two conversion factors:

$$\frac{1,000\,\text{cm}^3}{1\,\text{L}} \quad \text{and} \quad \frac{1\,\text{L}}{1,000\,\text{cm}^3}$$

that can be used to convert from either L to cm³, as in the first factor, or from cm³ to L, as in the second.

In the English system, 1 gallon (gal) is equivalent to 231 in³. As is typical with the English system of units, this is a fairly odd number. This leads to the two conversion factors:

$$\frac{231\,\text{in}^3}{1\,\text{gal}} \quad \text{and} \quad \frac{1\,\text{gal}}{231\,\text{in}^3}$$

that can be used to convert from either gal. to in^3, as in the first factor, or from in^3 to gal. as in the second.

To convert between the two systems, we have that 1 gal = 3785.412 cm^3, and that 1 L = 61.0237 in^3. Again, the conversions are fairly complicated. This leads to the two conversion factors:

$$\frac{3{,}785.412\,\text{cm}^3}{1\,\text{gal}} \quad \text{and} \quad \frac{1\,\text{gal}}{3{,}785.412\,\text{cm}^3}$$

that can be used to convert from either gal. to cm^3, as in the first factor, or from cm^3 to gal. as in the second.

EXAMPLE 1: Convert from 7 gal to cm^3.

We use the conversion factor above to convert from gal. to cm^3.

$$7\,\text{gal} \;=\; 7\,\cancel{\text{gal}} \times \frac{3{,}785.412\,\text{cm}^3}{1\,\cancel{\text{gal}}} \;=\; 7 \times 3{,}785.412\,\text{cm}^3 \;=\; 26{,}497.884\,\text{cm}^3$$

EXAMPLE 2: Convert 0.03L to cm^3.

We use the conversion factor above to convert from L to cm^3.

$$0.03\,\text{L} \;=\; 0.03\,\cancel{\text{L}} \times \frac{1{,}000\,\text{cm}^3}{1\,\cancel{\text{L}}} \;=\; 0.03 \times 1{,}000\,\text{cm}^3 \;=\; 30\,\text{cm}^3$$

EXAMPLE 3: Convert 0.75 gal to in^3.

We use the conversion factor above to convert from gal. to in.

$$0.75\,\text{gal} \;=\; 0.75\,\cancel{\text{gal}} \times \frac{231\,\text{in}^3}{1\,\cancel{\text{gal}}} \;=\; 0.75 \times 231\,\text{in}^3 \;=\; 173.25\,\text{in}^3$$

EXAMPLE 4: Convert 875in^3 to gal

We use the conversion factor above to convert from in. to gal.

$$875\,\text{in}^3 \;=\; 875\,\cancel{\text{in}^3} \times \frac{1\,\text{gal}}{231\,\cancel{\text{in}^3}} \;=\; \frac{875}{231}\,\text{gal} \;=\; 3.\overline{78}\,\text{gal}$$

All these conversion factors can be summarized in table form, as we did with the other types of conversions in Chapter 4.

Convert "from" "to"	Conversion Factor
in^2 to ft^2	$\dfrac{1\,ft^2}{144\,in^2}$
mm^2 to cm^2	$\dfrac{1\,cm^2}{100\,mm^2}$
in^3 to ft^3	$\dfrac{1\,ft^3}{1{,}728\,in^3}$
mm^3 to cm^3	$\dfrac{1\,cm^3}{1{,}000\,mm^3}$
cm^3 to m^3	$\dfrac{1\,m^3}{1{,}000{,}000\,cm^3}$
cm^3 to L	$\dfrac{1\,L}{1{,}000\,cm^3}$
in^3 to gal	$\dfrac{1\,gal}{231\,in^3}$
cm^3 to gal	$\dfrac{1\,gal}{3{,}785.412\,cm^3}$
in^3 to L	$\dfrac{1\,L}{61.0237\,in^3}$

Convert "from" "to"	Conversion Factor
ft^2 to in^2	$\dfrac{144\,in^2}{1\,ft^2}$
cm^2 to mm^2	$\dfrac{100\,mm^2}{1\,cm^2}$
ft^3 to in^3	$\dfrac{1{,}728\,in^3}{1\,ft^3}$
cm^3 to mm^3	$\dfrac{1{,}000\,mm^3}{1\,cm^3}$
m^3 to cm^3	$\dfrac{1{,}000{,}000\,cm^3}{1\,m^3}$
L to cm^3	$\dfrac{1{,}000\,cm^3}{1\,L}$
gal to in^3	$\dfrac{231\,in^3}{1\,gal}$
gal to cm^3	$\dfrac{3{,}785.412\,cm^3}{1\,gal}$
L to in^3	$\dfrac{61.0237\,in^3}{1\,L}$

EXERCISES 5.5

Perform the indicated conversions. If an exact answer is not possible, then round all answers to the thousandths place.

1. 4 ft^3 to in^3
2. 2.1 cm^2 to mm^2
3. 7.35 cm^3 to mm^3
4. 8.5 ft^2 to in^2
5. 2 gal to in^3
6. 0.05 L to cm^3
7. 3 gal to cm^3

8. 10 L to in^3
9. 500 in^2 to ft^2
10. 10368 in^3 to ft^3
11. 750000 cm^3 to m^3
12. 37 m^3 to cm^3
13. 1000 in^3 to gal
14. 0.58 L to mm^3

15. 15 mm^3 to L
16. 1 in^3 to 1 ft^3
17. 332 cm^3 to L
18. 3450 in^3 to gal
19. 36 ft^2 to in^2
20. 16 cm^3 to mm^3
21. 0.25 L to cm^3

5.6 π AND GEOMETRY

In some of the formulas above, we have encountered the number π. π is an irrational number whose value we can only approximate, which is why we have a special symbol to represent it. Although we know that it is irrational many have the misguided belief that 3.14 is equal to π. What we were actually taught is not that they are equal, but that you can sometimes use 3.14 in place of π and get a good **approximate** answer to the problem we might be solving, but

$$\pi \neq 3.14$$

The real $\pi = 3.1415926535\ 8979323846\ldots$, and requires an unending (infinite) process to define it.

A Brief History of π

π has its origins in geometry. It was known by the ancient Greeks – and Egyptians. However, they didn't know of it as an irrational number, nor was it known by the π symbol we use today. Instead it was known from the geometric observation that the circumference of any circle divided by its diameter, was always the same number. This was a number that they could not find an exact value for, which caused quite a bit of interest, and eventually, in the 1700's, it was given the current name of π. How could something so basic and fundamental, not have an exact answer?

Initially, during the 1600's, π was used as a symbol to represent the circumference or periphery of a circle. Then in 1706, the English writer William Jones used π to represent the ratio of the circumference to the diameter of a circle. Then in 1737 Euler adopted the notation and that sealed the deal (as Euler went, so went mathematics).

The first person to do some rigorous estimation of π was Archimedes (287–212 BCE). Various approximations for π existed in Ancient Egypt, Babylon, and Greece prior to Archimedes. The earliest estimates date from around 1900 BCE. The Egyptians would use the equivalent of $\frac{256}{81} \approx 3.16$, and the Babylonians used the equivalent of $\frac{25}{8} = 3.125$. Both these approximations were more than adequate for their basic calculations. Even today, using either 3.14 or $\frac{22}{7} \approx 3.14$, gives sufficiently accurate results for many applications.

Archimedes, however, took it a step further. Instead of looking for an approximation for π, he looked at trying to find out what the actual number equaled. Archimedes was definitely a mathematician before his time, which is why he is always on any list of the greatest mathematicians of all time. Archimedes created the method of exhaustion to estimate π. Again, this is another unending process. The idea goes as follows: We know that π is the ratio of the circumference to the diameter of a circle. If we choose a circle with a diameter, $D = 1$, then the circumference of this circle $C = \pi$.

$$\pi = \frac{C}{D} = \frac{C}{1} = C$$

So if we start with a circle with a diameter equal to one, then the circumference is equal to π. We then use regular polygons, whose perimeters we can find exactly, to estimate π as accurately as we desire. If we use inscribed polygons (polygons inside the circle), the length of its perimeter will be less than π, and using circumscribed polygons (the circle inside the polygon), the length of the perimeter will be greater than π.

Starting with a square, a polygon of four sides, we continue to double the number of sides of the polygon each time (see the picture below.) We see that after a very small number of doublings (4 to 8 to 16), the perimeters of the inscribed and circumscribed polygons get closer to the length of the circumference of the circle, which for this circle is equal to π.

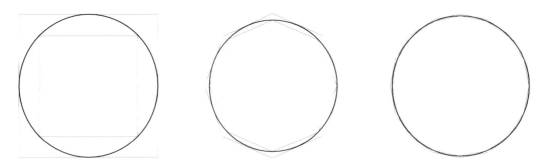

Four–Sided Polygons   Eight–Sided Polygons   Sixteen–Sided Polygons

Archimedes used the equivalent of a 96–sided polygon to come up with the estimate of π to be between $3\frac{10}{71} \approx 3.1408$ (the perimeter of the inner polygon), and $3\frac{1}{7} \approx 3.1429$ (the perimeter of the outer polygon.) If you take the average of these two values, the estimate is less than 1/100 of a percent off the actual value!

Theoretically, you could keep doubling the numbers of sides in the polygons, and both perimeters would get closer and closer to the actual value of π. This, however, is an unending process – the defining feature of an irrational number.

Since Archimedes' time, many other techniques have been discovered for calculating the value of π and as of today π has been calculated to over 5 trillion digits! Some people, obviously, have a lot of extra time on their hands.

What value should we use for π?

The most important point from our perspective is: what value should we use for π? When should we just write the symbol in our answer, and when should we use a decimal approximation? The answer is simple.

1. If you are not given any information about the accuracy you need, then you should leave your final answer in terms of the symbol representation of π.

2. If on the other hand, you are either given information that tells you something about the accuracy needed, or if you are given specific instructions: such as "use π = 3.14", then use a decimal approximation to the required or stated accuracy, i.e. the correct number of decimal places.

Again, for most ordinary calculations, using π = 3.14 will usually give you the accuracy you need. However, if you see numbers that are given to 4 or 5 places after the decimal, and this is required, then using π = 3.14 will not be accurate enough. Thus, in this case, you should always use the approximation for π that is at least one or two extra decimal places more than the number with the most places to the right of the decimal point in the problem. To be more specific:

    a. If you only have whole numbers in the problem, then use π = 3.14.

    b. If you have numbers with one place to the right of the decimal point (tenths position) then use π = 3.142.

    c. If you have numbers with two places to the right of the decimal point (hundredths position) then use π = 3.1416.

    d. If you have numbers with three places to the right of the decimal point (thousandths position) then use π = 3.14159.

       This can be continued as needed.

If there is any doubt in how much accuracy you need, and you are required to write a single number for your results with the pi symbol, then you can just use the value on your calculator, and round your result however you think is appropriate. Many calculators store at least 9 decimal places, which is probably accurate enough for any calculation you might encounter.

The π key is available on most calculators, either as a direct key or as a second function key. For example consider the TI-30X IIS below. This calculator uses a direct key for π.

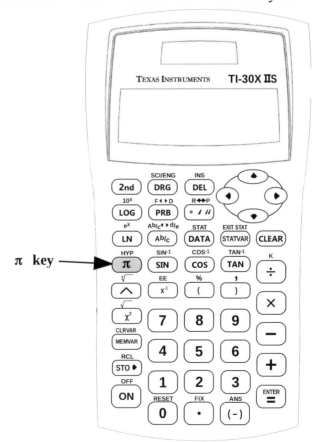

EXERCISES 5.6

Determine the number of decimal places needed for π in the following calculations.

1. $2\pi + 4$

2. 3π

3. $5\pi/2$

4. $\pi + 4.1958$

5. $\pi + 7.8$

6. $\pi + 2.718$

7. The volume of a cylinder with radius 4 inches and height 7 inches, approximated to 2 places after the decimal point

8. $\pi + \sqrt{11}$, approximated to 4 places after the decimal point

9. $\pi + \sqrt{49}$

10. The exact area of a circle with a radius of 2 cm

11. The volume of a sphere of radius 1 m, approximated to 3 places after the decimal point

12. The area of a circle where the dimension of the radius is know to 4 places to the right of the decimal point

13. The volume of a cylinder where both the height and radius of the cylinder are know to 5 places to the right of the decimal point

14. The circumference of a circle where the radius is 3.5 in

15. The circumference of a circle whose radius is 15 cm

16. The volume of a sphere with a radius of 30.0 mm

17. The volume of a sphere whose radius is 2 feet

18. The exact volume of a sphere whose radius is 1 foot

19. The exact volume of a cylinder whose height is 2 meters and whose radius is 0.75 meters

20. The exact circumference of a circle whose radius is 12 in

21. The area of a circle with a radius of 4.75 inches

Chapter 5 Practice Test

Determine the dimension of each object described below, and give a possible example of units to measure it.

1. The distance to the moon
2. Cargo space of an airplane
3. The windshield of an automobile
4. The distance around a house

Answer the following. If an exact answer is not possible, then round all answers to the thousandths place, unless stated otherwise.

5. Find the circumference of a circle that has a radius of 5.8 cm. Use $\pi = 3.14$.

6. Find the area of a rectangle that measures 48 inches in length by 2 feet width.

7. Find the area of a triangle with a height of 5 m and base of 3 m.

8. Find the volume of a rectangular solid that measures 7 cm by 13 cm by 20 cm

9. Find the volume of a cylinder with a height of 36.5 cm and a radius of 8 cm. Use $\pi = 3.14$.

Find the missing quantity/quantities

10. Find the third angle of a triangle if the first two angles are 36° and 41°.

11. Find the third angle of a right triangle if one angle is 20°.

12.

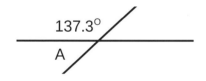

13.

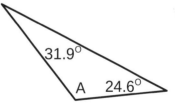

Perform the indicated conversions. If an exact answer is not possible, then round all answers to the thousandths place, unless stated otherwise.

14. 7 gal to in³

15. 2.25 ft² to in²

16. 38 mm³ to cm³

17. 54 cm³ to L

18. 36 in³ to cm³

19. 23.9 ft² to in²

Determine the number of decimal places needed for π in the following calculations.

20. $\pi + 7.89$

21. $\pi + 10$

22. To find the circumference of a circle that has a radius of 36.8 mm

23. To find the volume of a sphere with a radius of 74.874 cm

24. To find the exact volume of a cylinder that has a radius of 4 cm and height of 12 cm.

CHAPTER 6

Equations and Inequalities in One Variable

In the prior chapters we have presented some of the basics of the language we call algebra. We have introduced different types of algebraic expressions you will be exposed to, but at a simple level. We have also shown how to do arithmetic with different types of expressions. In this chapter, we begin to show how to apply the techniques to do something useful with the algebra. The first application we show is how to solve equations and inequalities involving polynomial expressions.

6.1 SOLVING FOR VARIABLES

To solve for a variable, means we are finding the value or values of the variable that make the equation a true statement. We can think of the equal sign as representing a scale that we must keep in balance. As with a scale we can add, subtract, multiply or divide whatever is on the scale, but to keep it in balance, we must do the same exact thing to both sides of the scale or equation.

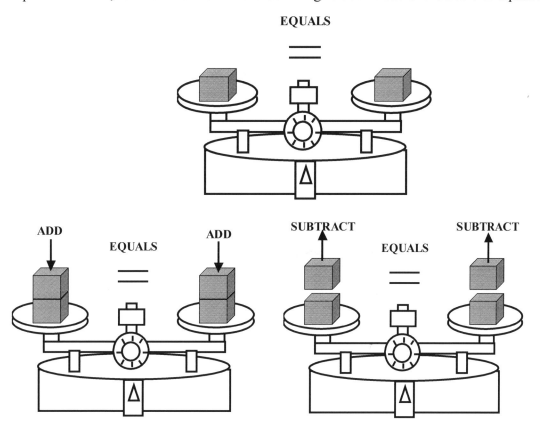

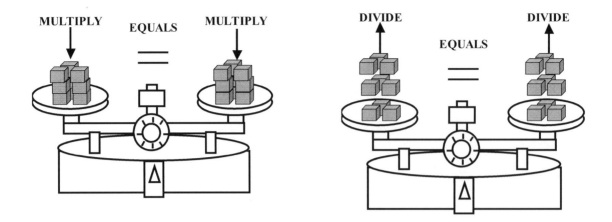

6.2 SOLVING LINEAR EQUATIONS

A linear equation is an equation where two linear expressions are set equal to each other. The objective in solving these types of equations is to find the value of the variable that makes the two expressions equal to each other. We accomplish this through a series of arithmetic operations that allow us to isolate the variable, with a unit coefficient, on one side of the equation, and a fixed number (not a variable) on the other.

In solving linear equations in a single variable, we are usually looking for a single solution. There are, however, special cases that will have no solution or that will have any value of the variable as a solution. These are more of an exception than the rule, so while we'll show them to you later in this chapter as unique cases, we will not focus on them, and their occurrence can be skipped without any loss of continuity in the material.

6.2a Solving One-Step Linear Equations

These are equations that only require one algebraic step to solve for the variable. Thus, we will either add or subtract a fixed number, or multiply or divide by a fixed number. Consider the following examples:

EXAMPLE 1: Solve the equation for the variable: $\qquad x - 7 = 5$

$$x - 7 = 5$$
$$\underline{\quad +7 \quad +7 \quad} \qquad \text{Step 1: Add 7 to both sides}$$
$$x = 12 \qquad\qquad \text{, or}$$

$$x - 7 = 5$$
$$x - 7 + 7 = 5 + 7 \qquad \text{Step 1: Add 7 to both sides}$$
$$x = 12$$

Notice, we can either add 7 beneath the equation, or inline with the equation. The result is the same.

EXAMPLE 2: Solve the equation for the variable: $\quad a + 2 = -3$

$$a + 2 = -3$$
$$\underline{-2 \quad -2} \qquad \text{Step 1: Subtract 2 from both sides}$$
$$a = -5$$

, or

$$a + 2 = -3$$
$$a + 2 - 2 = -3 - 2 \quad \text{Step 1: Subtract 2 from both sides}$$
$$a = -5$$

EXAMPLE 3: Solve the equation for the variable: $\quad 3y = 12$

$$\frac{\cancel{3}y}{\cancel{3}} = \frac{12}{3} \quad \text{Step 1: Divide both sides by 3}$$
$$y = 4$$

EXAMPLE 4: Solve the equation for the variable: $\quad \dfrac{p}{-5} = -7$

$$(-5)\frac{p}{-5} = -7(-5) \quad \text{Step 1: Multiply by } -5$$
$$\frac{\cancel{(-5)}}{1} \frac{p}{\cancel{-5}} = -7(-5)$$
$$p = 35$$

EXAMPLE 5: Solve the equation for the variable: $\quad -x = 4$

This equation is equivalent to $(-1)x = 4$, and can be solved in two different ways. The first approach is to multiply both sides of the equation by -1, and the second is to divide both sides of the equation by -1.

$$\underline{\text{Approach 1}}$$
$$(-1)(-x) = 4(-1) \quad \text{Step 1: Multiply by } -1$$
$$(-1)(-1)x = -4$$
$$x = -4$$

<u>Approach 2</u>

$$\frac{(-1)x}{(-1)} = \frac{4}{(-1)} \quad \text{Step 1: Divide by } -1$$

$$\frac{\cancel{(-1)}x}{\cancel{(-1)}} = \frac{4}{(-1)}$$

$$x = -4$$

Instead of adding or subtracting a fixed number we can also add or subtract a variable from both sides of an equation. Since the variable is the same, we are actually adding or subtracting an equivalent amount from both sides of the equation. This means we are not changing the solution to the equation. Consider the following example:

EXAMPLE 6: Solve the equation for the variable: $\quad 2w = w$

$$2w = w$$
$$2w - w = w - w \qquad \text{Step 1: Subtract w from both sides}$$
$$w = 0$$

Unique Cases (Optional): This section can be skipped without losing continuity.

EXAMPLE 7: Solve the equation for the variable

$$x + 3 = x + 5$$
$$x - x + 3 = x - x + 5 \quad \text{Step 1: Subtract x}$$
$$3 \neq 5 \qquad \qquad \text{Impossible result}$$

No Solution. The equation is a contradiction.

EXAMPLE 8: Solve the equation for the variable

$$w - 7 = w - 7$$
$$w - w - 7 = w - w - 7 \quad \text{Step 1: Subtract w}$$
$$-7 = -7 \qquad \qquad \text{Always true}$$

Any value of w will work, i.e. all Real numbers. The equation is always true (also called a Tautology in the language of logic).

If, when solving an equation, the variable vanishes, the answer is either no solution if you get a contradiction (both sides unequal), or all Real numbers if you obtain a true statement (both sides equal).

One-Step Equations

Solve each equation

1. $16 = 7 + x$
2. $5 + a = 9$
3. $w - 3 = 7$
4. $-3 = 16 + y$
5. $p - 8 = 13$
6. $n + 7 = -5$
7. $r - 15 = -7$
8. $5 + t = 31$
9. $-a = 3$
10. $-13 + v = 7$

11. $16x = 48$
12. $-72 = 9w$
13. $-y = -5$
14. $-8 = 4b$
15. $38 = -19a$
16. $-21s = 0$
17. $21 = -7p$
18. $\dfrac{t}{3} = -4$
19. $2x = 0$

20. $\dfrac{x}{5} = 10$
21. $17 = -c$
22. $-7 = \dfrac{m}{4}$
23. $27 = \dfrac{y}{3}$
24. $a + 11 = 11$
25. $-14 + b = -14$
26. $3y = 2$
27. $72 = -36v$

28. $-6 = \dfrac{x}{12}$
29. $-5 + k = 28$
30. $-b = -13$
31. $5x = 7$
32. $p - 26 = -28$
33. $x - 26 = x - 28$
34. $6 - y = -y + 6$
35. $3 - w = -w + 4$
36. $13 + z = z - 3$

6.2b Solving Two-Step Linear Equations

These are equations that require two algebraic steps to solve for the variable. Thus, we will either add or subtract a number, and then multiply or divide by a number, or vice-versa. Consider the following examples:

EXAMPLE 1: Solve the equation for the variable: $\qquad 3x + 5 = -7$

$$3x + 5 - 5 = -7 - 5 \quad \text{Step 1: Subtract 5}$$
$$3x = -12$$
$$\frac{\cancel{3}x}{\cancel{3}} = \frac{-12}{3} \quad \text{Step 2: Divide by 3}$$
$$x = -4$$

EXAMPLE 2: Solve the equation for the variable: $\qquad 5u - 6 = 15$

$$5u - 6 = 15 \qquad \text{Step 1: Add 6}$$

$$5u - 6 + 6 = 15 + 6$$

$$5u = 21$$

$$\frac{5u}{5} = \frac{21}{5} \qquad \text{Step 2: Divide by 5}$$

$$u = \frac{21}{5}$$

EXAMPLE 3: Solve the equation for the variable: $\qquad 2w + 3 = w - 5$

$$2w + 3 - 3 = w - 5 \ -3 \qquad \text{Step 1: Subtract 3}$$

$$2w = w - 8$$

$$2w - w = w - 8 - w \qquad \text{Step 2: Subtract w}$$

$$w = -8$$

EXAMPLE 4: Solve the equation for the variable: $\qquad 6s = 3s$

$$6s - 3s = 3s - 3s \qquad \text{Step 1: Subtract 3s}$$

$$3s = 0$$

$$\frac{3s}{3} = \frac{0}{3} \qquad \text{Step 2: Divide by 3}$$

$$s = 0$$

If we think about this last example further we can see that it makes sense. We are asking, what number when multiplied by 6 equals the same number multiplied by 3? The only possible number for which this is true is zero. Any other number would give a different result.

We should also note that while we might be tempted to divide both sides of this equation by the variable, s, that this would cause problems, since

$$6s = 3s$$

$$\frac{6s}{s} = \frac{3s}{s}$$

$$6 = 3$$

In this way, we get a contradiction and we might be tempted to say that there is no solution to this equation. However, the reason we get a contradiction is that the actual solution to this equation is s = 0. Thus, when we divided by s, we were actually dividing by zero. We should recall that we are never allowed to divide by zero. Since we performed an illegal operation, we got an incorrect result. We violated the rules of the game, so we can't expect the answer we get to be correct.

Multiplying or dividing by variables is one area where we have to be cautious. We always have to consider the possibility that the variable is equal to zero!

EXERCISES 6.2b

Two-Step Equations
Solve each equation
<u>Basic Problems</u>

1. $4 = 2a - 2$

2. $10 - 3x = -5$

3. $-16 + 5x = -9$

4. $9 = -3 + 4p$

5. $2x + 5 = 18$

6. $4p - 3 = -8$

7. $16 + u = 16$

8. $7 + 7c = 7$

9. $-10z + 1 = -99$

10. $5 - 3a = 12$

11. $13 - 3w = -7$

12. $1 - 4x = 17$

13. $2b - 13 = 13$

14. $7a + 24 = -4$

15. $-20 = 5p + 3$

16. $5x + 3 = 7$

17. $4c - 16 = 9$

18. $7 = 2x - 6$

19. $2p + 3 = 16$

20. $-7b - 5 = 21$

<u>Intermediate and Advanced Problems</u>

21. $-9 = \dfrac{n}{3} + 11$

22. $\dfrac{s}{21} - 3 = -4$

23. $13 + \dfrac{u}{-2} = 11$

24. $17 + \dfrac{y}{-5} = -3$

25. $\dfrac{p}{11} - 3 = 2$

26. $-5 = 5 + \dfrac{r}{4}$

27. $\dfrac{u}{-5} + 7 = -9$

28. $7 + \dfrac{x}{-5} = 3$

29. $5z + 3 = 4z - 2$

30. $-6 = \dfrac{v}{3} - 7$

31. $\dfrac{z}{7} + 5 = 7$

32. $7r = 2r$

6.2c Solving Multi-Step Linear Equations

EXAMPLE 1: Solve the equation for the variable: $2a - 3 + 5a + 1 = a - 7$

$$2a - 3 + 5a + 1 = a - 7 \qquad \text{Step 1: Combine like terms}$$
$$7a - 2 = a - 7$$
$$7a - 2 + 2 = a - 7 + 2 \qquad \text{Step 2: Add 2}$$
$$7a = a - 5$$

168

$$7a - a = a - a - 5 \quad \text{Step 3: Subtract a}$$
$$6a = -5$$
$$\frac{6a}{6} = \frac{-5}{6} \quad \text{Step 4: Divide by 6}$$
$$a = \frac{-5}{6}$$

EXAMPLE 2 Solve the equation for the variable: $\quad 1 - (2 - 3v) = 4(v - 7) - 6$

$$1 - 2 + 3v = 4v - 28 - 6 \quad \text{Step 1: Use Distributive Property}$$
$$-1 + 3v = 4v - 34 \quad \text{Step 2: Combine like terms}$$
$$-1 + 34 + 3v = 4v - 34 + 34 \quad \text{Step 3: Add 34}$$
$$33 + 3v = 4v$$
$$33 + 3v - 3v = 4v - 3v \quad \text{Step 4: Subtract 3v}$$
$$33 = v$$

EXERCISES 6.2c

Solving Multi-Step Equations
Solve each equation
<u>Basic Problems</u>

1. $8 = -8x + 2x - 4$

2. $3a - 4 = 2a - 3$

3. $1 - 2y = 3y - 9$

4. $4c - 3 = 2c - 11$

5. $p - 3 = p + 5$

6. $3 - z = 7 - z$

7. $9 = -3(-x + 4)$

8. $-(2x + 5) = 12$

9. $2(4y - 3) - 7 = y - 5$

10. $24 = -8(2s + 3)$

11. $-(2n - 3) = 18$

12. $2r + 3(r + 1) = -5$

13. $6(2m - 5) = -18$

<u>Intermediate and Advanced Problems</u>

14. $2(p + 2) - 4(2p - 1) = 18$

15. $2n - 6 = -3(n + 5)$

16. $4 - 3w = 6(2w + 5)$

17. $-6(u + 1) + 4(u - 1) = 3(u + 2)$

18. $-(a + 3) + 4(1 - a) = 7(2a + 1)$

19. $5 - 3(1 - 5a) = 1 - 3a$

20. $16k - 12 = -(2 - 3k)$

21. $13 - 21t = 3(4 - 2t)$

22. $7(x+4)-3(1-2x)=-36$

23. $32=-5(y+4)+3(y-2)$

24. $-(3-5p)=6-2p$

25. $16-3v=2-3(v+5)$

26. $-x+17+3x=5-2(1-2x)$

6.3 APPLICATIONS OF LINEAR EQUATIONS

In this section we demonstrate how to use linear equations to solve application problems. The problems themselves are fairly simple and they are mainly meant as an aid for learning how to set up and solve some beginning problems. Thus, we will focus more on the approach to setting up and solving the problem. The process loosely follows the following steps:

Step 1:   Read the problem thoroughly, a few times, not just once.
Step 2:   Write down what you are asked to find.
Step 3:   Determine what your variable (unknown you are solving for) is.
Step 4:   How is it related to itself or other numbers? This step gives us the equation that we have to solve.
Step 5:   Write down the equation, and reread the problem when looking at the equation to be sure that you have not missed or confused anything. Does the equation you have written represent what is being asked?
Step 6:   Solve the resulting equation.
Step 7:   Does the result make sense, or can it be tested or checked?

We illustrate the approach through examples:

EXAMPLE 1: Twelve more than six times a number is thirty-six. What is the number?

First, we notice that we are looking for a number, so we use "n" to represent the unknown number. Next, we convert the sentence above into an equation. We see that "twelve more than six times a number is 36" is equivalent to, "six times a number plus twelve equals thirty-six", or

$$6n+12=36$$

Solving this equation using the techniques developed above, we find that $n=4$.

Does it work? Since, $6(4)+12=24+12=36$, we see that we have found the correct solution.

EXAMPLE 2: Five less than four times a number is twenty-seven. Find the number.

Again we use "n" to represent the unknown number. Next, we convert the sentence above into an equation. We see that "five less than four times the number" is equivalent to $4n-5$. Setting this equal to 27 we have:

$$4n-5=27$$

Solving this equation using the techniques developed earlier, we find that $n=8$.

EXAMPLE 3: The length of one side of an equilateral triangle is 6 inches. What is the perimeter of the triangle?

First, we use "p" to represent the unknown number and ℓ to represent the length of its side. Next, we use the fact that an equilateral triangle has three equal sides. Thus, we are trying to find $p = 3\ell$, where ℓ is the length of one side of the triangle. Next, we substitute the known values into the perimeter equation to obtain:

$$p = 3\ell$$
$$= 3(6 \text{ inches})$$
$$= 18 \text{ inches}$$

EXAMPLE 4: A triangle has three interior angles such that one angle is twice the value of the first angle and the last angle is five times the size of the first angle. What are the values of all the angles?

Let "a" represent the value of the first angle. Then the values of the other two angles are $2a$ and $5a$ respectively. Next, we use the fact that the sum of the interior angles of a triangle equals 180 degrees. Thus, we are trying to find "a" such that:

$$a + 2a + 5a = 180°$$

Solving this we obtain

$$8a = 180°$$
$$a = \frac{180°}{8}$$
$$a = 22.5°$$

The first angle is $22.5°$; the values of the other two angles are: $(2)22.5° = 45°$, and $(5)22.5° = 112.5°$ respectively. Notice that thee three interior angles do sum to $180°$.

EXAMPLE 5: The perimeter of a rectangle is 360 feet. Its length is twice its width. What are the length and width of the rectangle?

The formula for the perimeter of a rectangle is: $p = 2\ell + 2w$

With w equal to the width of the rectangle, the length is given by: $\ell = 2w$. Replace ℓ in the perimeter equation with 2w, and the perimeter with 360 to obtain:

$$p = 2\ell + 2w$$
$$360 = 2(2w) + 2w$$
$$360 = 4w + 2w$$
$$360 = 6w$$
$$\frac{360}{6} = \frac{\cancel{6}w}{\cancel{6}}$$
$$60 = w$$

Thus, w = 60 feet and $\ell = 2(60)$ feet $= 120$ feet.

EXERCISES 6.3

Applications of Linear Equations

1. Six more than three times a number equals thirty-three. Find the number.

2. Five less than six times a number is seven. What is the number?

3. Twenty-five more than eight times a number is one. Find the number.

4. Seven less than two times a number is negative fifteen. Find the number.

5. The perimeter of a rectangle is sixty-four inches. Its length is three times as long as its width. What are the length and width of the rectangle?

6. The perimeter of a rectangle is 150 feet long. Its length is twice its width. What are the length and width of the rectangle?

7. The perimeter of an equilateral triangle is thirty-six inches. What is the length of each side?

8. The perimeter of an equilateral triangle is twelve feet. What is the length of each side?

9. If the interior angles of a triangle are such that the largest angle is three times as large as the smallest angle, and the middle angle is twice the smallest angle, what are the angles of the triangle?

10. If the interior angles of a triangle are such that the largest angle is five times as large as the smallest angle, and the middle angle is three times the smallest angle, what are the angles of the triangle?

11. The sum of two consecutive integers is twenty-three. What are the numbers?

12. The sum of two consecutive integers is thirty-nine. What are the numbers?

13. One number is twice another number, and their sum is thirty. What are the numbers?

14. One number is three times as large as another number. Their sum is eight. What are the two numbers?

15. You are given three numbers, where the largest number is three times the smallest number, and the middle number is two times as large as the smallest number. The sum of all three numbers is thirty-six. What are the three numbers?

16. The cost of two tables and three chairs is $745. If the table costs $60 more than the chair, find the cost of the table and the chair.

17. The length of a rectangle is five feet longer than its width. The perimeter of the rectangle is 38 feet around. What are the length and width of the rectangle?

18. The width of a rectangle is 12 meters shorter that its length. The perimeter of the rectangle is 156 meters around. What are the length and width of the rectangle?

19. The cost of a table and six chairs is $745. If the table costs $80 more than a chair, find the cost of the table and the chair.

20. The length of a rectangle is 2.5 times longer than its width. The perimeter is 189 cm. What are the length and width of the rectangle?

21. You have 15 feet of wood and you need to make 9 rungs (steps) for a ladder. How long will each rung be?

22. You are putting molding around the ceiling of a room whose width is 1/3 of its length. You have 24 meters of molding. What are the dimensions of the room?

6.4 SOLVING BASIC ABSOLUTE VALUE EQUATIONS

In Chapter 1 we introduced the absolute value operation. We said it was related to the distance a point is away from zero on the number line. Since distance is a common area of study in mathematics and the sciences, it makes sense that we would from, time-to-time, have to solve equations with the absolute value operation. In this section we show how to go about solving some basic absolute value equations.

The most basic absolute value equation is $|x|=1$. This equation is asking us to find all the values of x that are one unit away from zero on the number line. Obviously there are two such numbers, x = 1 and x = −1. When solving absolute value equations we have to always remember to solve for points both to the right and to the left of zero.

One potential problem, however, is sometimes the problem is ill-posed or misstated and you might be asked to solve $|x|=-1$. This is asking for us to find all points on the number line that are −1 units away from zero on the number line. Obviously this does not make any sense, because a distance cannot be negative. Thus, the answer to this problem is just to say no solution.

The approach to solving a basic absolute value equation is a follows.
- First, make sure the problem is not ill-posed, meaning you are trying to find a positive distance (the number on the other side of the absolute value symbol with a coefficient of one is not negative.)

- Next, set up and solve two separate linear equations. The first with one side of the equals sign with the number, and the second with it equal the same positive value, but now with a negative out front.

- Finally, you should always check your answers in the original equation.

We illustrate this approach through a few examples below.

EXAMPLE 1: Solve the absolute value equation $|2x|=4$

> Check that the problem makes sense to solve. Since the right hand side is positive we can move forward.
>
> Now, set up and solve two linear equations:

$$
\begin{array}{cc}
2x = 4 & 2x = -4 \\
\dfrac{2x}{2} = \dfrac{4}{2} & \dfrac{2x}{2} = \dfrac{-4}{2} \\
x = 2 & x = -2
\end{array}
$$

> State the solutions: x = 2. and x = −2
>
> Check the solutions:
> $$|2\times 2| = |4| = 4 \ \checkmark, \qquad |2\times -2| = |-4| = 4 \ \checkmark$$

EXAMPLE 2: Solve the absolute value equation $|-3x|=7$

Check that the problem makes sense to solve. Since the right hand side is positive we can move forward.

Now, set up and solve two linear equations:

$$-3x = 7$$
$$\frac{-3x}{-3} = \frac{7}{-3}$$
$$x = -\frac{7}{3}$$

$$-3x = -7$$
$$\frac{-3x}{-3} = \frac{-7}{-3}$$
$$x = \frac{7}{3}$$

State the solutions: x = 7/3. and x = −7/3

Check the solutions:

$$\left|-3\times\frac{-7}{3}\right| = \left|\frac{21}{3}\right| = 7 \ \sqrt{} \ ,$$

$$\left|-3\times\frac{7}{3}\right| = \left|-\frac{21}{3}\right| = 7 \ \sqrt{}$$

EXAMPLE 3: Solve the absolute value equation $|3+4x|=8$

Check that the problem makes sense to solve. Since the right hand side is positive we can move forward.

Now, set up and solve two linear equations:

$$3+4x = 8$$
$$4x = 8-3 = 5$$
$$\frac{4x}{4} = \frac{5}{4}$$
$$x = \frac{5}{4}$$

$$3+4x = -8$$
$$4x = -8-3 = -11$$
$$\frac{4x}{4} = \frac{-11}{4}$$
$$x = -\frac{11}{4}$$

State the solutions: x = 5/4. and x = −11/4

Check the solutions:

$$\left|3+4\times\frac{5}{4}\right| = |3+5| = 8 \ \sqrt{} \ ,$$

$$\left|3+4\times-\frac{11}{4}\right| = |3-11| = |-8| = 8 \ \sqrt{}$$

EXAMPLE 4: Solve the absolute value equation $|1-2x|=-6$

Check that the problem makes sense to solve. Since the right hand side is negative we can stop and simply state; No Solution..

EXAMPLE 5: Solve the absolute value equation $|1-2x|=5$

Check that the problem makes sense to solve. Since the right hand side is positive we can move forward.

174

Now, set up and solve two linear equations:

$$1-2x = 5$$
$$-2x = 5-1 = 4$$
$$\frac{-2x}{-2} = \frac{4}{-2}$$
$$x = -2$$

$$1-2x = -5$$
$$-2x = -5-1 = -6$$
$$\frac{-2x}{-2} = \frac{-6}{-2}$$
$$x = 3$$

State the solutions: $x = -2$. and $x = 3$

Check the solutions:
$$|1-2\times(-2)| = |1+4| = 5 \ \checkmark, \qquad |1-2\times3| = |1-6| = |-5| = 5 \ \checkmark$$

EXERCISES 6.4

Solve the absolute value equations.

1. $|-3x|=6$
2. $\left|\dfrac{x}{3}\right|=1$
3. $|5x|=2$
4. $|8x|=0$
5. $|-x|=-1$
6. $|-2x|=8$
7. $|2x-3|=3$
8. $|3x+1|=8$
9. $|x-2|=-9$
10. $|3-6x|=7$

6.5 SOLVING QUADRATIC EQUATIONS

We are now looking to find solutions to other forms of polynomial equations. The next level of difficulty is solving 2nd degree, or quadratic equations. Thus, we wish to solve equations of the form:

$$ax^2 + bx + c = 0$$

One way of approaching this problem is to see if we can write this equation in an alternative, but equivalent, form that might prove helpful in finding a solution.

We illustrate this approach with an example. Consider the following quadratic equation:
$$x^2 + x - 12 = 0$$
We factor this to obtain:

$$(x+4)(x-3) = 0$$

Now let's examine what we are trying to accomplish. We wish to find a value of x which when substituted into the above equation gives us zero. Let's try a few values, such as $x = 1$, $x = 2$, and $x = 3$.

We see that for $x = 1$, and $x = 2$ we get contradictory results,

$x = 1$

$$(1+4)(1-3) = 0$$
$$(5)(-2) = 0$$
$$-10 \neq 0$$

$x = 2$

$$(2+4)(2-3) = 0$$
$$(6)(-1) = 0$$
$$-6 \neq 0$$

so they are not solutions to the equation.

However, for $x = 3$ we get a true statement.

$$(3+4)(3-3) = 0$$
$$(7)(0) = 0$$
$$0 = 0$$

Thus, we see that $x = 3$ is a solution, and we can also verify this by substituting this value into the original equation.

$$x^2 + x - 12 = 0$$
$$(3)^2 + (3) - 12 =$$
$$9 + 3 - 12 =$$
$$0 = 0$$

We notice that the value of x that satisfied the original equation was the value of x that made one of the binomial factors zero. Thus, we only need to factor the quadratic and set each binomial factor equal to zero to find the solutions of this equation. What we have discovered is called the zero factor property.

Definition: The **zero factor property** states that if we have a product of two numbers that is equal to zero, then either one number, the other number, or both of the numbers must be zero, i.e.

$$a \cdot b = 0 ,$$

then either a = 0, or b = 0, or both are zero.

The factors can be written as monomial terms as above, or as binomial terms, as shown below:
$$(ax + b)(cx + d) = 0$$
,

then either $ax + b = 0$, or $cx + d = 0$, or both are zero.

Thus, to solve any quadratic equation, we first try and factor it, and if we can, we then set each factor equal to zero, and those new equations give us our solutions. We have been able to reduce solving some quadratic equations to solving either a one-step or a two-step linear equation. Again, we used the mathematics to transform a problem we don't know how to solve into one that we already know how to solve.

If we finish the above example we see that x = –4 is also a solution.

Let's consider several examples:

EXAMPLE 1: Solve the quadratic equation: $(w+6)(w-13)=0$

This equation is already in factored form, so we apply the zero factor property immediately.

Set each factor equal to zero and solve the linear equations.

$$w+6=0 \quad \text{and} \quad w-13=0$$
$$w=-6 \qquad\qquad w=13$$

Thus, this equation has two solutions: $w=-6$ and $w=13$

EXAMPLE 2: Solve the quadratic equation: $6y^2+7y-3=0$

Factor:

$$(3x-1)(2x+3)=0$$

Now use the zero factor property. Set each factor equal to zero and solve the linear equations.

$$3x-1=0 \qquad\qquad 2x+3=0$$
$$3x=1 \qquad\qquad 2x=-3$$
$$\text{and}$$
$$x=\frac{1}{3} \qquad\qquad x=\frac{-3}{2}$$

Thus, this equation has two solutions: $x=\frac{1}{3}$ and $x=-\frac{3}{2}$

EXAMPLE 3: Solve the quadratic equation: $a^2-25=0$

Factor, using the difference of squares formula:

$$(a-5)(a+5)=0$$

Now use the zero factor property. Set each factor equal to zero and solve the linear equations.

$$a-5=0 \quad \text{and} \quad a+5=0$$
$$a=5 \qquad\qquad a=-5$$

Thus, this equation has two solutions: $a=5$ and $a=-5$

EXAMPLE 4: Solve the quadratic equation: $\qquad n^2 = 3n + 4$

Before we can factor and apply the zero factor property, we must first have a zero on the right-hand-side (RHS) of the equation. Thus, we first get a zero on the RHS by subtracting 3n and 4.

$$n^2 - 3n - 4 = 0$$

Now we can factor: $(n-4)(n+1) = 0$

Use the zero factor property. Set each factor equal to zero and solve the linear equations.

$$\begin{array}{ccc} n - 4 = 0 & & n + 1 = 0 \\ & \text{and} & \\ n = 4 & & n = -1 \end{array}$$

Thus, this equation has two solutions: $n = 4$ and $n = -1$

EXAMPLE 5: Solve the quadratic equation: $\qquad (p-4)(p+3) = -6$

In this case we have the left-hand-side factored, but we do not have a zero on the right. Therefore, we will first have to get a zero on the right and then multiply the binomial terms, combine like terms and then factor the result, before we can apply the zero factor property.

$$(p-4)(p+3) + 6 = 0$$
$$p^2 - p - 12 + 6 = 0$$
$$p^2 - p - 6 = 0$$

Now we can factor:

$$(p-3)(p+2) = 0$$

Now use the zero factor property. Set each factor equal to zero and solve the linear equations.

$$\begin{array}{ccc} p - 3 = 0 & & p + 2 = 0 \\ & \text{and} & \\ p = 3 & & p = -2 \end{array}$$

Thus, this equation has two solutions: $p = 3$ and $p = -2$

EXERCISES 6.5

Solving Quadratic Equations by Factoring

Solve each equation by factoring.

Basic Problems

1. $(x-1)(x+2)=0$

2. $(a+4)(a-5)=0$

3. $(2b-1)(b+4)=0$

4. $(1+2y)(y-7)=0$

5. $-(p+12)(3p-5)=0$

6. $x^2-3x-10=0$

7. $y^2-6y+9=0$

8. $c^2+6c-16=0$

9. $w^2-8w+12=0$

10. $2x^2-5x-3=0$

11. $3m^2+11m+10=0$

12. $6p^2+p-12=0$

13. $5y^2-13y-28=0$

14. $4z^2-z-14=0$

15. $4x^2+12x+9=0$

Intermediate and Advanced Problems

16. $r^2-r-5=1$

17. $w^2+3w-13=-3$

18. $u^2-5u+8=4$

19. $c^2+10c-5=19$

20. $y^2-6y+3=19$

21. $(x+2)(x-3)=6$

22. $(w+5)(w+5)=4$

23. $(a-6)(a-5)=2$

24. $(p+2)^2=9$

25. $(u-3)^2=1$

26. $3z(z-5)=0$

27. $5a(2a+3)=0$

28. $x(x+3)=18$

29. $y(y-5)=14$

30. $2r(r-1)=5r-3$

31. $3a(a+4)=-2a-8$

32. $b^2=-3b-2$

33. $v^2=6v-9$

34. $x^2-3x+7=5x$

35. $m^2+8m-3=-19$

6.6 SOLVING LINEAR INEQUALITIES

We are often interested in finding a range of values that will work for a particular problem. For example, in business we may not want our costs to go above a certain value, or we may require a minimum profit on a selected product. In medicine we may want our body temperature to reside in a certain range of temperatures, or that certain levels of a particular chemical must remain within a certain range within our bodies. To handle problems of this type, we use inequalities instead of equations. The inequalities are represented using the inequality symbols, $<$, $\leq$, $>$, $\geq$, that were discussed earlier in the book.

Graphing Inequalities

The first thing we need to introduce is how to identify a set of numbers represented by an inequality. This is called graphing the inequality. We do this using the number line. For example to show the solutions to the inequality: $x > 3$, we draw a number line with the values of 0 and 3 identified on the graph, e.g.

We should point out that we have left an open circle around the number 3 on the number line. This is to illustrate that the value of 3 is not a solution. If our inequality has a greater than or equal to, or less than or equal to, we would shade in the circle to identify that the value of the number, 3 in this case, is a solution. Also, since x is greater than 3, we have drawn a thick dark line to the right of 3, showing all the values of the variable x that make this inequality true.

We illustrate a few variations of this type of problem with some of examples:

EXAMPLE 1: Graph the inequality: $w \le -5$

EXAMPLE 2: Graph the inequality: $y > 0$

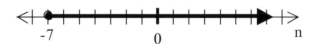

EXAMPLE 3: Graph the inequality: $n \ge -7$

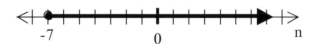

Solving Inequalities

Frequently the inequality does not have the variable isolated on one side of the inequality. So just like solving linear equations, we also have to do the same thing by solving the inequality. Solving inequalities is the same as solving equations with ONE EXCEPTION. If we either multiply or divide by a negative number then we must reverse the sense of the inequality. Here is why.

We start with a true statement: $2 < 7$

Now if we either multiply or divide this inequality by a negative number and don't change the sense (direction) of the inequality we have the following:

$$(-1)2 < 7(-1)$$
$$-2 < -7 \qquad \text{A contradiction}$$

or

$$\frac{2}{-1} < \frac{7}{-1}$$
$$-2 < -7 \qquad \text{A contradiction}$$

If, however, we reverse the sense of the inequality, we preserve the truth-value of the inequality, e.g.

$$(-1)2 > 7(-1)$$
$$-2 > -7 \qquad \text{A true statement}$$

or

$$\frac{2}{-1} > \frac{7}{-1}$$
$$-2 > -7 \qquad \text{A true statement}$$

This means that we solve inequalities the same way we solve equations, with the exception that if we multiply or divide by a negative number we must reverse the sense (direction) of the inequality.

The examples below follow the same approach we adopted when solving linear equations. We start with single-step inequalities, and then move on to two-step inequalities, and finally multi-step inequalities. In the following examples we show one of each type.

EXAMPLE 1: Solve and graph the inequality: $x - 5 \le 3$

We begin with the algebraic process.

$$x - 5 \le 3$$
$$x - 5 + 5 \le 3 + 5$$
$$x \le 8$$

Now we graph the inequality.

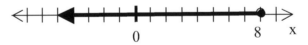

EXAMPLE 2: Solve and graph the inequality: $-3a + 5 \le 2$

We start with the algebraic part.

$$-3a + 5 \le 2$$
$$-3a + 5 - 5 \le 2 - 5 \qquad \text{Step 1: Subtract 5}$$
$$-3a \le -3$$
$$\frac{-3a}{-3} \ge \frac{-3}{-3} \qquad \text{Step 2: Divide by } -3, \text{ change the direction of the}$$
$$\qquad\qquad\qquad\qquad \text{inequality since we are dividing by a negative}$$
$$a \ge 1$$

Now we graph the inequality.

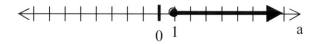

EXAMPLE 3: Solve and graph the inequality: $5y - 4 > 3y + 7$

We start with the algebraic part.

$$5y - 4 > 3y + 7$$
$$5y - 4 - 3y > 3y + 7 - 3y$$
$$2y - 4 > 7$$
$$2y - 4 + 4 > 7 + 4$$
$$2y > 11$$
$$\frac{\cancel{2}y}{\cancel{2}} > \frac{11}{2}$$
$$y > \frac{11}{2}$$

Now we graph the inequality.

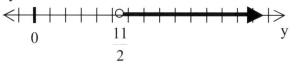

Recall that when solving equations we sometimes ended up with the variable on the right-hand-side of the equals sign, and this had no effect on the answer. When solving an inequality, however, the location of the variable in the end does impact the solution. $x < 3$ and $3 < x$ give two different solutions. If we take the second inequality we can move the x over to the left-hand-side using the rules developed above as follows:

$$3 < x$$
$$3 - x < x - x \qquad \text{Subtract x from both sides}$$
$$3 - x < 0$$
$$3 - 3 - x < 0 - 3 \qquad \text{Subtract 3 from both sides}$$
$$-x < -3$$

We have shown that $3 < x$ is equivalent to $x > 3$. Thus, when the variable is on the right-hand-side we can switch the variable and the constant (keeping the sign the same) to the other sides and reverse the direction of the inequality symbol, or we could also read the inequality from right-to-left, instead of from left-to-right.

$$(-1)(-x) > (-1)(-3) \quad \text{Multiply both sides by } -1 \text{ and change the direction}$$
$$x > 3 \qquad\qquad \text{of the inequality symbol}$$

$$3 < x \quad \text{Three is less than x}$$
→

$$3 < x \quad \text{x is greater than three}$$
←

EXAMPLE 4: Solve and graph the inequality: $-5 \leq 2 + b$

We begin with the algebraic process.

$$-5 \leq 2 + b$$
$$-5 - 2 \leq 2 - 2 + b$$
$$-7 \leq b$$

or

$$b \geq -7$$

Now we graph the inequality.

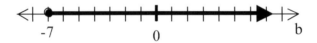
-7   0   b

Absolute Value Inequalities

Sometimes we wish to solve problems with a range of possible solutions that are between two numerical values. This is where the absolute value and its use in an inequality statement is useful.

For example, let's way we wish to find the values of a variable that are within 4 units (up to and including 4) of zero on the number line. This is equivalent to the absolute value inequality $|x| \leq 4$. This is equivalent to stating that x lies at or to the right of –4, and at or to the left of 4. Algebraically this is equivalent to $x \geq -4$ and $x \leq 4$. To be an acceptable x-value both of these inequalities must be true at the same time.

This can also be expressed using two inequality symbols in the same inequality statement. We could also state this as x lies between –4 and 4 (inclusive) on the number line, and this would be equivalent to the algebraic statement $-4 \leq x \leq 4$.

Another possibility is that we are looking to find a range of numbers that are greater than a certain distance from zero on the number line, such as $|x| > 8$. In this case, we need to write this as two separate inequalities, since either $x > 8$ or $x < -8$.

These two possibilities highlight some of the unique issues and challenges with solving absolute value inequalities. Then you can add the complexity of having a negative opposite the absolute value symbol, and just as in the case of the equation, this too becomes either an impossible or a trivial result. For example consider $|x| \leq -4$. This asks to find all the values of are that are less than a distance of –4 from zero on the number line. Obviously this is not possible, so the answer would be no solution. Alternatively you could consider $|x| \geq -4$. This asks the question of which numbers x are greater than –4 units away from zero on the number line. Obviously this is true for ANY x value, so the solution is simple state that x is any real number.

As you can see, unlike the other inequalities, absolute value inequalities require a great deal of thought before you set out to solve them. In what we present below, or focus will not be on the last two special cases we introduced above. We will only take a basic look at them in this presentation.

We now consider some of the more typical examples of solving absolute value inequalities.

EXAMPLE 1: Solve and graph the absolute value inequality $|x| > 7$

Solution: Since this is looking for values of x that are further than 7 units away from zero on the number line, we must solve two inequalities. Whenever we have a greater than symbol, we must solve to separate inequalities to find the solution to an absolute value inequality.

This states that either $x > 7$ or $x < -7$. We can represent this solution on a number line, as follows:

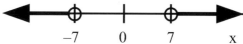

Notice in the graph above that we now have arrows representing the solution set going in two different directions.

EXAMPLE 2: Solve and graph the absolute value inequality $|x-1| \leq 4$

Solution: We are looking for values of the expression x –1 that are less than or equal to 4 units away from zero on the number line. To solve this, we have a choice, we can either solve a pair of inequalities or a single inequality statement with two inequality symbols. Whenever we have a less than (or less than or equal to) symbol, we either solve a pair of inequalities or a single inequality statement with two inequality symbols, to find the solution to an absolute value inequality. We'll show both approaches below.

Approach 1
This states that either $x-1 \leq 4$ or $x-1 \geq -4$. We first need to solve both of these inequalities.

$$
\begin{array}{ccc}
x-1 \leq 4 & & x-1 \geq -4 \\
x-1+1 \leq 4+1 & \text{and} & x-1+1 \geq -4+1 \\
x \leq 5 & & x \geq -3
\end{array}
$$

We can represent this solution on a number line, as follows:

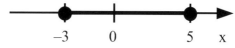

184

Notice in the graph above that we now have arrows representing the solution set going in two different directions.

Approach 2
Alternatively we can solve as a single inequality, since this is equivalent to:

$$-4 \leq x-1 \leq 4$$
$$-4+1 \leq x-1+1 \leq 4+1$$
$$-3 \leq x \leq 5$$

EXAMPLE 3: Solve and graph the absolute value inequality $|3-2x| \leq 5$

Solution: We are looking for values of the expression 3 –2x that are less than or equal to 5 units away from zero on the number line. To solve this, we have a choice, we can either solve a pair of inequalities or a single inequality statement with two inequality symbols. Whenever we have a less than (or less than or equal to) symbol, we either solve a pair of inequalities or a single inequality statement with two inequality symbols, to find the solution to an absolute value inequality. We'll show both approaches below.

Approach 1
This states that either $3-2x \leq 5$ or $3-2x \geq -5$. We first need to solve both of these inequalities.

$$3-2x \leq 5 \qquad\qquad 3-2x \geq -5$$
$$3-2x-3 \leq 5-3 \qquad\qquad 3-2x-3 \geq -5-3$$
$$-2x \leq 2 \qquad\qquad -2x \geq -8$$
$$\frac{-2x}{-2} \geq \frac{2}{-2} \qquad\qquad \frac{-2x}{-2} \leq \frac{-8}{-2}$$
$$x \geq -1 \qquad\qquad x \leq 4$$

We can represent this solution on a number line, as follows:

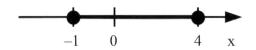

Approach 2
Alternatively we can solve as a single inequality, since this is equivalent to:
$$-5 \leq 3-2x \leq 5$$
$$-5-3 \leq 3-2x-3 \leq 5-3$$
$$-8 \leq -2x \leq 2$$
$$\frac{-8}{-2} \geq \frac{-2x}{-2} \geq \frac{2}{-2} \qquad \text{or} \quad -1 \leq x \leq 4$$
$$4 \geq x \geq -1$$

EXERCISES 6.6

Solving and Graphing Inequalities
Draw a graph for each inequality

1. $x > 5$

2. $b \leq -2$

3. $n < \dfrac{14}{2}$

4. $0 \leq p$

5. $s > -5$

6. $6 \geq w$

7. $x \geq -7$

8. $5 \leq z$

9. $-3 < c$

10. $b < 9$

11. $v \leq 0$

12. $t \geq -6$

13. $10 < y$

14. $x \leq -8$

One-Step Inequalities
Solve each inequality and graph the solution

15. $x - 3 > 2$

16. $b + 3 \leq -2$

17. $4 + n < 7$

18. $-2 \leq p + 3$

19. $-1 > r + 3$

20. $s + 2 \geq -5$

21. $4 + y \leq 4$

22. $6 \geq w + 7$

23. $0 \leq 2 + p$

24. $2x \geq -8$

25. $5 \leq 5z$

26. $\dfrac{t}{8} \geq -1$

27. $\dfrac{y}{-2} \geq 5$

28. $\dfrac{x}{-4} \leq -8$

Two-Step Inequalities
Solve each inequality and graph the solution

29. $5x - 10 > 5$

30. $3b + 1 \leq -2$

31. $4n - 9 < 7$

32. $20 \leq 4p + 6$

33. $-1 \geq r + 3$

34. $1 - s \geq -5$

35. $2 + 3y < 4$

36. $6 \geq 3w - 12$

37. $12 < 9 + p$

38. $2x - 3 \geq -7$

39. $5 \leq 3z - 7$

40. $4c - 3 > -3$

41. $-6 + 3b < 9$

42. $-5 > 2 - u$

Multi-Step Inequalities
Solve each inequality and graph the solution

43. $3x - x + 4 < -4$

44. $4b + 3 - b \leq -2$

45. $2n - 4 - 6n < 7$

46. $6 \leq p - 6p + 1$

47. $-1 > 6r - r + 2$

48. $2s - 8s + 19 \geq -5$

49. $4 - 2y - 3 < 4$

50. $6 - w \geq w + 4$

51. $4p - 1 \leq 3 - p$

52. $3(2x + 1) \geq -7 + x$

53. $5 \leq 3 - (z - 1)$

54. $4 + 2(c + 3) > -3$

Absolute Value Inequalities
Solve each inequality and graph the solution

55. $|4x| \leq 8$

56. $|2x - 5| > 5$

57. $|4 - x| < 7$

58. $|-3x| > 9$

59. $|4 + 5x| \geq 6$

60. $|3x + 2| \leq 3$

61. $|x - 3| > 2$

62. $|4x - 5| \leq 8$

6.7 APPLICATIONS OF LINEAR AND ABSOLUTE VALUE INEQUALITIES

In this section we introduce an array of different basic applications of linear and absolute value inequalities.

Linear Inequality Applications

EXAMPLE 1: You have saved $2,500 for a trip. Your airline ticket costs $825. Write and solve a linear inequality to find out how much you can spend for everything else on the trip. Also, write a complete sentence describing your solution.

Solution: Let x represents how much you can spend on everything else on the trip. Then we have that

$$x + \$825 \leq \$2,500$$

Solving we obtain:

$$x + \$825 \leq \$2,500 - \$825$$
$$x \leq \$1,675$$

I will have up to and including $1,675 left to spend on my trip.

EXAMPLE 2: John has a sales quota at work and must sell at least $10,000 worth of merchandise each week. On Monday he sold $1,500 and on Tuesday he sold $2,700. Write and solve a linear inequality to find out how much he has to sell for the remainder of the week to meet or exceed his quota. Also, write a complete sentence describing your solution.

Solution: Let x represents how much more John has to sell to meet or exceed his quota. Then we have that

$$x + \$1,500 + \$2,700 \geq \$10,000$$

Solving we obtain:

$$x + \$1,500 + \$2,700 \geq \$10,000$$
$$x + \$4,200 \geq \$10,000$$
$$x \geq \$10,000 - \$4,200$$
$$x \geq \$5,800$$

John needs to sell at least $5,800 for the remaining part of the week to meet his quota.

EXAMPLE 3: Sarah and her team are holding a car wash for a charity. She is charging $6 per car wash and wants to raise at least $750. Write and solve a linear inequality to find out how many cars she needs to wash to meet or exceed her goal. Also, write a complete sentence describing your solution.

Solution: Let x represents how many cars Sarah and her team needs to wash to raise at least $750.

Then we have that
$$\$6x \geq \$750$$

Solving we obtain:

$$\$6x \geq \$750$$
$$\frac{\$6x}{\$6} \geq \frac{\$750}{\$6}$$
$$x \geq 125$$

Sarah and her team need to wash at least 125 cars to meet her donation goal.

EXAMPLE 4: A construction elevator can safely hold up to 5,600 pounds. Angela needs to transport skids of bricks that weigh 1,500 pounds each. Write and solve a linear inequality to find out how many skids she can load on the elevator safely. Also, write a complete sentence describing your solution.

Solution: Let x represents how many skids of bricks Angela can safely load on the construction elevator. Then we have that

$$x \times 1,500 \text{ pounds} \leq 5,400 \text{ pounds}$$

Solving we obtain:

$$x \times 1,500 \text{ pounds} \leq 5,400 \text{ pounds}$$
$$\frac{x \times 1,500 \text{ pounds}}{1,500 \text{ pounds}} \leq \frac{5,400 \text{ pounds}}{1,500 \text{ pounds}}$$
$$x \leq 3.6$$

Angela can load no more that 3.6 skids of bricks on the construction elevator safely.

Absolute Value Inequality Applications

EXAMPLE 1: The average starting salary at a company for a factory worker is $32,000. Depending on the experience of the applicant there is a salary spread of plus or minus $2,800. Write and solve an absolute value inequality to find out the range of salaries for starting employees. Also, write a complete sentence describing your solution.

Solution: Let x represents the salary an employee in the factory is hired at. Then we have that
$$|x - \$32,000| \leq \$2,800$$

Solving we obtain:

$$-\$2,800 \leq x - \$32,000 \leq \$2,800$$
$$-\$2,800 + \$32,000 \leq x - \$32,000 + \$32,000 \leq \$2,800 + \$32,000$$
$$\$29,200 \leq x \leq \$34,800$$

A new factory employee will earn between $29,200 and $34,800 at this company.

188

EXAMPLE 2: The average height of the female members of a fashion club is 65 inches with a standard deviation (the plus or minus spread of 68% of the female members) of 2.8 inches. Write and solve an absolute value inequality showing the height range of the female members that are plus or minus two standard deviations (the plus or minus spread of 95% of the female members) from the average height. Also, write a complete sentence describing your solution.

Solution: Let x represents the height of some female club member. Then we have that

$$|x - 65\,\text{inches}| \leq 2 \times 2.8\,\text{inches}$$

$$|x - 65\,\text{inches}| \leq 5.6\,\text{inches}$$

Solving we obtain:

$$-5.6\,\text{inches} \leq x - 65\,\text{inches} \leq 5.6\,\text{inches}$$

$$-5.6\,\text{inches} + 65\,\text{inches} \leq x - 65\,\text{inches} + 65\,\text{inches} \leq 5.6\,\text{inches} + 65\,\text{inches}$$

$$59.4\,\text{inches} \leq x \leq 70.6\,\text{inches}$$

The range of heights of women in a fashion club that are within two standard deviations of the average height is from 59.4 inches up to 70.6 inches.

EXAMPLE 3: A cereal manufacturer tries to put 17.5 ounces of cereal in each box. The box can be off by no more than 0.75 ounces. Write an absolute value inequality that displays the acceptable amount of cereal in a box. Also, write a complete sentence describing your solution.

Solution: Let x represents the actual amount of cereal in a box. Then we have that

$$|x - 17.5\,\text{ounces}| \leq 0.75\,\text{ounces}$$

Solving we obtain:

$$-0.75\,\text{ounces} \leq x - 17.5\,\text{ounces} \leq 0.75\,\text{ounces}$$

$$-0.75\,\text{ounces} + 17.5\,\text{ounces} \leq x - 17.5\,\text{ounces} + 17.5\,\text{ounces} \leq 0.75\,\text{ounces} + 17.5\,\text{ounces}$$

$$16.75\,\text{ounces} \leq x \leq 18.25\,\text{ounces}$$

The amount of cereal in a box that should contain 17.5 ounces of cereal will range between 16.75 ounces and 18.25 ounces.

EXERCISES 6.7

Linear Inequality Applications
1. Javier saved $3,700 for a vacation. His airline ticket costs $1,025, and his hotel costs are $689. Write and solve a linear inequality to find out how much he can spend for everything else on the trip. Also, write a complete sentence describing your solution.

2. A Class 5 truck has a maximum carrying capacity of 19,500 pounds. The driver needs to transport skids of building materials that weigh 2,250 pounds each. Write and solve a linear inequality to find out how many skids she can load on the truck safely. Also, write a complete sentence describing your solution.

3. A quality control inspector has a quota of 875 products during a 7.5 hr shift. She has inspected 325 products during the first 3 hours. Write and solve a linear inequality to find out how many more she has to inspect in the remaining time to meet or exceed her quota? Also, write a complete sentence describing your solution.

4. To cover their weekly costs a bakery must sell $6,500 worth of baked goods. During the first 3 days of a 7 day work week they sold $3,800 of goods. Write and solve a linear inequality to find out how much more they need to sell to make a profit for the week. Also, write a complete sentence describing your solution.

Absolute Value Inequality Applications

5. The average hourly wage at a construction company is $28. There is a range of plus or minus $7.5. Write an absolute value inequality that displays the problem, then solve the problem. Also, write a complete sentence describing your solution.

6. A cereal manufacturer tries to put 14.75 ounces of cereal in each box. The box can be off by no more than 0.65 ounces. Write an absolute value inequality that displays the acceptable amount of cereal in a box. Also, write a complete sentence describing your solution.

7. The average height of the male members in a basketball club is 74 inches with a standard deviation (the plus or minus spread of 68% of the male members) of 2.9 inches. Write and solve an absolute value inequality showing the height range of the male members that are plus or minus two standard deviations (the plus or minus spread of 95% of the male members) from the average height. Also, write a complete sentence describing your solution.

8. The average amount of money spent on back-to-school supplies is $122 per student with a standard deviation (the plus or minus spread in costs of 68% of the students) of $10.50. Write and solve an absolute value inequality showing the cost range per student that are plus or minus two standard deviations (the plus or minus spread in costs of 95% of the students) from the average cost. Also, write a complete sentence describing your solution.

Chapter 6 Practice Test

Solve each linear equation

1. $z - 3 = -3$

2. $5 - x = -6$

3. $3 + 2x = -9$

4. $5y - 2 = -17$

5. $\frac{1}{2}x - 6 = 8$

6. $-5 = \frac{y}{8}$

7. $5 - x = 2x - 3$

8. $2 + 4(3 - x) = -4x + 9$

Solve the applications

9. Five less than three times a number is negative seventeen. Find the number.

10. The perimeter of a rectangle is ninety-six inches. Its length is three times as long as its width. What are the length and width of the rectangle?

11. The perimeter of a rectangle is 120 feet long. Its length is five times its width. What are the length and width of the rectangle?

12. The length of a rectangle is 5 cm longer than its width. The perimeter is 154 cm. What are the length and width of the rectangle?

13. The cost of a table and four chairs is $525. If the table costs $150 more than a chair, find the cost of the table and the chair.

Solve the absolute value equations.

14. $|2x - 9| = 15$

15. $|-3x| = 14$

16. $|1 + 2x| = -3$

Solve the quadratic equations.

17. $(x + 2)(3x - 1) = 0$ 18. $x^2 - 5x + 6 = 0$

19. $x^2 - 3x - 10 = 0$

20. $x^2 - 8x + 16 = 0$

Solve the inequalities.

21. $3x + 5 > -2$

22. $1 - 2x \leq -4$

23. $|x + 3| < 4$

24. $|1 - 2x| \geq 7$

Solve the inequality applications

25. Kaitlyn has a sales quota at work and must sell at least $6,500 worth of merchandise each week. On Monday she sold $500, on Tuesday she sold $2,100, and on Wednesday she sold $1,400. Write and solve a linear inequality to find out how much she has to sell for the remainder of the week to meet or exceed her quota. Also, write a complete sentence describing your solution.

26. A flour produces tries to put 36 ounces of flour in each package. The amount can be off by no more than 0.75 ounces. Write an absolute value inequality that displays the acceptable amount of flour in a package. Also, write a complete sentence describing your solution.

CHAPTER 7

Ratios, Percents, Rates, and Proportions

In this chapter we introduce three related ideas. That of a ratio, a rate and a proportion. These important ideas help us connect the concept of fractions to division, and to expand upon the meaning of a fraction.

7.1 WHAT IS A RATIO?

The dictionary definition of ratio is: "the quantitative relation between two amounts showing the number of times one value contains or is contained within the other." For example, the ratio of 35 to 5, asks how many fives are in thirty-five, or how many copies of 5 are contained in 35?

Mathematically we are asking the question $? \times 5 = 35$, which is equivalent to to the division statement $35 \div 5 = ?$. This connects the idea of a ratio to the process of division. Thus, more simply, a ratio can be viewed as dividing two numbers.

$$16 \text{ to } 2 \text{ is equivalent to } 16 \div 2$$

$$34 \text{ to } 17 \text{ is equivalent to } 34 \div 17$$

We now connect the idea of division, which is a verb or an action, to that of a fraction, which is a noun or a name. Most likely you have been doing this all along in your math courses, but never gave it any thought, because you were taught to just do it. Now, however, we want to examine it more clearly, and hopefully you will understand that they are actually two different concepts that can be seen to give equivalent results.

Division of whole numbers is defined as follows:

The **division of a whole number m by a whole number n > 0**, written as $m \div n$, is the whole number k that satisfies $m = k \times n$. Where the number k is called the **quotient**, n the **divisor**, and m the **dividend**.

For example, $12 \div 4 = 3$ is equivalent to $12 = 3 \times 4$, or $28 \div 4 = 7$ is equivalent to $28 = 7 \times 4$.

This definition can be extended to include cases where k is not a whole number.

The **division of a whole number m by a whole number n > 0**, written as $m \div n$, is by definition the length of one part when a segment of length m is divided into n equal parts. This new definition requires the introduction of unit fractions, from Appendix A.1, as the n equal parts. This idea, coupled with the commutative property of multiplication, then gives a similar multiplication statement, $m = n \times k$, but now k is a unit fraction.

For example, $5 \div 3$ is equivalent to the multiplication statement $5 \times \dfrac{1}{3} = \dfrac{5}{3}$, which is essentially the definition of a fraction. This means that $5 \div 3$ is equivalent to the fraction $\dfrac{5}{3}$, or from now on we can replace one with the other because $5 \div 3 = \dfrac{5}{3}$.

This means that any ratio, with is one number divided by another, can also be thought of as a fraction.

Using this concept of ratios we can compare the sizes of groups of the same objects.

EXAMPLE 1: Compare the two groups using ratios. 10 people to 7 people.

$$\frac{10 \text{ people}}{7 \text{ people}} = \frac{10 \ \cancel{\text{people}}}{7 \ \cancel{\text{people}}} = \frac{10}{7}$$

The units of people cancel, so this is equivalent to saying that the group of 10 people to 7 people is equivalent to writing, $\dfrac{10}{7}$.

Stated another way, the ratio of 10 people to 7 people is equivalent to the fraction $\dfrac{10}{7}$. Thus, for every 7 people in group 2 there are 10 people in group 1.

EXAMPLE 2: Compare the two groups using ratios. 24 cars to 6 cars.

$$\frac{24 \text{ cars}}{6 \text{ cars}} = \frac{24 \ \cancel{\text{cars}}}{6 \ \cancel{\text{cars}}} = \frac{24}{6} = 4$$

The units of cars cancel, so this is equivalent to saying that the group of 24 cars to 6 cars is equivalent to writing, $\dfrac{24}{6} = 4$. This means that the group with 24 cars is 4 times as large as the group with 6 cars.

Stated another way, the ratio of 24 cars to 6 cars is equivalent to the fraction $\dfrac{24}{6} = \dfrac{4}{1} = 4$. Thus for every 6 cars in group 2 there are 24 cars in group 1, or equivalently, for every 1 car in group 2 there are 4 cars in group 1.

Notice that in the two examples above the final number has no units, it is just a number. All ratios are without units. They simply mean we are comparing the relative size of two groups of the same types of objects.

Another common type of ratio is the percentage. Percentages are ratios out of a fixed number, 100. A percentage can always be written as a fraction over 100, e.g.; $20\% = \dfrac{20}{100}$, $7\% = \dfrac{7}{100}$, $93\% = \dfrac{93}{100}$, etc. Percentages also compare the same objects, only they compare it to a fixed number of 100 objects. We can extend what we do for ratios to include percentages.

EXAMPLE 3: Compare the two groups using ratios. 10 people to 7 people. Then rewrite as a decimal rounded to the nearest hundredths place, then as a fraction over 100, and finally as a percent.

$$\frac{10 \text{ people}}{7 \text{ people}} = \frac{10 \text{ }\cancel{\text{people}}}{7 \text{ }\cancel{\text{people}}} = \frac{10}{7} = 1.42857\ldots \approx 1.43 = \frac{143}{100} = 143\%$$

EXAMPLE 4: Compare the two groups using ratios. 8 boats to 15 boats. Then rewrite as a decimal rounded to the nearest hundredths place, then as a fraction over 100, and finally as a percent.

$$\frac{8 \text{ boats}}{15 \text{ boats}} = \frac{8 \text{ }\cancel{\text{boats}}}{15 \text{ }\cancel{\text{boats}}} = \frac{8}{15} = 0.5333\ldots \approx 0.53 = \frac{53}{100} = 53\%$$

EXERCISES 7.1

Compare the number of objects in the second group to the number in the first, and then explain what the resulting ratio means when written as a fraction. Also rewrite as a decimal approximated to the hundredths place, then as a fraction over a 100, and finally as a percentage.

1. $32 to $25
2. 36 cases of books to 12 cases of books.
3. 16 bottles to 19 bottles
4. 45 meters to 10 meters
5. 175 cm to 25 cm
6. 28 tickets to 56 tickets
7. 120 square feet to 10 square feet.
8. 4 boats to 3 boats

7.2 WHAT IS A RATE, OR RATE-OF-CHANGE?

A rate is simply a ratio where the two quantities we are comparing have different units. In particular, consider the common rate called speed. This is a comparison of a distance (for example miles) to time (for example hours). This is called the speed which is characterized by the rate of miles per hour.

In this case the rate, which is the speed, is characterized by a distance traveled (or a change in distance), over an elapsed time (or a change in time). A rate is simply a measure of how the quantity in the numerator (top) changes as the quantity in the denominator (bottom) changes. A rate is also called a rate-of-change. We will use the terms interchangeably. Now, we frequently think of a rate as something that changes with time, such as:

$$\text{Rate} = \frac{\text{Change in Miles}}{\text{Change in Hours}}$$

$$\text{Rate} = \frac{\text{Change in volume}}{\text{Change in time}}$$

$$\text{Rate} = \frac{\text{Number of heartbeats}}{\text{Number of minutes}}$$

$$\text{Rate} = \frac{\text{Change in the Earth's temperature}}{\text{Change in the number of years}}$$

In all these rates we are considering how the quantity on top (numerator) changes with a changing quantity of time on the bottom (denominator).

However, we can go beyond the "time" interpretation and consider other possible rates-of-change. For example, consider the following rates:

$$\text{Rate} = \frac{\text{Profit from an item}}{\text{Cost of the item}} \text{ , or, how the profit changes with a changing cost.}$$

$$\text{Rate} = \frac{\text{Temperature}}{\text{Altitude}} \text{ , or, how the temperature changes with a changing altitude.}$$

Thus, we can think of a rate more generally, as how changing one quantity, changes another quantity.

This is an extremely important concept in both mathematics as well as the sciences. With this in mind, we consider some basic examples that involve finding rates.

We should, however, point out that we are actually finding an **average rate** in all the quantities above. For example, we know that it is possible to drive from location A to location B, and change your speed as you travel along the way. However, if we only consider the total distance traveled divided by the total time to travel it, this will only give us the average speed in which we traveled. It says nothing about how our speed might have changed along the way. This is also true for rates not involving time. Thus, in this section we are only considering average rates and assuming that our speed never changes. We realize that the last assumption is not realistic, but it is necessary at this level of mathematics.

EXAMPLE 1: A plane flies 1,500 miles in 3 hours. What is its average rate of change or speed?

$$\text{rate} = \frac{1500 \, \text{miles}}{3 \, \text{hours}} = 500 \frac{\text{miles}}{\text{hours}} = 500 \, \text{mph}$$

This means that on average, the plane flew at 500 mph.

EXAMPLE 2: As we climb 2,000 feet in altitude, the temperature goes down by 35 °F. What is its average rate-of-change?

$$\text{rate} \ = \ \frac{-35\,^{O}F}{2,000\,\text{feet}} \ = \ -0.0175\,\frac{^{O}F}{\text{foot}}$$

This means that on average, the temperature decreased by 0.0175 °F per foot of increased altitude.

EXAMPLE 3: In a glycolysis reaction that takes 3.25 hours, the volume of CO_2 produced is 3.5 mL. What is the rate of CO_2 production?

Note: Since we are looking for a rate of production, that implies that we are looking for how much "stuff" is produced over time. Therefore, the rate should be expressed in units of $\frac{\text{volume}}{\text{time}}$.

$$\text{rate} \ = \ \frac{3.5\,\text{mL}}{3.25\,\text{hours}} \ \approx \ 1.077\,\frac{\text{mL}}{\text{hour}}$$

This means that the CO_2 production was at an average rate of about 1.077 mL per hour.

Sometimes we are given one set of units, but are asked to find the rate in a different set of units. The example below shows ways in which to handle this.

EXAMPLE 4: A competitive high school swimmer takes 54 sec to swim 100 yards. What is her rate in yards per second? What is her rate in feet per minute? Round your answer to the hundredths place.

We can compute the rate directly by placing 100 yards over 54 seconds.

$$\text{rate} \ = \ \frac{100\,\text{yards}}{54\,\text{sec}} \ \approx \ 1.85\,\frac{\text{yards}}{\text{sec}}$$

To change the rate to feet per minute we have two options. The first is more direct, and usually a bit easier for students. The second converts from the yards per second rate to the feet per minute rate. We shall show both, but the second approach involves a deeper knowledge of complex fractions and should only be used if you are comfortable with it, as is can cause confusion, mistakes and loss of accuracy.

Approach 1: We start by changing the yards into feet and the seconds into minutes.

$$100\,\text{yards} \ = \ 100\,\text{yards} \times \frac{3\,\text{feet}}{1\,\text{yard}} \ = \ 100 \times 3\,\text{feet} \ = \ 300\,\text{feet}$$

$$54\,\text{sec} \ = \ 54\,\text{sec} \times \frac{1\,\text{minute}}{60\,\text{sec}} \ = \ \frac{54}{60}\,\text{minutes} \ = \ 0.9\,\text{minutes}$$

Then we take the ratio of the results:

$$\text{rate} = \frac{300 \text{ feet}}{0.9 \text{ minutes}} \approx 333.33 \frac{\text{feet}}{\text{minute}}$$

<u>Approach 2</u>: We start with the already calculated rate of yards per sec and convert to feet per minute.

$$\text{rate} \approx 1.85 \frac{\text{yds}}{\text{sec}} = \frac{1.85 \, \cancel{\text{yds}} \times \frac{3 \text{ ft}}{1 \, \cancel{\text{yd}}}}{\cancel{\text{sec}} \times \frac{1 \text{ min}}{60 \, \cancel{\text{sec}}}} = \frac{1.85 \times 3 \text{ ft}}{\frac{1}{60} \text{ min}} = \frac{1.85 \times 3}{\frac{1}{60}} \frac{\text{ft}}{\text{min}}$$

$$= \frac{5.55}{\frac{1}{60}} \frac{\text{ft}}{\text{min}} = \frac{555 \times 60}{1} \frac{\text{ft}}{\text{min}} = 333 \frac{\text{ft}}{\text{min}}$$

Notice how we lose accuracy with this approach as opposed to the previous Approach 1. This is because we approximated early on in the process in Approach 2 and not in the end, as we did with Approach 1. As you can also see, it is a far more complicated way to get the end result. Thus, we recommend Approach 1 for these types of problems.

Finally, an important idea related to the rate concept is the distance, constant rate (speed), and time relationship.

A constant rate of speed is really just a rate of the form

$$\text{Constant Rate (or Speed)} = \frac{\text{Total Distance Traveled}}{\text{Total Time to Travel it}}$$

If we multiply both sides of this relation by the "Time to Travel it", we can rearrange the result and obtain the following important relationship.

$$\text{Distance Traveled} = \text{Constant Rate (or Speed)} \times \text{Time to Travel it}$$

or

$$\text{Distance} = \text{Speed} \times \text{Time}$$

We can now use this relationship to find the total distance traveled provided we know both the average speed, and the time we traveled at that speed.

EXAMPLE 5: You travel at a speed of $25 \frac{\text{miles}}{\text{hour}}$ for 8 hours. How far have you traveled?

We simply replace our speed and time in the formula

$$\text{Distance} = \text{Speed} \times \text{Time}$$

to obtain:

$$\text{Distance} = 25\,\frac{\text{miles}}{\text{hour}} \times 8 \text{ hours}$$

Units follow the same algebraic principles as numbers, thus, we can rewrite the equation as follows:

$$\text{Distance} = (25 \times 8) \left(\frac{\text{miles}}{\text{hour}} \times \text{hours} \right)$$

$$= (200) \left(\frac{\text{miles}}{\cancel{\text{hour}}} \times \cancel{\text{hours}} \right)$$

$$= 200 \text{ miles}$$

Again, we use the concept of dimensional analysis, and cancel the units in an appropriate way to get the required final unit of miles.

EXAMPLE 6: You travel at a speed of $30\,\frac{\text{feet}}{\text{second}}$ for 10 seconds. How far have you traveled?

We simply replace our speed and time in the formula

$$\text{Distance} = \text{Speed} \times \text{Time}$$

to obtain:

$$\text{Distance} = 30\,\frac{\text{feet}}{\text{second}} \times 10 \text{ seconds}$$

Units follow the same algebraic principles as numbers, thus, we can rewrite the equation as follows:

$$\text{Distance} = (30 \times 10) \left(\frac{\text{feet}}{\text{second}} \times \text{seconds} \right)$$

$$= (300) \left(\frac{\text{feet}}{\cancel{\text{second}}} \times \cancel{\text{seconds}} \right)$$

$$= 300 \text{ feet}$$

Again, we use the concept of dimensional analysis, and cancel the units in an appropriate way to get the required final unit of feet.

These examples should also make sense logically, if you think about it. Example 4 says that if you go at a constant speed of 25 miles/hour for 8 hours, you would travel a total distance of 200 miles. In Example 5, it says that if you travel at a speed of 30 feet/second for 10 seconds, you would travel a total distance of 300 feet.

All these types of problems will require that you are fully versed with time conversion problems, as we show in the next example. Thus, you should know that there a 60 seconds in a minute, 60 minutes in an hour and 24 hours in a day. Conversions between different measures of time, are also achieved using conversion factors as done in Chapter 4. A table of the common conversion factors are shown below.

Convert "from" "to"	Conversion Factor
seconds to minutes	$\dfrac{1\,\text{min}}{60\,\text{sec}}$
minutes to hours	$\dfrac{1\,\text{hour}}{60\,\text{min}}$

Convert "from" "to"	Conversion Factor
minutes to seconds	$\dfrac{60\,\text{sec}}{1\,\text{min}}$
hours to minutes	$\dfrac{60\,\text{min}}{1\,\text{hour}}$

EXAMPLE 7: You travel at a speed of $60\,\dfrac{\text{miles}}{\text{hour}}$ for 45 minutes. How far have you traveled?

We simply replace our speed and time in the formula

$$\text{Distance} = \text{Speed} \ \times \ \text{Time}$$

to obtain:

$$\text{Distance} = \ 60\,\frac{\text{miles}}{\text{hour}} \ \times \ 45\ \text{minutes}$$

We now notice a problem. The units of time and the speed units for time are not the same type. In our speed we measure it using hours, but in our time we are given minutes. Thus, we will first need to covert our time in minutes to time in hours before we carry out the arithmetic in the formula.

$$
\begin{aligned}
45\,\text{minutes} \ &= \ 45\,\text{minutes} \times \frac{1\ \text{hour}}{60\ \text{minutes}} \\
&= \ \frac{45}{60}\ \frac{\cancel{\text{minutes}} \times \text{hour}}{\cancel{\text{minutes}}} \\
&= \ 0.75\ \text{hours}
\end{aligned}
$$

Now replacing 45 minutes with 0.75 hours we obtain:

$$
\begin{aligned}
\text{Distance} \ &= \ 60\,\frac{\text{miles}}{\text{hour}} \times 0.75\ \text{hours} \\
&= \ (60 \times 0.75)\left(\frac{\text{miles}}{\text{hour}} \times \text{hours}\right) \\
&= \ (45) \ \left(\frac{\text{miles}}{\cancel{\text{hour}}} \times \cancel{\text{hours}}\right) \\
&= \ 45\,\text{miles}
\end{aligned}
$$

EXERCISES 7.2

Given the following, find the rates. Be sure to include the appropriate units. If an exact answer is not possible, then round all answers to the hundredths place, unless stated otherwise.

1. A car travels 345 miles in 6 hours, What is its average rate-of-change in miles per hour?
2. A car travels 745 miles in 12 hours, What is its average rate-of-change in miles per hour?
3. As we climb 3,000 feet in altitude, the temperature goes down by 50 $^{\circ}$F. What is its average rate-of-change of $^{\circ}$F per foot?
4. As we climb 15,000 feet in altitude, the temperature goes down by 75 $^{\circ}$F. What is its average rate-of-change of $^{\circ}$F per foot?
5. On an electrocardiogram, you observe that a patient's normal heartbeat occurs every 0.84 seconds. What is the patient's heart rate in heartbeats per minute?
6. On an electrocardiogram, you observe that a patient's normal heartbeat occurs every 0.75 seconds. What is the patient's heart rate in heartbeats per minute?
7. In a glycolysis reaction that takes 2 hours, the volume of CO_2 produced is 3.25 mL. What is the rate of CO_2 production?
8. In a glycolysis reaction that takes 2.5 hours, the volume of CO_2 produced is 4.75 mL. What is the rate of CO_2 production?
9. In a glycolysis reaction that takes 90 minutes, the volume of CO_2 produced is 2.75 mL. What is the rate of CO_2 production?
10. A competitive high school swimmer takes 52 s to swim 100 yards. What is his rate in yards per second? What is his rate in feet per minute?
11. The *Alvin*, a submersible research vessel, can descend into the ocean to a depth of approximately 4500 m in just over 5 hr. Determine its average rate of submersion in meters per hour. What is its average rate in kilometers per second?
12. A competitive college runner ran a 5K (5 km) race in 15 min, 25 s. What was her pace in km per min? Meters per second?
13. A compressor will compress 20 liters of air in 15 minutes. What is the rate of compression in liters per hour?
14. A weather balloon rises 275 feet in 15 seconds. How fast is it rising in feet per minute?
15. A parachute descends 255 meters in 8 minutes. At what rate, in meters per minute is the parachute descending?
16. In taking a person's pulse you count 15 heartbeats in 10 seconds. What is the heart rate in beats per minute?
17. In taking a person's pulse you count 8 beats in 5 seconds. What is the heart rate in beats per minute?
18. A plane descends 25 miles in 5 minutes. What is the rate is it descending in miles per hour?
19. A runner can run 320 feet in 10 seconds. At what rate in miles per hour are they running?
20. A bird can fly 250 meters in 20 seconds. How fast are they traveling in meters per hour?
21. You ski down a hill and in 15 seconds you travel 110 meters. How fast are you traveling in meters per minute?
22. The potassium permanganate dye in a gel is diffusing at 18 mm every 2 minutes. What is the rate of diffusion in centimeters per hour (cm/hr)?
23. The dye in a gel is diffusing at 12 mm every 3 minutes. What is the rate of diffusion in centimeters per hour (cm/hr)?

Find the distance traveled, given the following. Be sure to show how you cancel the units appropriately.

24. speed = 5 m/second, for 12 seconds

25. speed = 55 miles/hour, for 3.5 hours

26. speed = 250 cm/minute, for 15 minutes

27. speed = 32 feet/second, for 24 seconds

28. speed = 32 m/minute, for 40 minutes

29. speed = 560 miles/hour, for 4 hours

30. speed = 112 km/hour, for 1.25 hours

31. speed = 550 miles/hour, for 45 minutes

32. speed = 1.73 feet/second, for 3 minutes

33. speed = 36.7 km/hour, for 9.81 seconds

34. speed = 80 km/hour, for 5 hours

35. speed = 1800 m/hour, for 2.5 hours

36. speed = 900 cm/second, for 45 seconds

37. speed = 35 feet/minute, for 12 minutes

38. speed = 2.5 m/second, for 25 seconds

39. speed = 60 miles/hour, for 4.5 hours

40. speed = 325 cm/second, for 15 seconds

41. speed = 72 feet/minute, for 30 minutes

42. speed = 215.5 m/second, for 22 seconds

43. speed = 75 miles/hour, for 3.75 hours

44. speed = 1550 miles/hour, for 90 minutes

45. speed = 23.4 feet/second, for 5 minutes

7.3 WHAT IS A PROPORTION?

The term proportion can either be viewed as a noun or as a verb. For our purposes we shall view it as a verb, meaning there is an action to be performed. More specifically, a proportion is when we set two ratios or rates equal to each other.

For example:

$$\frac{16\,\text{miles}}{3\,\text{hours}} = \frac{48\,\text{miles}}{9\,\text{hours}}\,, \quad \frac{25}{100} = \frac{1}{4}\,, \quad \frac{\$220}{1\,\text{week}} = \frac{\$550}{2.5\,\text{weeks}}\,, \quad \frac{33}{70} = \frac{5}{7}$$

Now, just because we set two ratios or rates equal to each other, does not imply that they are indeed equal. The last proportion in the examples above does not have equal ratios. Thus, the first step we must perform is a verification process, where we determine if the two ratios or rates are equal to each other.

This then leads to the more important application of proportions, which is solving a proportion when part of one of the ratios or rates is unknown, and we wish to determine the value of that unknown part which will make the proportion true, or verified.

Proportions either require that we verify that the proportion is true, or that we find a value for an unknown in the proportion that makes it a true statement. The latter operation is called solving a proportion. In the next two sections we see how to first verify and then solve proportions.

7.4 VERIFYING PROPORTIONS

When working with a proportion involving two ratios, the verification process is already familiar to us. This is essentially verifying that we have equivalent fractions. The process to verify if two fractions are equivalent, is the cross-multiplication algorithm. The cross-multiplication algorithm states that two fractions (ratios) are equal if, when we cross-multiply (the denominator of one fraction times the numerator of the other fraction), we get equal numbers on both sides of the equal sign. If this is true then the fractions or ratios are equal.

EXAMPLE 1: Verify the proportion of the two ratios, i.e. determine if the proportion is a true statement or not.

$$\frac{25}{100} = \frac{1}{4}$$

cross multiplying

$$\frac{25}{100} \bowtie \frac{1}{4}$$

then verifying

$$4 \times 25 \stackrel{?}{=} 1 \times 100$$

$$100 = 100 \checkmark$$

This proportion is true.

EXAMPLE 2: Verify the proportion of the two ratios, i.e. determine if the proportion is a true statement or not.

$$\frac{33}{70} = \frac{5}{7}$$

cross multiplying

$$\frac{33}{70} \bowtie \frac{5}{7}$$

then verifying

$$7 \times 33 \stackrel{?}{=} 70 \times 5$$

$$231 \neq 350$$

This proportion is not true.

The same is true when we look at proportions involving rates. Proportions with rates, however, have the added complexity of units, since the units do not cancel like they do for ratios. The end result, using cross-multiplication to verify equality, is still the same as it is for ratios, as we now show.

EXAMPLE 3: Verify the proportion of the two rates, i.e. determine if the proportion is a true statement or not.

$$\frac{16\,\text{miles}}{3\,\text{hours}} = \frac{48\,\text{miles}}{9\,\text{hours}}$$

cross multiplying

$$\frac{16\,\text{miles}}{3\,\text{hours}} \bowtie \frac{48\,\text{miles}}{9\,\text{hours}}$$

202
then verifying

$$9 \times 16 \frac{\text{miles}}{\text{hour}} \overset{?}{=} 3 \times 48 \frac{\text{miles}}{\text{hour}}$$

$$144 \frac{\text{miles}}{\text{hour}} = 144 \frac{\text{miles}}{\text{hour}} \checkmark$$

This proportion is true. Note, however, that the result of 144 miles/hour is meaningless. Nothing is actually going at this rate. This is just the verification process stating that the two original rates are equal!

Now the alternative is to write this cross-multiplication as:

$$9 \text{ hours} \times 16 \text{ miles} \overset{?}{=} 3 \text{ hours} \times 48 \text{ miles}$$

$$144 \,(\text{mile})(\text{hours}) = 144 \,(\text{mile})(\text{hours}) \checkmark$$

However, since the units of (mile)(hours) has no physical meaning we chose the other alternative. The real point is that the numerical coefficients in front of the units are the same. If we want we can ignore the units altogether and simply write:

$$9 \times 16 \overset{?}{=} 3 \times 48$$

$$144 = 144 \checkmark$$

This is the approach we will choose to use in the future.

EXAMPLE 4: Verify the proportion of the two rates, i.e. determine if the proportion is a true statement or not. A compressor will compress 30L of air in 15 minutes, or 20L of air in 10 minutes. Are they proportionally equivalent?

$$\frac{30\,\text{L}}{15\,\text{minutes}} = \frac{20\,\text{L}}{10\,\text{minutes}}$$

cross multiplying

$$\frac{30\,\text{L}}{15\,\text{minutes}} \times \frac{20\,\text{L}}{10\,\text{minutes}}$$

then verifying

$$10 \times 30 \overset{?}{=} 15 \times 20$$

$$300 = 300 \checkmark$$

This proportion is true.

Verify the following proportions

1. $\dfrac{36}{60} = \dfrac{54}{90}$ Are the two ratios equivalent?

2. $\dfrac{40}{90} = \dfrac{8}{18}$ Are the two ratios equivalent?

3. $\dfrac{15}{75} = \dfrac{3}{24}$ Are the two ratios equivalent?

4. $\dfrac{120}{45} = \dfrac{24}{7}$ Are the two ratios equivalent?

5. $\dfrac{345}{85} = \dfrac{21}{5}$ Are the two ratios equivalent?

6. $\dfrac{12}{16} = \dfrac{3}{4}$ Are the two ratios equivalent?

7. A car travels 235 miles in 5 hours. Another car travels 188 miles in 4 hours. Are the two rates proportionally equivalent?

8. A car travels 371 miles in 7 hours. Another car travels 212 miles in 4 hours. Are the two rates proportionally equivalent?

9. A car travels 640 miles in 10 hours. Another car travels 330 miles in 6 hours. Are the two rates proportionally equivalent?

10. A climber climbs 2,000 feet in altitude, and the outside air temperature goes down by 42 $^\circ$F. Another climber climbs 1600 feet in altitude and the temperature goes down 35 $^\circ$F. Are the two rates proportionally equivalent?

11. A climber climbs 1,000 feet in altitude, and the outside air temperature goes down by 25 $^\circ$F. Another climber climbs 800 feet in altitude and the temperature goes down 20 $^\circ$F. Are the two rates proportionally equivalent?

12. In one glycolysis reaction 3.7mL of CO_2 is produced in 2 hours. In a second glycolysis reaction 8.325mL of CO_2 is produced in 4.5 hours. Are the two rates proportionally equivalent?

13. In one glycolysis reaction 9.75mL of CO_2 is produced in 5.4 hours. In a second glycolysis reaction 2.35mL of CO_2 is produced in 1.7 hours. Are the two rates proportionally equivalent?

14. A compressor compresses 36 liters of air in 21 minutes. A second compressor compresses 48 liters of air in 28 minutes. Are the two rates proportionally equivalent?

15. A compressor compresses 115 liters of air in 60 minutes. A second compressor compresses 175 liters of air in 1.5 hours. Are the two rates proportionally equivalent?

7.5 SOLVING PROPORTIONS

To solve a proportion means we have to verify the proportion containing a variable. We do this by cross-multiplying the proportion as shown above and then finding the value of the variable that makes the proportion true. We first illustrate this process through proportion examples below with ratios with their units removed, and then later we show the same results for proportions involving rates.

EXAMPLE 1: Solve the proportion of ratios: $\dfrac{x}{3} = \dfrac{5}{15}$

To solve for the variable x, we can **cross-multiply** the equation by 3 and 15. To **cross-multiply** means we multiply the x by fifteen (15) and the five (5) by three (3). In general, we cross-multiply the denominators times the **opposite** numerators.

This results in the following:

$$15 \times x = 3 \times 15$$

$$15 \times x = 15$$

This tells us that we need to find a number x that when multiplied by 15 gives us 15. We can see that number is simply x = 1. That is the value that solves our proportion, or makes it a true statement, i.e.

$$\frac{1}{3} = \frac{5}{15}$$

EXAMPLE 2: Solve the proportion of ratios: $\dfrac{8}{x} = \dfrac{7}{5}$

To solve for the variable x, we cross-multiply the equation by the x and the 5.

$$\frac{8}{x} = \frac{7}{5}$$

$$8 \times 5 = 7 \times x$$

$$40 = 7x$$

We now want to find a number that when multiplied by 7 gives us 40. Since no whole number will work, we try the fraction $x = \dfrac{40}{7}$ as shown below.

$$7 \times \frac{40}{7} = \frac{7 \times 40}{7} = \frac{\cancel{7} \times 40}{\cancel{7}} = 40$$

Notice this is equivalent to simply dividing both sides of $40 = 7x$ by 7. Thus, that will be our approach, to divide both sides of the equation by the coefficient in front of the variable as a last step.

The solution to the proportion is $x = \dfrac{40}{7}$

$$\frac{8}{\frac{40}{7}} = 8 \div \frac{40}{7} = 8 \times \frac{7}{40} = \frac{56}{40} = \frac{56 \div 8}{40 \div 8} = \frac{7}{5}$$

EXAMPLE 3: Solve the proportion of ratios: $\dfrac{w-3}{5} = \dfrac{2}{5}$

To solve for the variable w, we cross-multiply the equation by 5 and 5. Then we solve the resulting linear equation.

$$5(w-3) = 2\cdot 5$$
$$5w - 15 = 10$$
$$5w - 15 + 15 = 10 + 15$$
$$5w = 25$$
$$\frac{5w}{5} = \frac{25}{5} = \frac{5\cdot 5}{5}$$
$$w = 5$$

EXAMPLE 4: Solve the proportion: $\dfrac{2}{5} = \dfrac{1}{a-2}$

To solve for the variable a, we cross-multiply the equation by 5 and $(a-2)$. Then we solve the resulting linear equation.

$$2(a-2) = 5\cdot 1$$
$$2a - 4 = 5$$
$$2a - 4 + 4 = 5 + 4$$
$$2a = 9$$
$$\frac{\cancel{2}a}{\cancel{2}} = \frac{9}{2}$$
$$a = \frac{9}{2}$$

EXAMPLE 5: Solve the proportion: $\dfrac{3}{b+2} = \dfrac{2}{2b-3}$

To solve for the variable b, we cross-multiply the equation by $(b+2)$ and $(2b-3)$. Then we solve the resulting linear equation.

$$3(2b-3) = 2(b+2)$$
$$6b-9 = 2b+4$$
$$6b-2b-9 = 2b-2b+4$$
$$4b-9 = 4$$
$$4b-9+9 = 4+9$$
$$4b = 13$$
$$\frac{4b}{4} = \frac{13}{4}$$
$$b = \frac{13}{4}$$

> We should point out that in all the above problems we have not been concerned about the possible division by zero. This can occur because the variable is in the denominator. It is a potential problem, however, we will not address it here. It is something that we discuss in detail in following courses on algebra.

In the next example we introduce a very common rate which is speed, $\frac{distance}{time}$, and show how to solve proportions related to speed.

EXAMPLE 6: Solve the proportion of rates.
$$\frac{440\,\text{miles}}{8\,\text{hours}} = \frac{320\,\text{miles}}{x\,\text{hours}}$$
Round your answer to the nearest tenths of an hour

To solve for the variable x, we ignore the units, and cross-multiply the equation to obtain:
$$440 \times x = 320 \times 8$$
$$440\,x = 2,560$$

To solve for x, divide both sides of the proportion by 440 and write you answer as a decimal using your calculator. Then round to the nearest tenths place.

$$\frac{440\,x}{440} = \frac{2,560}{440}$$
$$x = 5.818\ldots$$
$$x = 5.8$$

This means it would take approximately 5.8 hours.

$$\frac{440\,\text{miles}}{8\,\text{hours}} \approx \frac{320\,\text{miles}}{5.8\,\text{hours}}$$

EXERCISES 7.5

Solving Proportions

Solve each ratio proportion.

Basic Problems

1. $\dfrac{x}{3} = \dfrac{4}{5}$

4. $\dfrac{4}{a} = \dfrac{3}{10}$

7. $\dfrac{n}{4} = \dfrac{3}{8}$

10. $\dfrac{s}{7} = \dfrac{3}{14}$

2. $\dfrac{2}{7} = \dfrac{b}{6}$

5. $\dfrac{7}{12} = \dfrac{5}{d}$

8. $\dfrac{6}{x} = \dfrac{3}{2}$

11. $\dfrac{w}{2} = \dfrac{9}{16}$

3. $\dfrac{n}{12} = \dfrac{3}{8}$

6. $\dfrac{7}{y} = \dfrac{9}{8}$

9. $\dfrac{5}{r} = \dfrac{2}{7}$

12. $\dfrac{p}{5} = \dfrac{8}{9}$

Intermediate and Advanced Problems

13. $\dfrac{x+2}{3} = \dfrac{4}{7}$

17. $\dfrac{w-5}{3w+5} = \dfrac{1}{7}$

21. $\dfrac{n-5}{7} = \dfrac{2n}{3}$

24. $\dfrac{b+3}{15} = \dfrac{b}{4}$

14. $\dfrac{m-3}{8} = \dfrac{6}{5}$

18. $\dfrac{4}{r+3} = \dfrac{2}{r-1}$

22. $\dfrac{p+13}{p-3} = \dfrac{7}{2}$

25. $\dfrac{2y}{3} = \dfrac{7}{2}$

15. $\dfrac{5}{3} = \dfrac{1}{a+5}$

19. $\dfrac{6}{v-4} = \dfrac{5}{2v+3}$

23. $\dfrac{c-2}{c+3} = \dfrac{9}{8}$

16. $\dfrac{p+3}{2p-1} = \dfrac{3}{8}$

20. $\dfrac{x}{2} = \dfrac{4x-5}{9}$

Solve each rate proportion and round your answer to the nearest tenths place as needed.

26. $\dfrac{200\,\text{miles}}{3\,\text{hours}} = \dfrac{250\,\text{miles}}{x\,\text{hours}}$

28. $\dfrac{x\,\text{km}}{3\,\text{hours}} = \dfrac{415\,\text{km}}{2\,\text{hours}}$

30. $\dfrac{x\,\text{m}}{5\,\text{sec}} = \dfrac{1575\,\text{m}}{3\,\text{sec}}$

27. $\dfrac{565\,\text{miles}}{9\,\text{hours}} = \dfrac{x\,\text{miles}}{4\,\text{hours}}$

29. $\dfrac{1500\,\text{km}}{3\,\text{hours}} = \dfrac{x\,\text{km}}{4.5\,\text{hours}}$

31. $\dfrac{4440\,\text{m}}{x\,\text{sec}} = \dfrac{532\,\text{m}}{8\,\text{sec}}$

7.6 MORE APPLICATIONS OF PROPORTIONS

Proportions can be used to solve a variety of problems across a number of disciplines. In this section we highlight examples from a few areas. We use the cross-multiplication algorithm we introduced in the previous sections to solve proportions, and apply it to some real-life examples.

The examples we consider in this section are called direct proportions. This means that as the value of one object increases the other object's value also increases. If, on the other hand, the value of the object decreases the other object's value also decreases. We call this a direct proportion.

We provide a range of problems in this section to highlight the utility of the proportion concept. In the first three examples we choose to set up and solve our proportions as ratios. In the last three examples we solve the same three problems, but instead write them as rates. Both techniques work equally well, and you can choose whichever approach you feel most comfortable with.

EXAMPLE 1: If 2 pencils cost $1.50, how many pencils can you buy with $9.00?

Since the cost of the pencils varies directly with their number meaning if we increase the number of pencils the cost also increases, this is a direct proportion.

To solve, we set up the proportion as a ratio by placing like objects above like objects in our proportion, so that our units cancel. We also place related objects directly across from each other in the proportion.

$$\frac{2 \, \text{pencils}}{x \, \text{pencils}} = \frac{\$1.50}{\$9.00}$$

To solve for the x pencils, we cancel the units, cross-multiply and solve for x.

$$(2)(9.00) = (1.50)x$$
$$18.00 = 1.50x$$
$$\frac{18.00}{1.50} = \frac{\cancel{1.50}x}{\cancel{1.50}x}$$
$$12 = x$$

This means we can buy 12 pencils with $9.00.

EXAMPLE 2: If Dimitri ran 200 meters at a constant speed in 28 seconds, how long did he take to run 1 meter.

Since the distance varies directly with time, this is a direct proportion.

To solve we set up the direct proportion equation:
$$\frac{28 \, \text{sec}}{x \, \text{sec}} = \frac{200 \, \text{m}}{1 \, \text{m}}$$

To solve for the time, we cancel the units and cross-multiply and solve for x.

$$28 \times 1 = 200 \times x$$
$$28 = 200x$$
$$\frac{28}{200} = \frac{\cancel{200}x}{\cancel{200}}$$
$$0.14 = x$$

It would take Dimitri 0.14 seconds to run 1 meter.

EXAMPLE 3: If $\frac{2}{7}$ of a tank can be filled in 5 minutes, how many minutes will it take to fill the entire tank?

Since the portion of the tank filled varies directly with time this is a direct proportion. To solve we set up the direct proportion equation:

$$\frac{5\,\text{minutes}}{x\,\text{minutes}} = \frac{\frac{2}{7}\,\text{tank}}{1\,\text{tank}}$$

We cancel the units, and cross-multiply

$$1 \times 5 = \frac{2}{7} \times x$$

$$5 = \frac{2}{7}x$$

To solve for x we need to divide both sides of the proportion by the fraction $\frac{2}{7}$. This gives us:

$$5 \div \frac{2}{7} = \frac{2}{7}x \div \frac{2}{7}$$

$$5 \times \frac{7}{2} = \frac{2}{7}x \times \frac{7}{2}$$

$$\frac{35}{2} = \frac{2 \times x \times 7}{7 \times 2}$$

$$17.5 = \frac{\cancel{14}\,x}{\cancel{14}} = x$$

It would take 17.5 minutes to fill the entire tank.

EXAMPLE 4: Solve the proportion of rates. You traveled 380 miles in 6 hours, how long did it take you to travel 275 miles given that your speed was a constant value. Round your answer to the nearest tenths of an hour

Set up the two rates as miles per hour, and set them equal to each other.

$$\frac{380\,\text{miles}}{6\,\text{hours}} = \frac{275\,\text{miles}}{x\,\text{hours}}$$

To solve for the variable x, we ignore the units, and cross-multiply the equation to obtain:

$$380 \times x = 275 \times 6$$

$$380\,x = 1{,}650$$

To solve for x, divide both sides of the proportion by 380 and write you answer as a decimal using your calculator. Then round to the nearest tenths place.

$$\frac{\cancel{380}\,x}{\cancel{380}} = \frac{1,650}{380}$$

$$x = 4.3421\ldots$$

$$x = 4.3$$

This means it would take approximately 4.3 hours.

$$\frac{380\,\text{miles}}{6\,\text{hours}} \approx \frac{275\,\text{miles}}{4.3\,\text{hours}}$$

CHEMISTRY MASS CALCULATIONS

Proportions can also be used in chemistry to solve problems related to the amount (mass) of chemicals desired or present in solutions and other forms. We illustrate its utility through two examples.

EXAMPLE 5: You have a container with 4000 g of water in it. How many grams of oxygen are in the container? You are given the chemical make-up of water as H_2O, and that the gram atomic weight of hydrogen is 1 and oxygen is 16.

The gram atomic weight of water is 18 (two hydrogen atoms and one oxygen atom, $2 + 16 = 18$). Let x represent the gram atomic weight of oxygen. Then we can set up the proportion:

$$\frac{x\,\text{grams of oxygen}}{4000\,\text{grams}} = \frac{16\,\text{gram atomic weight of oxygen in water}}{18\,\text{gram molecular weight of water}}$$

On the left-hand-side of the proportion we have the rate (proportion) of the gram amount of oxygen for a fixed gram amount of water, and on the right-hand-side we have the rate (proportion) of the gram molecular weight of oxygen in water for a fixed total gram molecular weight of water.

Canceling the gram units, but keeping the meaning of the x quantity in mind, we have the mathematical proportion:

$$\frac{x}{4000} = \frac{16}{18}$$

Solve the proportion by cross multiplying and solving for x.

$$18\,x = 16(4000)$$

$$x = \frac{16(4000)}{18} = \frac{64000}{18} \approx 3555.6$$

There are approximately 3,556 g of oxygen in 4,000 g of water.

EXAMPLE 6: You have a container with 1700 g of ammonia. How many grams of nitrogen are in the container? You are given the chemical formula of ammonia as NH_3, and that the gram atomic weight of hydrogen is 1, and nitrogen is 14.

The gram molecular weight of ammonia is 17 (three hydrogen atoms and one nitrogen atom, 3 + 14 = 17). Let x represent the gram atomic weight of nitrogen in the 1700 g. Then we can set up the proportion:

$$\frac{x \text{ grams atomic weight of nitrogen}}{1700 \text{ grams}} = \frac{14 \text{ gram atomic weight of nitrogen in ammonia}}{17 \text{ gram molecular weight of ammonia}}$$

removing descriptions as in the previous example, we have the mathematical proportion:

$$\frac{x}{1700} = \frac{14}{17}$$

Solve the proportion by cross multiplying and solving for x.

$$17x = 14(1700)$$

$$x = \frac{14(1700)}{17} = \frac{23800}{17} = 1400$$

There are 1,400 g of nitrogen in 1,700 g of ammonia.

EXERCISES 7.6

If an exact answer is not possible, then round all answers to the hundredths place, unless stated otherwise.

1. If 5 pencils cost $1.75, how many pencils can you buy with $10.00?

2. If 7 batteries cost $11.50, how many batteries can you buy with $34.50?

3. If 2 pounds of nails cost $2.75, how many many pounds can you buy with $12.00?

4. If 12 pounds of chlorine cost $13.75, how many pounds can you buy with $45.00?

5. If Shirley drove 65 miles per hour constantly for 6 hours. How long did it take her to drive 300 miles.

6. If Juan ran 400 meters at a constant speed in 54 seconds. How long did he take to run 10 meters.

7. If Jessica ran 1 km at a constant speed in 280 seconds. How long did he take to run 20 meters.

8. If Justin drove 58 miles per hour constantly for 4 hours. How long did it take him to drive 150 miles.

9. If $\dfrac{3}{10}$ of a tank can be filled in 15 minutes, how many minutes will it take to fill the entire tank?

10. If $\dfrac{3}{8}$ of a tank can be filled in 1 hours, how long will it take to fill the entire tank?

11. If $\dfrac{5}{9}$ of a tank can be filled in 3 hours, how long will it take to fill the entire tank?

12. If $\dfrac{7}{12}$ of a tank can be filled in 7 minutes, how many minutes will it take to fill the entire tank?

13. On an electrocardiogram, you observe that a patient's normal heartbeat occurs every 21 squares on the graph. If a square represents 0.04 seconds, how many seconds does it take for a heartbeat to occur?

14. On an electrocardiogram, you observe that a patient's normal heartbeat occurs every 25 squares on the graph. If a square represents 0.04 seconds, how many seconds does it take for a heartbeat to occur?

Chemistry Mass Calculations

15. You have a container with 2500 g of water in it. How many grams of oxygen are in the container? You are given the chemical make-up of water as H_2O, and that the gram atomic weight of hydrogen is 1 and oxygen is 16.

16. You have a container with 500 g of hydrochloric acid (HCl) in it. How many grams of chlorine are in the container? You are given that the gram atomic weight of hydrogen is 1 and chlorine is 35.45.

17. You have a container with 340 g of ammonia. How many grams of nitrogen are in the container? You are given the chemical formula of ammonia as NH_3, and that the gram atomic weight of hydrogen is 1, and nitrogen is 14.

18. You have a container with 3200 g of potassium chloride (KCl) in it. How many grams of potassium are in the container? You are given the gram atomic weight of potassium is 39 and chlorine is 35.45.

19. You have a container with 500 mg of barium chloride (BaCl) in it. How many grams of chlorine are in the container? You are given the gram atomic weight of barium is 137.3 and chlorine is 35.45.

20. You have a container with 4500 mg of Ethylene (C_2H_4) in it. How many grams of carbon are in the container? You are given the gram atomic weight of carbon is 12 and hydrogen is 1.

21. You have a container with 1300 mg of Ethylene (C_2H_4) in it. How many grams of hydrogen are in the container? You are given the gram atomic weight of carbon is 12 and hydrogen is 1.

7.7 PERCENTS AS PROPORTIONS

As we showed in 9.1 a percent is just a ratio whose denominator is 100, i.e. $27\% = \dfrac{27}{100}$. Then in 9.3 we showed proportions can come from setting two ratios equal to each other. In this section, we consider the case where the two ratios are percentages. This will allow us to apply what we learned about solving proportions in 9.5 and 9.6 to percent problems. For example, consider the following types of problems related to a percent increase or decrease of an amount.

1. Madeline's salary increased from $14 an hour to $15. What percent increase is that?
2. The movie was cut from 120 minutes down to 105 minutes. What percent decrease is that?
3. The cost of the new coat was reduced by 22%, saving you $15. What was the original cost of the coat?

Problems of this type can be confusing to set up, unless you remember this simple fact. The percent increase or decrease of anything, written as a decimal, is equal to the amount of the increase or decrease, divided by the original amount. If we write the decimal as a percentage, then the following direct proportion is equivalent to that statement.

$$\frac{\text{Amount of Increase (or Decrease)}}{\text{Original Amount}} = \frac{\text{Percent Increase (or Decrease)}}{100\,\text{Percent}}$$

Using this proportional formulation we can solve all the problems introduced above.

EXAMPLE 1: Madeline's salary increased from $14 an hour to $15. What percent increase is that?

Before using the direct proportion equation from above, we first need to find the Amount of the Increase = $1, and the Original Amount = $14. We substitute these into the formula and with x representing the percent increase, we have:

$$\frac{\$1}{\$14} = \frac{x\%}{100\%}$$

$$\frac{1}{14} = \frac{x}{100}$$

$$(1)\times(100) = (14)\times x \qquad \text{Cross multiply}$$

$$100 = 14\,x$$

$$\frac{100}{14} = x \qquad \text{Divide by 14}$$

$$7.14\ldots = x$$

This means she received a little over 7% for her raise.

EXAMPLE 2: The movie was cut from 120 minutes down to 105 minutes. What percent decrease is that?

Before using the direct proportion equation from above, we first need to find the Amount of the Decrease = 120 minutes – 105 minutes = 15 minutes, and the Original Time = 120 minutes. We substitute these into the formula, with x representing the % decrease and obtain:

$$\frac{15 \text{ minutes}}{120 \text{ minutes}} = \frac{x\%}{100\%}$$

$$\frac{15}{120} = \frac{x}{100}$$

$$(15) \times (100) = (120) \times x \qquad \text{Cross multiply}$$

$$1500 = 120\,x$$

$$\frac{1500}{120} = x \qquad \qquad \text{Divide by 120}$$

$$12.5 = x$$

This means the length of the movie was reduced by 12.5%.

EXAMPLE 3: The cost of the new coat was reduced by 22%, saving you $15. What was the original price of the coat?

Before using the direct proportion equation from above, we first need to find the Amount of the decrease = $15, and the Percent Decrease = 22%. We substitute these into the formula, and letting x equal the price of the original coat, we obtain:

$$\frac{\$15}{\$x} = \frac{22\%}{100\%}$$

$$\frac{\$15}{\$x} = \frac{22\%}{100\%}$$

$$(15) \times (100) = (22) \times x \qquad \text{Cross multiply}$$

$$1500 = 22\,x$$

$$\frac{1500}{22} = x \qquad \qquad \text{Divide by 22}$$

$$68.18 = x$$

This means the original price for the coat was $68.18, and you actually paid $53.18 for it.

EXAMPLE 4: The cost of a new dress is $149. The sales tax is 8.625%. How much do you pay in sales tax? What is the final price of the dress? Round your answer to the nearest cent.

In this example we are given the original cost of $149 and the percent increase of 8.625%, and are being asked to find the amount of the increase. Thus we have:

$$\frac{\$\,x}{\$\,149} = \frac{8.625\,\%}{100\,\%}$$

$$\frac{\$\,x}{\$\,149} = \frac{8.625\,\%}{100\,\%}$$

$$(100)\times(x) = (8.625)\times 149 \qquad \text{Cross multiply}$$

$$100\,x = 1285.125$$

$$x = 12.85125 \qquad \text{Divide by 100}$$

$$x = 12.85 \qquad \text{Round to the nearest cent}$$

The price of the dress is $149 plus the $12.85 for the sales tax, so the total is $161.85.

EXERCISES 7.7

Solve the following direct proportions involving a percent increase or decrease. If not an exact answer, round all answers to the hundredths place, unless stated otherwise.
1. Aaron's yearly salary increased from $45,000 to $48,000. What percent increase is that?
2. The train ride to the city was reduced from 48 minutes down to 44 minutes. What percent decrease is that?
3. The cost of a new pair of boots was reduced by 15%, saving you $12. What was the original cost of the boots? Round your answer to the nearest dollar amount.
4. The price of sliced ham was originally $4.99 a pound, but it is now $3.99 a pound. What percent decrease is that? Round your answer to the nearest percent.
5. A bridge toll is being increased from $5.54 to $5.76. What percent increase is that? Round your answer to the nearest percent.
6. The price for a weekly unlimited subway card increased by $1 to $32. What percent increase is that? Round your answer to the nearest percent.
7. A particular Stradivarius violin was sold in 2005 for $5.6 million. This was a 250% increase from is original price 1990. What was the price of the violin in 1990? Round your answer to the nearest dollar.
8. A brand new car is sold for $25,000. Two years later, the car is being sold used for $21,000. What percent decrease is that? Round your answer to the nearest percent.

9. Today's stock price of a departmental clothing store decreased from yesterday by $1.60 (a 7% decrease). What was the stock price yesterday? Round your answer to the nearest cent.

10. Rick's monthly apartment rent has increased from $1200 to $1275. What percent increase is that?

11. Jaya's medical insurance co-payment has increased by 15% from the original amount of $20. How much is she paying now?

12. During a plane flight, the pilot announces that she will have to make a route adjustment to avoid some turbulent weather. The distance of the new route will be 1728 kilometers, an increase of 8%. What was the distance of the original route?

13. One of the purposes of cashless tolling for bridges and tunnels is to reduce commuting times. Malik notices that, after cashless tolling was introduced, his commute time has decreased from 75 minutes to 66 minutes. What percent decrease is that?

14. To attract new customers, a mechanic shop announces a promotion for first-time customers: 7% off their first service bill. If you are a first-time customer with a service bill of $120, how much would you pay?

15. At the cashier's register, you discover that the coat you selected is part of a 60% discount clearance sale, so you only have to pay $50 for it. What was the original price of the coat?

16. The cost of gasoline per gallon last month at a certain gasoline station was $2.59. The cost of gasoline per gallon at that same station this month is $2.19. What is the percent decrease from last month to this month?

17. I bought a food processor for $89.99. It was originally priced at $129.99. What is the percent discount?

18. The population in a town grew from 4000 to 8800 in 10 years. What percent increase was the population in this town in the last 10 years?

19. John scored 70% on his first test and was determined to score better on his second test. He studied hard and scored 91% on the second test. What percent increase did John receive in the second test?

20. A sweater that is originally $98 is on sale for $35. Calculate approximately the percent discount of this sweater. Round your answer to the nearest percent

21. The price of a new suit was $289. The sales tax is 8.625%. What is the amount paid in sales tax? What is the final cost of the suit? Round your answer to the nearest cent.

22. The price of a coat is $89. The sales tax is 8.625%. How much sales tax do you have to pay? What will be the total cost of the coat? Round your answer to the nearest cent.

Chapter 7 Practice Test

Find the average rates-of-change. Be sure to include the appropriate units. If an exact answer is not possible, then round all answers to the hundredths place, unless stated otherwise.

1. A car travels 384 miles in 8 hours, What is its average rate-of-change in miles per hour?

2. As we descend 1,750 feet in altitude, the temperature goes up by 35 $^{\circ}$F. What is its average rate-of-change of $^{\circ}$F per foot?

3. A weather balloon rises 75 meters in 10 seconds. What is the average rate of change in meters per second?

4. On an electrocardiogram, you observe that a patient's normal heartbeat occurs every 0.65 seconds. What is the patient's heart rate in heartbeats per minute?

5. In a glycolysis reaction the volume of CO_2 produced is 2.95 mL in 45 minutes. What is the average rate of change in mL per minute?

Find the average speed for the given information. If an exact answer is not possible, then round all answers to the hundredths place.

6. 3200 feet in 12 minutes

7. 490 miles in 9 hours

Find the distance traveled, given the following. Be sure to show how you cancel the units appropriately. If an exact answer is not possible, then round all answers to the hundredths place, unless stated otherwise.

8. speed = 38 mi/hour, for 1.2 hours

9. speed = 36 feet/second, for 30 seconds

10. speed = 53 mi/hour, for 30 minutes

11. speed = 238 feet/minute for 45 seconds

Verify the following proportions

12. $\dfrac{12}{20} = \dfrac{3}{5}$ Are the two ratios equivalent?

13. $\dfrac{35}{42} = \dfrac{5}{7}$ Are the two ratios equivalent?

14. A car travels 235 miles in 5 hours. Another car travels 188 miles in 4 hours. Are the two rates proportionally equivalent?

15. A car travels 371 miles in 7 hours. Another car travels 212 miles in 4 hours. Are the two rates proportionally equivalent?

Solve each proportion of ratios exactly. If an exact answer is not possible, then round all answers to the hundredths place, unless stated otherwise.

16. $\dfrac{x}{5} = \dfrac{2}{7}$

17. $\dfrac{2}{r} = \dfrac{5}{4}$

18. $\dfrac{3}{4} = \dfrac{y}{2}$

19. $\dfrac{x+2}{4} = \dfrac{8}{3}$

20. If Stanley ran 2,000 meters at a constant speed in 350 seconds, how long did he take to run 300 meters?

21. If Rachael drove 648 miles in 5 hours, how long did it take her to drive 175 miles?

22. If $\dfrac{4}{15}$ of a tank can be filled in 30 minutes, how many minutes will it take to fill the entire tank?

Solve the following proportions involving a percent increase or decrease. If an exact answer is not possible, then round all answers to the hundredths place, unless stated otherwise.

23. The price of the new car was reduced from $27,500 to $25,000. What percent decrease is that?

24. The height of the plant grew from 34 cm to 46 cm. What is the percent increase?

25. The cost of an item is $135. The sales tax is 8.625%. How much do you pay in sales tax, and what is the final cost of the item. Round your answer to the nearest cent.

26. Your salary was increased from$12.50 an hour to $13.75 per hour. What percent raise did you receive?

CHAPTER 8

Linear Equations in Two Variables

Up until now we have studied Algebra with only a single variable. It has been the Algebra of the real number line. The power of Algebra, however, is unleashed when we start to look at how variables are related, or how changing one variable affects another. For example, how changing temperatures or pressures change the weather. How a changing acceleration affects the force on a body. How changing the public perception changes the economy. How the changing of time impacts the amount of a radioactive isotope left in an object. How changing the price of a commodity impacts its sales. How changing a person's weight impacts their health. We could go on and on. Algebra provides us with the language that allows us to analyze these and other important questions by constructing mathematical models to approximate these phenomena. In this chapter, we begin to lay the foundation for some fundamental analysis techniques that we will continue to build upon in future mathematics courses, so that we can answer some important questions regarding patterns in the world.

8.1 CARTESIAN COORDINATE SYSTEM

When we worked with equations involving a single variable earlier in this textbook, we could show the solutions to these equations as a point or a set of points on a real number line. Equations involving two variables are more complicated, and require a new way to visualize the solution. René Descartes, a French philosopher and mathematician, came up with a new way to visualize solutions to equations in two variables. Descartes, who lived from about 1598 to 1650, made many important contributions to philosophy and mathematics during his life. His major contribution to Algebra was to show that Algebra could be closely linked with Geometry. He gave Algebra a new visual aspect that had not existed before.

He showed us that algebraic equations, such as:

$$y = 2x + 3, \quad y = x^2 + 3x - 1, \quad y = 1 - 3x^3, \text{ or } \quad y = \frac{x+3}{x^2 - 1},$$

all have a picture (graph) that can be associated with them. Descartes "Geometrized" Algebra and provided an explosion of new potential.

Descartes used a two dimensional grid (two perpendicular, or right-angled, number-lines) to show the relationship between two variables. Today this **grid** bears his name, and is called the Cartesian Coordinate System (also called the Cartesian Grid, and the Rectangular Grid or Coordinate System). The basic idea is to use two perpendicular numbers lines, one for each of the variables, as shown below.

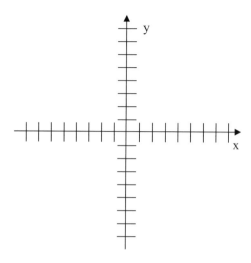

Now extend perpendicular grid lines along the horizontal number line, also called the horizontal or x-axis. Do the same along the vertical number line, or the vertical or y-axis, and obtain the following Cartesian or Rectangular Grid.

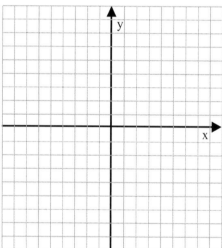

Useful Terminology

x-axis: the horizontal number line

y-axis: the vertical number line

quadrants: The x- and y-axes divide the plane into four separate regions called quadrants. The quadrants are labeled in a counter-clockwise fashion, starting in the upper right of the graph, as quadrants I, II, III, and IV.

points: a location of any place on the grid

coordinates of the point: points are located by specifying how many units along the horizontal and the vertical axes you must move to reach the point on the grid.

x-coordinate: how far the point is away from the y-axis. If it is to the right it is positive, and to the left it is negative.

y-coordinate: how far the point is away from the x-axis. If it is above it is positive, and below it is negative.

ordered pair: the x- and y-coordinates written as (x-coordinate, y-coordinate), identifies a point on the graph.

origin: the point where the horizontal and the vertical axes intersect designated by the ordered pair (0,0)

See the figure below, which highlights some of these new terms:

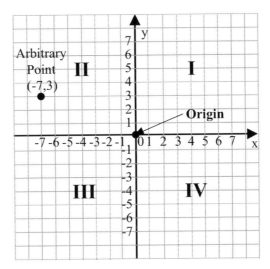

In scientific problems we seldom use x and y as the variables. Instead we choose variables that more closely describe what we are analyzing, such as t for time, d for distance, m for mass, etc. When this happens we need to better define the two variables so we know which variable goes with the horizontal axis, and which goes with the vertical axis.

We use the terms dependent and independent, since it establishes a causal relationship between the two variables that is part of our experiment. For example, as an object moves in time we say time is the independent variable and the position or distance is dependent, since the position an object is at, depends upon the time we observe it.

independent variable: the variable along the horizontal-axis (in our case the x-axis).

dependent variable: the variable along the vertical-axis (in our case the y-axis).

Later on in the chapter we'll discuss this further, but for now we'll develop the concepts we need using x and y. However, this is an important topic that needs to be fully understood, especially when you are in a laboratory science class.

If we continue on with this new geometric way to "view" Algebra, we can now create an image of algebraic expressions. We can actually answer the question – what does $y = 2x + 3$ look like?

In mathematics, the word graph means you are going to create an image of the equation in two variables, or more precisely, an image of all the points (ordered pairs) that make the equation true. To make the equation in two variables true means we have found one number for x and one number for y that when substituted into the equation gives a true (equal) result.

EXAMPLE 1: $x = 3$, and y = –4 when substituted into the equation, $y = 2x - 10$ gives a true statement, since $2(3) - 10 = -4$. We say that the pair of values $x = 3$, and y = –4 is a solution to the equation, $y = 2x - 10$. We can also write this solution using the ordered pair notation, $(x, y) = (3, -4)$.

EXAMPLE 2: $x = -5$, and y = –20, when substituted into $y = 2x - 10$ also gives a true statement.

We could find an unending (infinite) number of x-y combinations (ordered pairs) that will work for each of these examples. Drawing a line through all the points creates a visual image of the solutions, which we call a "graph of the equation."

The graph of an equation in two variables is a picture of all the solutions to that equation, i.e. all the x-y pairs of numbers that solve the equation. Thus, graphing the solutions of an equation in two variables is a picture of the set of ordered pairs, and not a single x- value, or set of, x- values, as we saw in Chapter 7, that solve the equation.

As a further example we see that the ordered pairs: $(0,3)$, $(1,5)$, and $(-1,1)$, all satisfy the equation: $y = 2x + 3$, since if we insert the appropriate x and y values into the equation, we get a true statement, e.g.

$$y = 2x + 3 \qquad\qquad y = 2x + 3 \qquad\qquad y = 2x + 3$$
$$(3) = 2(0) + 3, \qquad (5) = 2(1) + 3, \quad \text{and} \quad (1) = 2(-1) + 3$$
$$3 = 3 \qquad\qquad\quad 5 = 5 \qquad\qquad\qquad 1 = -2 + 3$$
$$1 = 1$$

However, the point $(3, 2)$ does not work since:

$$y = 2x + 3$$
$$(2) \neq 2(3) + 3$$
$$2 \neq 9$$

Thus, the points $(0,3)$, $(1,5)$, and $(-1,1)$, are on the graph, since they are solutions, but the point, (3, 2), is not, since it is not a solution.

EXAMPLE 3: Determine which ordered pairs $(-1,4)$, $(0,6)$, or $(1,2)$ are solutions to the equation $y = -x + 3$.

Solution: Test each ordered pair separately.

For $(-1,4)$ set $x = -1$ and see if $y = 4$ comes out of the equation.

$$y = -x + 3$$
$$y = -(-1) + 3$$
$$y = 1 + 3$$
$$y = 4$$

Since $y = 4$ when $x = -1$, we say that $(-1,4)$ is a solution.

For $(0,6)$ set $x = 0$ and see if $y = 6$ comes out of the equation.

$$y = -x + 3$$
$$y = 0 + 3$$
$$y = 3$$

Since $y \neq 6$ when $x = 0$, we say that $(0,6)$ is NOT a solution.

For $(1, 2)$ set $x = 1$ and see if $y = 2$ comes out of the equation.

$$y = -x + 3$$
$$y = -(1) + 3$$
$$y = -1 + 3$$
$$y = 2$$

Since $y = 2$ when $x = 1$, we say that $(1,2)$ is a solution.

EXAMPLE 4: Find three ordered pairs that are solutions to the equation $y = 3x - 4$.

Solution: Pick any value for x and solve for y

<u>Pick</u> $x = 0$

$y = 3x - 4$
$y = 3(0) - 4$ Substitute $x = 0$ into the equation
$y = 0 - 4$
$y = -4$

Thus, $(0,-4)$ is a solution.

<u>Pick</u> $x = 1$

$y = 3x - 4$
$y = 3(1) - 4$ Substitute $x = 1$ into the equation
$y = 3 - 4$
$y = -1$

Thus, $(1,-1)$ is a solution.

<u>Pick</u> x = 2

$$y = 3x - 4$$

$$y = 3(2) - 4 \quad \text{Substitute } x = 2 \text{ into the equation}$$

$$y = 6 - 4$$

$$y = 2$$

Thus, $(2, 2)$ is a solution. Can you find others? Having these solutions, we could them plot them on a Cartesian grid.

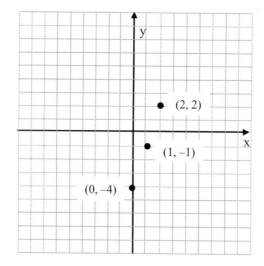

We see from the plot that these solutions all fall on a straight line connecting them. If we found more solutions and plotted them we would see that all the solutions for this equation would fall on a straight line. This is not always the case, but for this type of equation it is. We'll talk more about this type of equation later on in this chapter and about the graph we can expect. The graph of all solutions to this equation can be represented by a straight line through all these points.

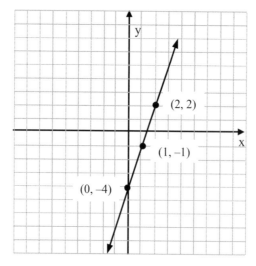

Cartesian Coordinate Systems

1. Identify the quadrants of the Cartesian Grid

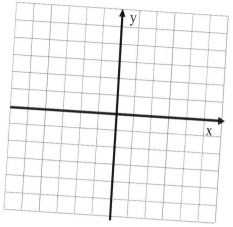

2. Identify the coordinates of the points on the graph

A =   B =   I =

C =   D =

E =   F =

G =   H =

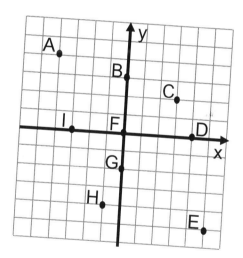

3. Plot the following points on the grid to the right:

A = $(-2,3)$ B = $(4,0)$ C = $(-3,-4)$

D = $(-1, 0)$

E = $(2, 5)$ F = $(5,-1)$

G = $(0,0)$

H = $(0,2)$ I = $(-5,-3)$

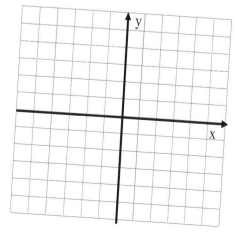

Determine which ordered pairs are solutions to the given equation

4. $(2,3)$, $(-1,1)$, $(0,1)$; $y = -2x - 1$

5. $(-1,-1)$, $(0,3)$, $(2,11)$; $y = 4x + 3$

6. $(0,3)$, $(3,-3)$, $(-3,-7)$; $y = \dfrac{2}{3}x - 5$

7. $(2,3)$, $(0,2)$, $(-2,-1)$; $y = -\dfrac{1}{2}x - 2$

8. $(2,5)$, $(-1,0)$, $(0,9)$; $y = 7x - 9$

9. $(0,6)$, $(-1,-8)$, $(2,28)$; $y = 14x + 6$

Find three ordered pairs that solve the given equation, and plot them on a Cartesian grid. Connect the points with a straight line.

10. $y = x - 6$

11. $y = -3x + 7$

12. $y = -x + 3$

13. $y = 2x - 1$

14. $y = -2(x - 3)$

15. $y = 5(2 - x)$

8.2 PROPERTIES OF LINEAR EQUATIONS IN TWO VARIABLES

We begin by looking at graphs of equations, starting with one of the most basic relationships in Algebra, and that is the linear relationship. The simplest relationship between two variables is a linear relationship. Recall that the first basic polynomial is a linear polynomial. Thus, the most elementary example of a two-variable relationship can be written as an equation in two linear variables of the form:

$$Ax + By = C$$

This is called the **standard form** of a linear equation. Here A, B, and C are unspecified numerical coefficients (constants), and x and y are the two variables.

Examples of Linear Equations:

$$5x - 3y = 7, \qquad \frac{2}{3}x + 7y = \frac{1}{8}, \qquad -3x + y = 0, \qquad 2x = 13$$

What Descartes enabled us to do, was to visualize all the x- and y-values that make these equations true, i.e. the x- and y-coordinates, or points that satisfy the given equation. The name linear equation is not a fluke. It is called a linear equation because its graph is that of a straight line. We have already shown this at the end of the previous section, but we'll highlight this further with a few examples here as well.

EXAMPLE 1: Find three solutions to the linear equation: $2x + y = 6$. Plot the solutions on a grid and then connect with a straight line.

Solution: Pick values for either x or y and solve for the other variable.

<u>Pick</u> $x = 0$

$2x + y = 6$

$2(0) + y = 6$ Substitute $x = 0$ in the equation

$0 + y = 6$ Solve for y

$y = 6$

Thus, $(0,6)$ is a solution.

<u>Pick</u> $x = 3$

$2x + y = 6$

$2(3) + y = 6$ Substitute $x = 3$ in the equation

$y = 0$ Solve for y

Thus, $(3,0)$ is a solution.

<u>Pick</u> $x = 1$

$2x + y = 6$

$2(1) + y = 6$ Substitute $x = 1$ in the equation

$2 + y = 6$ Solve for y

$y = 4$

Thus, $(1,4)$ is a solution. Now plot the solutions and connect them with a straight line.

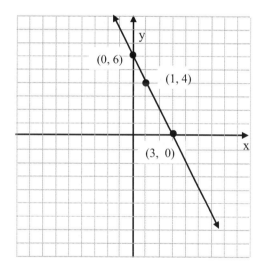

EXAMPLE 2: Find three solutions to the linear equation: $4x - y = 8$. Plot the solutions on a grid and then connect with a straight line.

Solution: Pick values for either x or y and solve for the other variable.

<u>Pick</u> $x = 0$

$4x - y = 8$

$4(0) - y = 8$ Substitute $x = 0$ in the equation

$0 - y = 8$ Solve for y

$y = -8$

Thus, $(0, -8)$ is a solution.

<u>Pick</u> $y = 0$

$4x - y = 8$

$4x - 0 = 8$ Substitute $y = 0$ in the equation

$4x = 8$ Solve for x

$x = 2$

Thus, $(2, 0)$ is a solution.

228

Pick $y = 4$

$4x - y = 8$

$4x - 4 = 8$ Substitute $y = 4$ in the equation

$4x = 12$ Solve for x

$x = 3$

Thus, $(3, 4)$ is a solution. Now plot the solutions and connect them with a straight line.

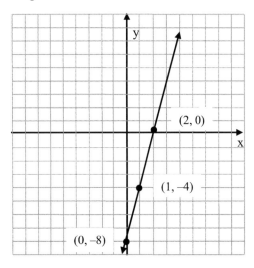

In the exercises you'll try a number of examples that should convince you that all graphs of equations of this form are straight lines.

Some additional examples of linear graphs are shown below:

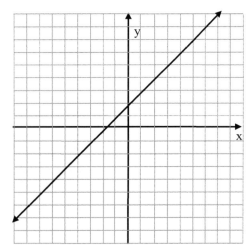
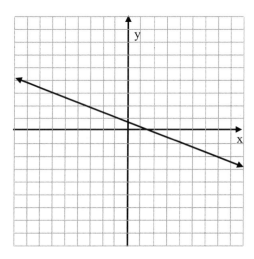

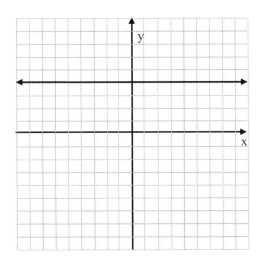
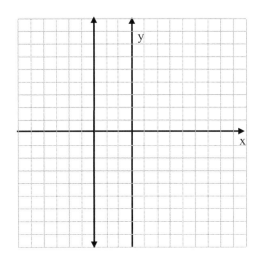

The last two graphs from above are special cases of lines; horizontal and vertical. We will discuss them in more detail later in the book. If we look at a variety of lines we see that there are two important features that completely characterize any line – its steepness, and where it crosses either the x- or the y-axis (its intercepts).

Intercepts

The point where a graph intercepts (crosses) either the x- or y-axis is important to know. The point at which a graph crosses the y-axis is called the y-intercept, and the point where it crosses the x-axis is called the x-intercept. See the graph to the right:

To obtain the x-intercept we set the y-value equal to zero and solve for x. To obtain the y-intercept we set the x-value equal to zero and solve for y.

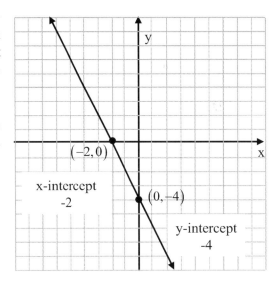

EXAMPLE 1: Find the x- and y-intercepts of the following equation: $3y + 5x = 15$

To find the y-intercept we set x =0 and solve for y, i.e.,

$$3y + 5(0) = 15$$
$$3y = 15$$
$$\frac{3y}{3} = \frac{15}{3} = 5$$
$$y = 5$$

To find the x-intercept we set y = 0 and solve for x, i.e.,

$$3(0) + 5x = 15$$
$$5x = 15$$
$$\frac{5x}{5} = \frac{15}{5} = 3$$
$$x = 3$$

Thus, the x-intercept is (3,0) and the y-intercept is (0, 5).

EXAMPLE 2: Find the x- and y-intercepts of the following equation: $2y - 3x = 30$

To find the y-intercept we set x =0 and solve for y, i.e.,

$$2y - 3(0) = 30$$
$$2y = 30$$
$$\frac{2y}{2} = \frac{30}{2} = 15$$
$$y = 15$$

To find the x-intercept we set y = 0 and solve for x, i.e.,

$$2(0) - 3x = 30$$
$$-3x = 30$$
$$\frac{-3x}{-3} = \frac{30}{-3} = -10$$
$$x = -10$$

Thus, the x-intercept is (−10, 0) and the y-intercept is (0, 15).

EXAMPLE 3: Find the x- and y-intercepts of the following equation: $5y + 7x = 40$

To find the y-intercept we set x =0 and solve for y, i.e.,

$$5y + 7(0) = 40$$
$$5y = 40$$
$$\frac{5y}{5} = \frac{40}{5} = 8$$
$$y = 8$$

To find the x-intercept we set y = 0 and solve for x, i.e.,

$$5(0)+7x=40$$
$$7x=40$$
$$\frac{7x}{7}=\frac{40}{7}=\frac{40}{7}$$
$$x=\frac{40}{7}$$

Thus, the x-intercept is ($\frac{40}{7}$, 0) and the y-intercept is (0, 8).

Thus, we see that the intercept does not have to equal an integer. It can be any real number.

The Slope of a Line

The numerical measure of the **steepness** of a line is called its slope. We use the letter "m" to designate the slope of a line. Nobody is really sure why we use "m." Some speculate that it is from the term modulus of slope, but it is something we will probably never know. Other countries around the world use different designations such as "k," "a," or "s." In the US, however, we have adopted the symbol "m" to represent the slope of a straight line.

To construct a numerical measure of slope, we use the observation that the slope can be characterized by how high up or down we go, as we move towards the right in the coordinate system. Some have used the words "rise" and "run" to describe these motions. These words are related to the risers and runners of a staircase. The slope then is a measure of how steep the "stairs" are.

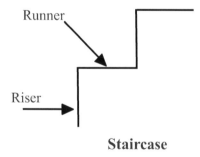

Staircase

Thus, the slope can be defined as the ratio of the "rise" over the "run." We can also say it is "the change-in-y" over "the change-in-x." In mathematics we sometimes use the capital Greek letter delta, Δ, to represent change.

Consequently, we may refer to the slope in any of the following ways:

$$m = \frac{rise}{run} = \frac{change\text{-}in\text{-}y}{change\text{-}in\text{-}x} = \frac{\Delta y}{\Delta x}$$

Using this as our descriptive definition, we can now develop a numerical measure of the slope of a line, by looking at its graph.

Given the graph below, we find the slope as follows. We pick two points on the graph, and then find the distance we move vertically, either up (positive) or down (negative), and divide that by the distance we move in the horizontal direction, as we move from the left-most point to the right-most point.

We now have a quantitative way of identifying the slope. In this case we say the line has a slope equal to $\frac{7}{9}$. This means that for every 9 units we move to the right, we move up 7 units.

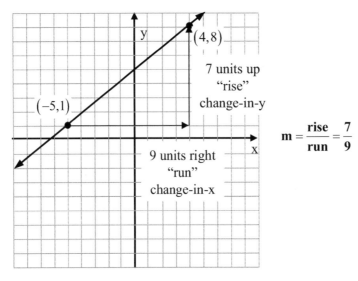

Consider another line.

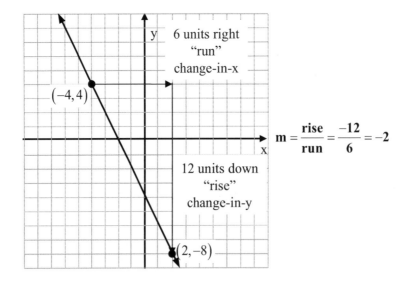

The slope of this line is -2. Notice that a negative slope corresponds to a line sloping downward as we move to the right, and a positive slope rises as we move to the right.

Instead of relying on the graphs to compute the slope, we can also use the coordinates of the two points directly. As we said, the slope is the change-in-y over the change-in-x. The change-in-y is just the difference of the final y-location and the initial y-location. A similar result holds for the change-in-x as well. Thus, if we identify y_f with y final, and y_i with y initial, and do the same for x, we can write an equation for the slope as:

$$m = \frac{y_f - y_i}{x_f - x_i}$$

Or, if we identify the two points as: (x_1, y_1), and (x_2, y_2), we can express the slope of a line passing through two points as:

$$m = \frac{y_2 - y_1}{x_2 - x_1}$$

We should note that it can be shown that it does not matter which point you choose as point 1 and which is point 2. When you do the computation, the formula automatically accounts for this variation. Thus, you don't always have to choose the point furthest to the left as point 1.

The slope can also be thought of as a constant rate-of-change. It tells us how the y-variable changes with-respect-to the changing x-variable.

EXAMPLE 1: Find the slope of the line through the given points
$$(1, -1), \text{ and} (3, -7)$$

We determine that $x_1 = 1$, $x_2 = 3$, $y_1 = -1$, and $y_2 = -7$

$$m = \frac{y_2 - y_1}{x_2 - x_1} = \frac{-7 - (-1)}{3 - 1} = \frac{-7 + 1}{2} = \frac{-6}{2} = \frac{-3}{1} = -3$$

Interpretation: If x is increased by 1, y will be decreased by -3

EXAMPLE 2: Find the slope of the line through the given points
$$(-2, 0), \text{ and} (-1, 4)$$

We see that $x_1 = -2$, $x_2 = -1$, $y_1 = 0$, and $y_2 = 4$

$$m = \frac{y_2 - y_1}{x_2 - x_1} = \frac{4 - 0}{-1 - (-2)} = \frac{4}{-1 + 2} = \frac{4}{1} = 4$$

Interpretation: If x is increased by 1, y will be increased by 4

EXAMPLE 3: Find the slope of the line through the given points
$$(-2,-3), \text{ and } (5,2)$$

We see that $x_1 = -2$, $x_2 = 5$, $y_1 = -3$, and $y_2 = 2$

$$m = \frac{y_2 - y_1}{x_2 - x_1} = \frac{2-(-3)}{5-(-2)} = \frac{2+3}{5+2} = \frac{5}{7}$$

Interpretation: If x is increased by 7, y will be increased by 5.

Special Cases to Consider
Sometimes either the x- or y-coordinates of the two points will be the same. We'll now consider when this happens. First let's consider two points with the same y-coordinate.

EXAMPLE 4: Find the slope of the line through the given points
$$(-5,7), \text{ and } (8,7)$$

We see that $x_1 = -5$, $x_2 = 8$, $y_1 = 7$, and $y_2 = 7$

$$m = \frac{y_2 - y_1}{x_2 - x_1} = \frac{7-7}{8-(-5)} = \frac{0}{8+5} = \frac{0}{13} = 0$$

A line with a zero slope is a horizontal line. This means that as we change x, the y value does not change. We stay at the same height, y. Anytime the y-coordinates of our two points are the same and the x-coordinates are different, we will always have a zero slope, which simply means we are finding the slope of a horizontal line. This can be see by graphing the two points from the example and connecting them with a straight line.

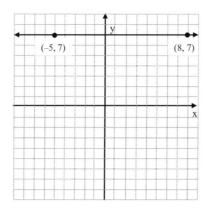

The other possibility is if the x-coordinates of the two points are the same, but the y-coordinates are different.

EXAMPLE 5: Find the slope of the line through the given points
$$(3,5), \text{ and}(3,-7)$$

We see that $x_1 = 3, \ x_2 = 3, \ y_1 = 5, \text{ and } y_2 = -7$

$$m = \frac{y_2 - y_1}{x_2 - x_1} = \frac{-7-5}{3-3} = \frac{-12}{0} = \text{undefined}$$

A line with an undefined slope is a vertical line. We can see this if we plot the two points and connect them with a straight line.

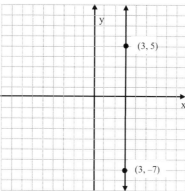

Thus, anytime the x-coordinates are the same and the y-coordinates are different, we will always have an undefined slope, which means we have a vertical line. The slope is not defined, since in looking at the picture we cannot say if the line is sloping upwards or downwards.

In addition to finding the slope of a line given two points which it passes through, as the graphs above show, we can also find the slope of a line directly from its graph, and points in the graph. Consider the following examples:

EXAMPLE 1: Find the slope of the line through the given points using the graph below.

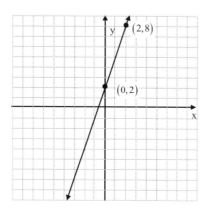

To find the slope we recall that the slope is $\frac{\text{rise}}{\text{run}}$. Starting at the point furthest to the left in the graph

(0, 2), we see that to get to the other point (2, 8), we'd have to move two "steps" to the right (a "run" of two), and six steps upwards (a "rise" of six), or we could move up 6 units (rise of 6) then go the right 2 units (a run of two).

The slope of this line is

$$\frac{\text{rise}}{\text{run}} = \frac{6}{2} = \frac{3}{1} = 3$$

This means that if we were to move to the right along the line we would increase our height by three units for every one unit we move to the right.

EXAMPLE 2: Find the slope of the line through the given points using the graph below.

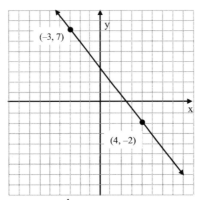

To find the slope we recall that the slope is $\frac{\text{rise}}{\text{run}}$. Starting at the point furthest to the left in the graph (–3, 7), we see that to get to the other point (4, –2), we'd have to move seven "steps" to the right (a "run" of seven), and nine steps downwards (a "rise" of negative nine).

The slope of this line is

$$\frac{\text{rise}}{\text{run}} = \frac{-9}{7} = -\frac{9}{7}$$

This means that if we were to move to the right along the line we would decrease our height by nine units for every seven units we move to the right.

EXERCISES 8.2

Finding Solutions to Equations in Two Variables

Given the following equations, find at least three solutions. Plot the solutions on a grid and then connect with a straight line.

1. $y = 3x + 5$
2. $y = -x - 2$
3. $y = x - 7$
4. $y = -2x + 11$
5. $y = 5x - 8$
6. $2y + 3x = 6$
7. $y - x = 4$
8. $-y + 4x = -8$

Finding Intercepts

Given the following equations, find the x- and y-intercepts.

9. $y = 3x + 5$
10. $y = x - 7$
11. $2y + 3x = 6$
12. $y - x = 4$
13. $-y + 4x = -8$
14. $y = -2x + 11$
15. $-3y + 5x = 15$
16. $2y - 7x = -14$

Finding Slopes from Points

Find the slope of the line through the given points.

17. $(1, 3)$, and $(2, 5)$
18. $(5, 8)$, and $(6, 3)$
19. $(10, 4)$, and $(1, 7)$
20. $(6, 5)$, and $(5, 6)$
21. $(-1, -3)$, and $(2, 5)$
22. $(-5, -4)$, and $(3, 8)$
23. $(-5, 6)$, and $(2, -5)$
24. $(3, -4)$, and $(4, 9)$

25. $(4, -8)$, and $(1, 1)$
26. $(8, 4)$, and $(-8, -4)$
27. $(2, 5)$, and $(7, 15)$
28. $(-3, -3)$, and $(-5, 2)$
29. $(1, 3)$, and $(2, 3)$
30. $(2, -5)$, and $(-3, -5)$
31. $(1, 3)$, and $(1, 5)$
32. $(-3, 4)$, and $(-3, 7)$

Finding Slopes from Graphs

Find the slope of the line through the given points

33.

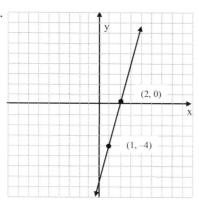

34.

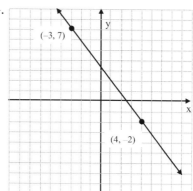

238

35.

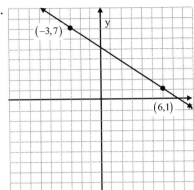

36.

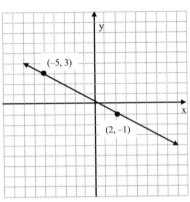

37.

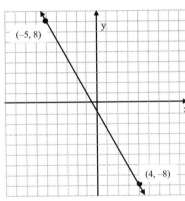

38.

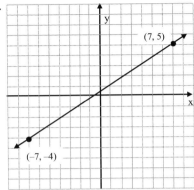

39.

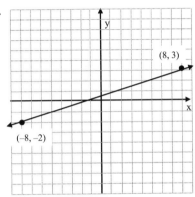

8.3 SLOPE-INTERCEPT FORM OF A LINE

A linear equation in two variables can be written in numerous ways. For example,

$$2x + 3y = 7 \text{ is equivalent to,}$$

$$3y = 7 - 2x \text{, as well as,}$$

$$y = \frac{-2x + 7}{3} \text{, and}$$

$$y = -\frac{2}{3}x + \frac{7}{3}$$

We have already introduced the Standard Form: Ax + By = C of a linear equation. We now wish to introduce another form called the slope-intercept form of a line. The slope-intercept form of a line is a linear equation that is solved for the y variable, i.e.

$$y = mx + b$$

When we do this, it turns out that the coefficient in front of the x term is the slope of the line, and the constant term on the right, b, is the y-coordinate of the y-intercept. Hence, we see where the name, slope-intercept form, comes from.

We will find it useful to be able to rewrite our equation into either form. A critical skill, is to change our equation from standard form to slope-intercept form.

To do this we first move the x-term over by subtracting it from both sides of the equation. We then divide both sides of the new equation by the coefficient in front of the y-term. This will require dividing each term on the right-hand-side of the equation by that coefficient separately, and rewriting the coefficient of the x-term as a fraction times the x-variable. This is shown in the example below:

EXAMPLE 1: Rewrite the equation 4x – 3y = 9 into the slope-intercept form of a line.

$$4x - 3y = 9$$

$$\cancel{4x} - \cancel{4x} - 3y = -4x + 9 \qquad \text{subtract 4x from both sides and cancel the 4x term}$$

$$-3y = -4x + 9$$

$$\frac{-3y}{-3} = \frac{-4x + 9}{-3} \qquad \text{divde by the coefficient in front of the y term, } -3$$

$$y = \frac{-4x}{-3} + \frac{9}{-3} \qquad \text{rearrange}$$

$$y = \frac{4}{3}x + (-3)$$

$$y = \frac{4}{3}x - 3 \qquad \text{change adding a negative number into subtraction}$$

The final slope-intercept form is: $y = \frac{4}{3}x - 3$

With the equation written in this form, we now have an easy means for identifying the slope of the equation. It is just the coefficient in front of the x-term. Thus, in the equation above, the slope is:

$$m = \frac{4}{3}$$

EXAMPLES: Find the slopes of the following linear equations:

1. $y = 2x - 4, \quad m = 2$

2. $y = -\frac{3}{4}x - 4, \quad m = -\frac{3}{4}$

3. $y = 7, \quad m = 0$ (Note: no x term)

4. $2y = -5x + 8$, We must first solve this equation for y.

$$2y = -5x + 8, \quad \text{solve for y}$$

$$\frac{2y}{2} = \frac{-5x + 8}{2} \quad \text{divide by 2}$$

$$y = -\frac{5}{2}x + 4, \quad m = -\frac{5}{2}$$

EXERCISES 8.3

Finding Slopes from Equations

Find the slopes of the following linear equations:

<u>Basic Problems</u>

1. $y = 8x + 7$

2. $y = -3x - 4$

3. $y = -\dfrac{3}{5}x + 9$

4. $y = \dfrac{4}{3}x - 3$

<u>Intermediate Problems</u>

5. $2y = 4x + 8$

6. $y = 4 - x$

7. $x = y + 3$

8. $x = 9$

9. $y = 10 - 12x$

10. $y = 14$

11. $3x - 2y = 8$

12. $x = -1$

13. $y - x = -5$

14. $x = \dfrac{1}{2}y + 7$

<u>Advanced Problems</u>

15. $2y + 4x - 8 = 0$

16. $5x + 2y = -4$

17. $3x = 2y - 5$

18. $3y + 2 = -3x$

19. $2y - 4 = x$

20. $2y = -\dfrac{4}{3}x - 5$

8.4 GRAPHING LINEAR EQUATIONS

Now that we have learned how to write and manipulate linear equations, as well as what a slope and intercept are, we can now show two different techniques for constructing the graph of a linear equation.

Graphing by Plotting a Table of Points

The first approach is to create a table of values. In this approach you pick an arbitrary value for one of the variables, usually x and then use the equation to find the other value that makes the equation true. This provides the x–,y–coordinates of the ordered pair $(x-\text{coordinate}, y-\text{coordinate})$. We do this several times (two is sufficient, but it is advisable to use three or four points, just to make sure you did not make any mistakes), and then plot the points associated with thee ordered pairs. We then draw a straight line through the points. Note, that if the points are not coincident (can be connected by a straight line), then we have made a mistake, and should re-check our work.

EXAMPLE 1: Graph by plotting points: $y = 3x + 2$

We begin by creating a table, and picking the x-values to substitute into the equation. For this problem we chose: $x = 0, \ 1, \ -1, \text{ and } 2$.

x	$y = 3x + 2$	Ordered Pairs (x, y)
0	$3(0) + 2 = 2$	$(0, 2)$
1	$3(1) + 2 = 5$	$(1, 5)$
−1	$3(-1) + 2 = -1$	$(-1, -1)$
2	$3(2) + 2 = 8$	$(2, 8)$

We now plot the ordered pairs and the straight line connecting them as shown below.

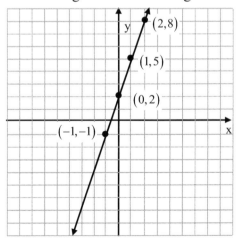

EXAMPLE 2: Graph by plotting points: $y = -\dfrac{2}{3}x + 5$

We begin by creating a table, and picking the x-values to substitute into the equation. For this problem we chose: $x = 0, 3, -3,$ and 6. Note: we have chosen these specific numbers, since they are all divisible by three, and will also give an integer solution for the y-coordinate.

x	y	Ordered Pairs $(\mathbf{x}, \mathbf{y})$
0	$-\dfrac{2}{3}(0) + 5 = 5$	$(0, 5)$
3	$-\dfrac{2}{3}(3) + 5 = -2 + 5 = 3$	$(3, 3)$
−3	$-\dfrac{2}{3}(-3) + 5 = 2 + 5 = 7$	$(-3, 7)$
6	$-\dfrac{2}{3}(6) + 5 = -4 + 5 = 1$	$(6, 1)$

We now plot the ordered pairs and the straight line connecting them:

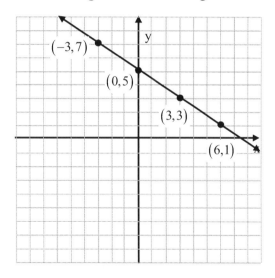

Graphing Using the Slope-Intercept Form of a Line

The second technique we show for graphing a line is useful when the equation is given in the slope-intercept form: $y = mx + b$. If the equation is written in this form, or if we decide to rewrite it into this form, then we can readily find the y-coordinate of the y-intercept, b, and the slope, m. With these two pieces of information we can then construct the graph of the line.

We start by plotting the y-coordinate of the y-intercept, b, and then use the slope to obtain another point on the graph. For example, consider the following equation:

$$y = \frac{4}{5}x - 1$$

244

From this we see that the slope $m = \dfrac{4}{5}$, and the y-intercept $b = -1$. We first plot the y-intercept, and then use the slope which is the $\dfrac{\text{rise}}{\text{run}}$ to obtain another point, as illustrated in the graph below:

After using the slope to move from $(0,-1)$, up 4 and right 5, we are at a new point $(5,3)$ (see graph to the right).

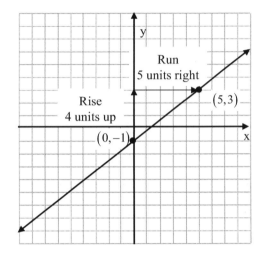

Then we draw the line that passes through the two points.

EXAMPLE 1: Graph the equation using the slope-intercept technique:

$$y = -\dfrac{6}{7}x + 5$$

We find the slope $m = -\dfrac{6}{7}$, which we can write as $m = \dfrac{-6}{7}$, and the y-coordinate of the y-intercept $b = 5$.

Plot the y-intercept and then use the information about the slope to move six units down and seven units to the right, as shown in the graph below.

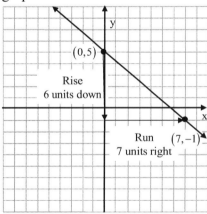

Connect these two points with a straight line, and this is the associated graph.

EXAMPLE 2: Graph the equation using the slope-intercept technique.
$$y = 4x - 2$$
We find the slope m = 4, which we can write as $m = \dfrac{4}{1}$, and the y-coordinate of the y-intercept b = –2.

Plot the y-intercept and then use the information about the slope to move four units up and one unit to the right, as shown in the graph below:

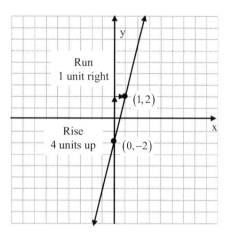

EXAMPLE 3: Graph the equation using the slope-intercept technique.
$$y = -6$$
We find the slope m = 0, since there is no x term, and the y-coordinate of the y-intercept b = –6.
Plot the y-intercept and then use the information about the slope to draw a horizontal line (zero slope), as shown in the graph below:

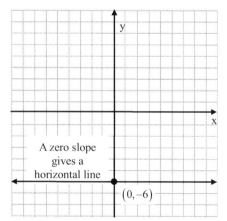

EXERCISES 8.4

Graph by Plotting Points

1. Graph the equation: $y = 2x + 1$

2. Graph the equation: $y = -x + 5$

3. Graph the equation: $y = -4x + 7$

4. Graph the equation: $y = x - 5$

5. Graph the equation: $y = \dfrac{2}{3}x + 2$

6. Graph the equation: $y = -\dfrac{1}{2}x - 1$

7. Graph the equation: $y = \dfrac{3}{5}x - 4$

8. Graph the equation: $y = 4x$

9. Graph the equation: $y = -\dfrac{5}{2}x + 4$

10. Graph the equation: $y = 3x - 3$

11. Graph the equation: $x - y = 4$

12. Graph the equation: $2x - 3y = 6$

13. Graph the equation: $4x - y = -8$

14. Graph the equation: $8x + 9y = 72$

15. Graph the equation: $6x = -24$

16. Graph the equation: $-2x = 14$

17. Graph the equation: $4y = 3x - 8$

18. Graph the equation: $7y = -14$

Graph Using the Slope Intercept Form

19. Graph the line by finding the slope and y-intercept: $y = 3x - 4$

20. Graph the line by finding the slope and y-intercept: $y = x - 6$

21. Graph the line by finding the slope and y-intercept: $y = -3x + 2$

22. Graph the line by finding the slope and y-intercept: $y = -4x + 1$

23. Graph the line by finding the slope and y-intercept: $y = \dfrac{3}{2}x - 5$

24. Graph the line by finding the slope and y-intercept: $y = \dfrac{1}{2}x + 2$

25. Graph the line by finding the slope and y-intercept: $y = -\dfrac{2}{5}x + 7$

26. Graph the line by finding the slope and y-intercept: $y = -\dfrac{1}{4}x + 2$

27. Graph the line by finding the slope and y-intercept: $2y - 5x = 8$

28. Graph the line by finding the slope and y-intercept: $-3y + x = -9$

29. Graph the line by finding the slope and y-intercept: $2y - x = 10$

30. Graph the line by finding the slope and y-intercept: $6y + 9x = -18$

31. Graph the line by finding the slope and y-intercept: $y + 2x = 5$

32. Graph the line by finding the slope and y-intercept: $y - 7x = -3$

33. Graph the line by finding the slope and y-intercept: $x - y = -3$

8.5 GRAPHING REAL–WORLD DATA

What we have shown above is "perfect data." All our points fall precisely on a straight line. This, however, is not practical in the real world. We typically obtain our data from measurements performed during experiments, and this usually leads to approximations and variations in the measured data. The phenomena under study may follow a linear pattern, but the points may not be coincident with a straight line. This is the real world, and in what follows we will show how to work with and handle this sort of data.

There are four aspects when working with real data that we now consider. The first is to determine which of the variables are dependent, and which are independent. Next, we must adjust to the fact that most of our data values are positive numbers, the third is focused on variations in the measured data, and finally, the scales of the x- and y-values may be dramatically different for various data sets. We'll discuss each of these separately, and show how to handle each of these situations.

When running experiments we measure many different things. It could be a pH-value, a mass, time, a concentration amount, a temperature, etc. Typically we wish to determine how one variable changes relative to another variable. For example in a Physics lab we may want to see how the position of an object changes as time changes, or in a Biology lab how the temperature or pH-value affects the rate of an enzyme catalyzed reaction. The question that arises is which variable do we place along the x-axis and which goes along the y-axis? Or, in other words, which is the independent and which is the dependent variable?

Independent variables are factors that can be controlled or changed by the experimenter.

Dependent variables are factors that are measurable and change because of the independent variable. (The dependent variable depends on the independent variable.) These are the effects measured by the experimenter.

For example:
1. Photosynthesis is related to the amount of light, so the amount of light is the independent variable and photosynthesis is the dependent variable, since photosynthesis depends upon the amount of light. Light does not depend upon photosynthesis.

2. The thickness of the annual rings in trees is related to annual rainfall, so the amount of annual rainfall is the independent variable and the thickness of the annual tree ring is the dependent variable, since the ring thickness depends upon the rainfall.

3. The acceleration of a vehicle is related to the mass of the vehicle. Thus, the mass is the independent variable and the acceleration is the dependent variable, since the acceleration depends upon the mass of the vehicle.

Another factor that needs to be addressed is the occurrence of only positive data values, but this is not really a problem. The only adjustment we make is that we only draw the first quadrant of our grid. Since data values do not fall in the other quadrants, it makes no sense to include these regions when graphing. See the figure below. We should note that this does not happen all the time. Sometimes you will be asked to graph data with negative values, such as a graph of negative temperatures versus time.

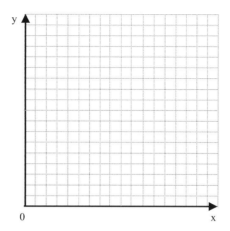

The next difference is data that does not fall exactly on a straight line.

EXAMPLE 1: Consider the following set of data:

x	y	Ordered Pairs (x, y)
1	1	(1, 1)
3	1.75	(3, 1.75)
4	3.5	(4, 3.5)
6	5	(6, 5)
8	5.5	(8, 5.5)
9	7	(9, 7)

If we plot this data on a graph it looks like:

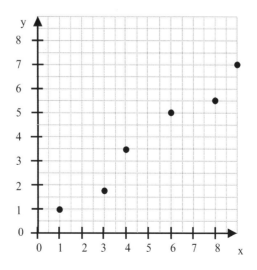

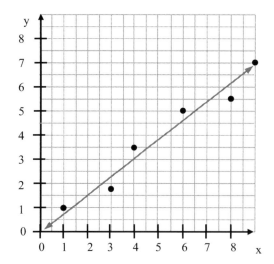

This graph is called a scatter-plot or scatter-diagram. As we can see, the data does not fall on a straight line, but it does appear to be increasing, as we move left to right, in a trend similar to a straight line. In fact, we can use our eyes to fit a straight line to the trend of the points, as shown in the following graph:

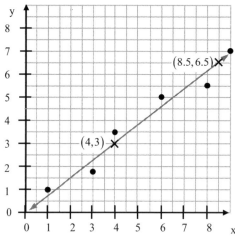

The line we have drawn by eye, follows the increasing linear trend of the data, and can be used to extract some useful information. In particular, we can find the slope of this line, by seeing where it intersects the grid and computing the rise over the run as we go from the left-most point to the right-most point.

In this graph we have identified two points on the grid, marked with $\times$'s, that locate points where the line crosses the grid. These points are $(4,3)$ and $(8.5, 6.5)$. We can either use these two points or the graph itself to approximate the slope. Either way, the rise is 3.5 and the run is 4.5. Thus, the slope is:

$$m = \frac{6.5-3}{8.5-4} = \frac{3.5}{4.5} = 0.\overline{7}$$

We should point out that this number will differ slightly from person to person, as we all may place the line at slightly different locations and orientations. However, the values of the slope should be relatively close to each other. We also note that there is a more consistent method of finding this line, but it is beyond the scope of the course. It is called regression analysis, and is sometimes covered in a basic statistics course.

We can use this same approach to identify either the x- or y-intercept, as they may have a useful physical meaning. In the graph above the x- and y-intercepts are at (0,0).

EXAMPLE 2: Approximate the slope and the x- and y-intercepts of the following scatter-plot.

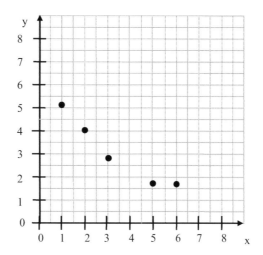

There are five data points that appear to follow a downward linear trend. We take a straight edge and look for the line that appears to approximate the general trend on the data. Usually we want the same number of points above and below the line, or we try and visually balance the net distance the points are above and below, the line we draw, equally.

We show our visual placement of this line in the graph below:

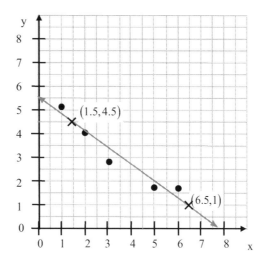

Using this line as a guide, we pick two points on the line from which to determine the slope. We have chosen $(1.5, 4.5)$ and $(6.5, 1)$. From this we can compute an estimate for the slope of:

$$m = \frac{1 - 4.5}{6.5 - 1.5}$$
$$= \frac{-3.5}{5}$$
$$= -0.7$$

We can also extrapolate (go beyond) and estimate the y-coordinate of the y-intercept to be at 5.5 and the x-coordinate of the x-intercept to be approximately at 7.75.

Another situation we must account for, are data sets where the x- and y-coordinate values are on different scales, so that if we plotted them with equal distances the data would be obscured. In the next example we show how to handle this, and we also introduce some **real world** data to show how this is really used.

EXAMPLE 3: Consider the data from a biology experiment that identifies the amount of DNA present, based upon the absorption of the color blue as measured on a spectrograph. Here is some sample data collected from a lab experiment showing the µg of DNA in the sample with respect to the Absorbance:

µg of DNA	Absorbance (A)
0	0
100	0.050
200	0.090
300	0.175
400	0.238
600	0.310
700	0.375
800	0.404

Note: In this set of data the µg of DNA is a known value from calibration specimens and the Absorbance is calculated from the know samples. This is why the Absorbance is a dependent variable and the µg of DNA is the independent variable.

If we plot this data on a graph where the x- and y-axes had the same scale, the graph would look like the graph below. Notice that we plotted the Absorbance along the y-axis, since that is the dependent variable, and the DNA amount along the x-axis, as that is the independent variable. We also note the other issue, in that all the data points seem to fall on the horizontal axis and we can't see any variations.

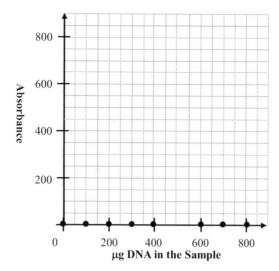

Obviously, we cannot see any variation in the data. However, if we create a different scale along the vertical Absorbance axis, an obvious pattern emerges.

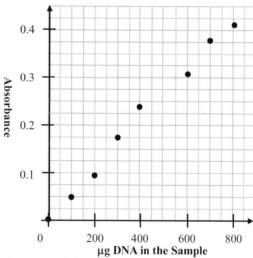

Thus, we always need to take into consideration the range of values for both the horizontal and the vertical axes when setting up the scales.

The range is calculated as:

Range = Largest value – Smallest value

The range of values, in the example above, for the horizontal axis is $800 - 0 = 800$, and the range of values, in the example above, for the vertical axis is $0.404 - 0 = 0.404$

We cannot plot these data using the same scales. Instead we have to pick different ranges for both the vertical and horizontal axes. Thus, we chose to make our grid from 0 to 900 on the horizontal axis, and from 0 to 0.45 along the vertical axis (other values could be chosen and would give the same results, as long as the smallest value chosen is at least as small as the minimum data value and the largest value chosen is at least as large as the maximum data value.)

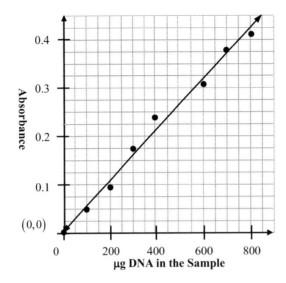

We can now estimate the slope of this line just as we did above. We fit a line to the data.

Next, we identify two points on this line from which to calculate the rise and the run.

The points we have chosen are the origin $(0,0)$ and (800, 0.425).

We should state here that the further away the points are from each other, the more accurate the approximation will be. So, you should not choose points that are too close to each other, if it can be avoided.

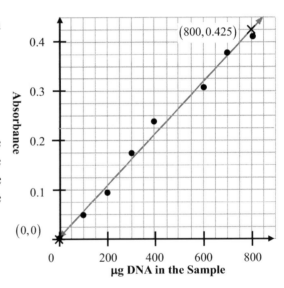

We estimate the slope of this line to be:

$$m = \frac{0.425 - 0}{800 - 0} \frac{\text{Absorbance}}{\mu\,\text{g of DNA}}$$

$$= \frac{0.425}{800} \frac{\text{Absorbance}}{\mu\,\text{g of DNA}}$$

$$= 0.00053125 \frac{\text{Absorbance}}{\mu\,\text{g of DNA}}$$

With real data, we can assign a physical interpretation to the slope. As we said earlier, the slope is a rate of change. What this value means, is that for every µg (micro-gram, or one millionth of a gram) increase in DNA there is a 0.00053125 (or 0.053125%) increase in the absorbance of the sample. The slope tells us how the Absorbance changes with increasing DNA amount.

EXAMPLE 4: Consider the data for a given class and their final test scores on a Regent's exam

Study Hours	Regent's Score
1	35
2	60
3	70
4	85
6	85
7	95

We first plot the data with appropriate scales for the vertical and horizontal axes. Then fit the "best" straight line to the data. Again, we are trying to visually balance the distance the points are away from the line, both above and below. The result will differ slightly from person to person.

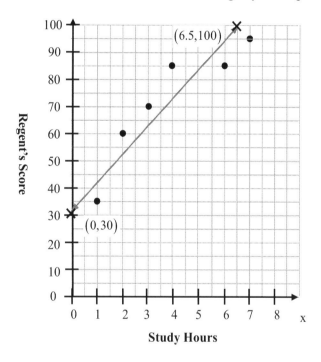

From this graph we can determine the slope,

$$m = \frac{100-30}{6.5-0} \frac{\text{Regent's Score}}{\text{Study hours}} = \frac{70 \text{ Regent's score}}{6.5 \text{ Study hours}} \approx 10.8 \frac{\text{Regents's Score}}{\text{Study hours}} \ ,$$

and the y-coordinate of the y-intercept is 30.

The x-intercept has no meaning, since, if extended to the left, the graph would cross the x-axis when x is negative, and you can't study a negative number of hours.

The y-intercept corresponds to the grade someone would receive without studying, and the slope means that for every hour you study, you will increase your grade by roughly 10.8 points.

EXERCISES 8.5

Find the Independent and the Dependent Variables

1. The relationship between pigments responsible for fall colors of leaves and the chlorophyll by-products in the leaves.
2. Determining the effect on moths to light brightness.
3. The relationship between the vapor pressure of a liquid and temperature.
4. The relationship between temperature and the rate of photosynthesis.
5. The relationship between animal metabolism and the ambient temperature.
6. The relationship between the effectiveness of disinfectants and exposure time.
7. The height of a metal ball on a ramp is measured at different times.
8. The relationship between transpiration (water evaporation from a plant) and humidity.
9. The relationship between how far you can go in your car based upon the amount of gas used.
10. Root growth and its relation to gravity.
11. The temperature of the hydrochloric acid and how it affects the rate of carbon dioxide production when reacting with calcium carbonate.
12. The relationship between the hatching of brine shrimp and the salinity of the water.
13. Diffusion efficiency is related to the ratio of surface to volume, thus decreasing the S/V (increasing volume of a cell mass) will decrease the rate of diffusing into a cellular mass.
14. How your vision is affected by the amount of carrots you eat.

15. Redwood trees are kept at different humidity levels inside a greenhouse for 12 weeks. One tree is left outside in normal conditions. The height of the tree is measured once a week.

Find the Best Fit Line

Find the "best-fit" line using a ruler or a straight edge, and then compute the slope of the line.

16.

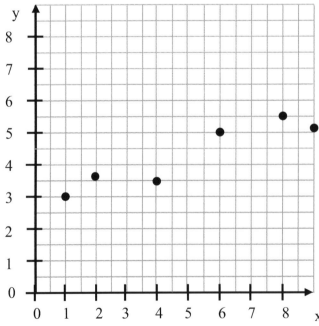

17.

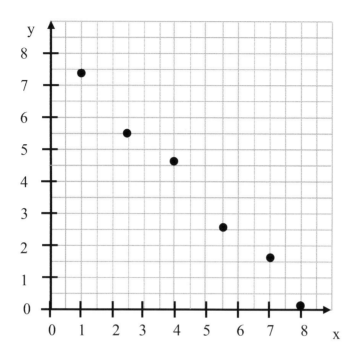

18.

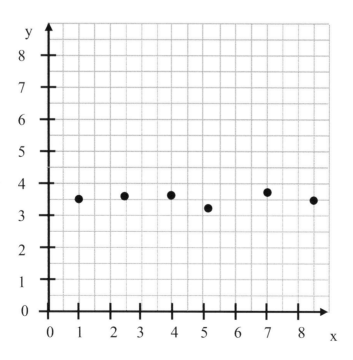

19.

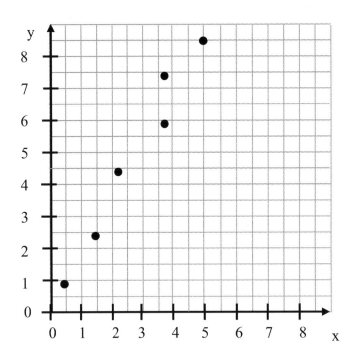

20.

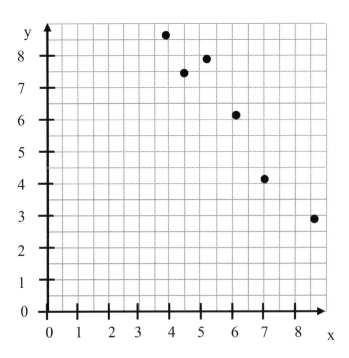

21.

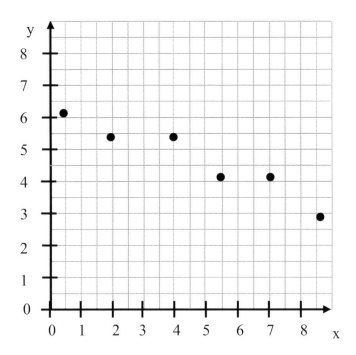

22.

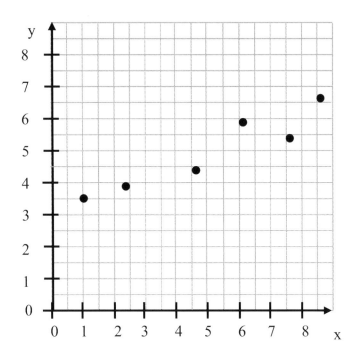

23.

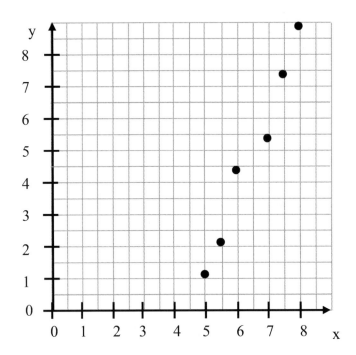

24.

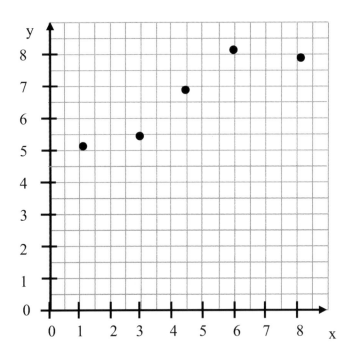

25.

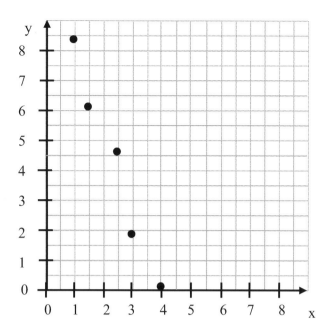

Plot the data points, construct a "best-fit" line with a ruler or straight edge, and then approximate the slope.

26.

x	y
1.0	1.5
2.5	3.5
4.0	4.0
5.5	6.5
6.5	8.0
7.0	9.0

27.

x	y
1.0	8.0
2.0	6.0
3.5	5
4.5	4.5
5.5	3.0
6.0	2.5

28.

x	y
2.5	4.5
3.0	3.5
4.5	4.0
5.5	5.0
6.0	4.5
6.5	4.0

29.

x	y
0.5	7.5
2.5	5.5
4.5	3.0
6.0	2.5
7.5	1.0

30.

x	y
0.0	6.5
1.5	5.0
3.0	4.0
4.0	4.5
5.5	3.0

31.

x	y
2.0	0.5
2.5	1.5
4.0	3.0
4.5	4.0
5.5	4.5
6.5	7.0

32.

x	y
2.0	8.5
4.0	7.0
6.0	4.5
8.0	3.0

33.

x	y
3.5	4.5
4.5	5.5
5.0	6.0
5.5	6.5
6.0	7.0
7.0	8.0

34.

x	y
4.5	8.5
5.5	7.0
6.0	4.0
7.5	3.0
8.0	1.0
8.5	0.0

262

35.

x	y
1.0	0.5
3.0	4.5
5.0	4.5
7.0	6.5
9.0	7.0

Applications

If an exact answer is not possible, then round all answers to the thousandths place, unless stated otherwise.

36. The scientist Edwin Hubble made some calculations that showed the universe is expanding. Hubble measured the recession speed of galaxies (based upon the red shift phenomena) in km/sec and plotted that against the estimated distance from earth and found the following data (this data is not really Hubble's, but is similar. It has been updated.)

Distance (mpc, Mega parsecs)	Speed Km/sec
50	3,000
100	5,000
200	14,000
250	16,000
300	18,000
350	24,000

Plot this data and estimate how the speed changes with distance. This rate of change is a measure of how fast the universe is expanding.

37. The trend in US infant mortality rate per 1,000 live births is given by the table below. Plot the data, construct a "best-fit" line, and estimate the rate of change. When will it be zero? Is that possible? Why or why not?

Year	Mortality
1950	29.2
1960	26
1970	20
1980	12.6
1990	9.2
2000	6.9

Note: The year data does not start at zero. You can avoid this by using the x-axis to represent the years since 1950, or you could label the origin of the x-axis at 1950.

38. Sunspot activity on the sun goes through cycles. Below is some sample sunspot data. Plot the data and find the "best-fit" line for the data. What is the value of the slope? (see the Note regarding the years in Problem 37 above), estimate the rate of change. What is happening?

Year	Sunspots
2000	110
2001	60
2002	100
2003	75
2004	50
2005	30
2006	25
2007	25
2008	10
2009	3

39. Health care costs have been rising. Here are the data for health care costs as a percent of the total economy called the GDP (Gross Domestic Product).

Year	% of GDP
1960	5.2
1965	5.9
1970	7.2
1975	8.1
1980	9.1
1985	10.4
1990	12.3
1995	13.7
2000	13.8
2007	16.2

Plot the data (see the Note regarding the years in Problem 37 above), construct a "best-fit" line, and estimate the rate of change. What is happening?

40. Is global warming real? Here are some data. See if the temperature is changing or not.

Year	Average Land Temp increase in °F
1880	0
1900	0.2
1920	0.3
1940	0.5
1960	0.35
1980	0.5
2000	0.9
2005	1.1

Plot the data (see the Note regarding the years in Problem 37 above), construct a "best-fit" line, and estimate the rate of change. What is happening?

8.6 INTERPRETING GRAPHS OF DATA

An important skill that needs to be learned is how to interpret and understand what graphs of data are telling us about how the dependent and independent variables are related to each other. This is really just an extension of what we learned in the previous section regarding real-world data. This is also a skill that is important in our science courses, where the focus is on how the various quantities are related to each other. These relationships are often expressed as laws, or calibration curves. In what follows we explore this important concept of graph interpretation further using a variety of examples from science courses. The focus will not be on any particular law or curve, but rather on what we can learn from any graph or curve.

EXAMPLE 1: In chemistry there is something called Beer's Law. The law describes the linear relationship between absorbance and concentration of a particular solution. A graphical example illustrating the law is shown below.

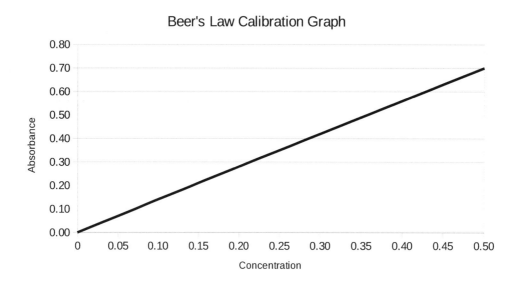

This graph shows us how something called the absorbance of a quantity, is related to its concentration. Now, at this point we really don't have to know what the absorbance or the concentration are related to, to still be able to provide an interpretation of what is happening. For example, we can read from the graph, that an absorbance of 0, corresponds to a concentration of 0, since the point (0,0) is on the graph. We also see that an absorbance of 0.7 corresponds to a concentration of 0.5, since the point (0.5, 0.7) is also on the graph as shown below.

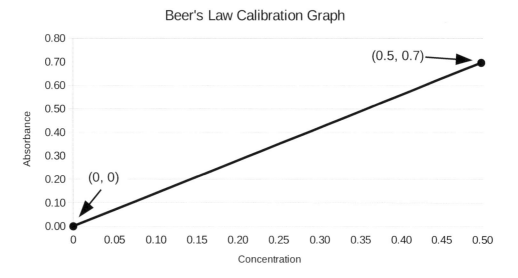

However, we can also estimate other concentrations or absorbance values. For example, what is the concentration for an absorbance value of 0.3. We see from the graph below that this is approximately 0.217, as read from the graph. This is, however, just an estimate of where we think the point falls along the Concentration Axis.

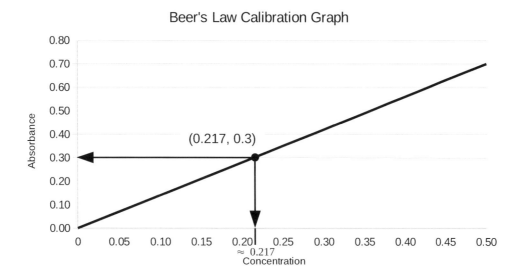

EXAMPLE 2: In Chapter 4 on units, we showed a formula for converting from Celsius to Fahrenheit. If we look closely at the formula, we see that this is just a linear relationship whose graph looks like this:

Relationship between Fahrenheit and Celsius Temperatures

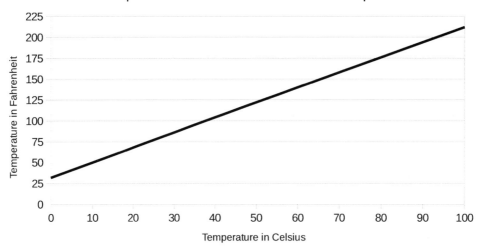

We can use this graph, instead of the formula, to approximate a temperature in ^{O}C as a temperature in ^{O}F. For example, what temperature is 40 ^{O}C associated with? To find an approximate value we simply read up from the 40 ^{O}C line until we hit the line, and then read the approximate value to the left, as shown in the graph below.

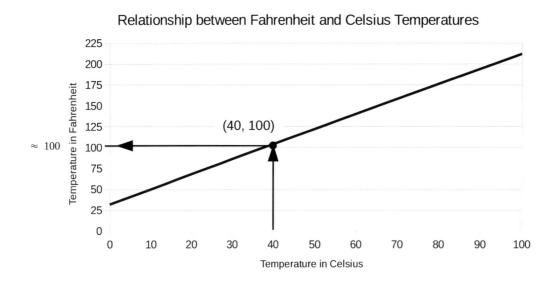

From the graph we read off a value that is slightly over 100 ^{O}F (the exact answer is 104 ^{O}F). For many situations, approximate answers can be almost as useful as exact answers, especially if we don't have access to a precise formula.

EXERCISES 8.6

Answer the following. If an exact answer is not possible, then round all answers to the thousandths place, unless stated otherwise.

1. The Beer's Law describes the linear relationship between absorbance and concentration of a particular solution. A graphical example illustrating the law is shown below.

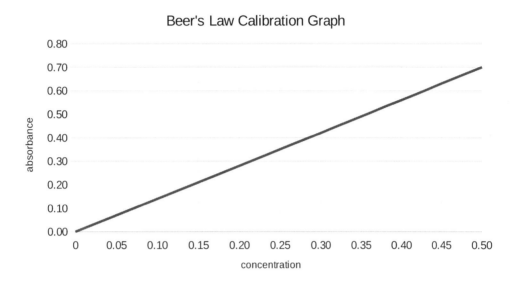

Beer's Law Calibration Graph

   Use this graph to answer the following questions:
   a) Approximately what is the concentration of the solution when the absorbance is 0.2?
   b) Approximately what is the concentration of the solution when the absorbance is 0.25?
   c) Approximately what is the absorbance when the concentration of the solution is 0.3?
   d) Approximately what is the absorbance when the concentration of the solution is 0.45?

2. There is a linear relationship between Celsius and Fahrenheit temperatures. This relationship is shown in the graph below.

Relationship between Fahrenheit and Celsius Temperatures

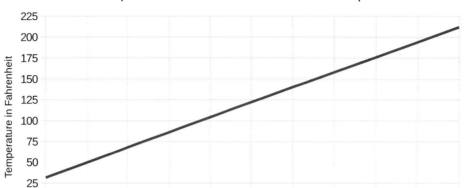

Use this graph to answer the following questions:
a) What is the temperature of 50°C in Fahrenheit, approximately?
b) What is the temperature of 70°C in Fahrenheit, approximately?
c) What is the temperature of 50°F in Celsius, approximately?
d) What is the temperature of 150°F in Celsius, approximately?

3. In Meteorology the temperature profile through the lower atmosphere follows graphs similar to the one below.

Altitude vs. Temperature Curve

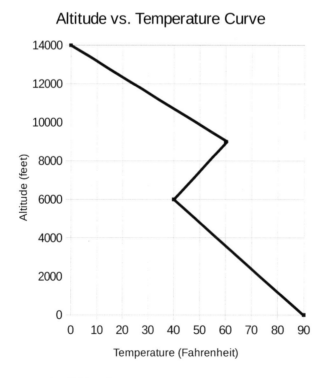

From this graph answer the following:

a) What is the approximate temperature at 12000 feet?
b) What is the approximate temperature at 8000 feet?
c) What is the approximate temperature at 2000 feet?
d) What is the approximate temperature at 14000 feet?
e) What is the approximate temperature at 0 feet?
f) As the altitude drops from 9000 feet to 6000 feet, what happens to the temperature?
g) As the altitude drops from 14000 feet to 9000 feet, what happens to the temperature?
h) As the altitude drops from 6000 feet to 0 feet, what happens to the temperature?

4. The depth of the bottom of the ocean from the shoreline on the beach chart is shown below. The horizontal axis represents the distance (in meters) from the shoreline on a beach while the vertical axis represents the depth of the bottom of the ocean (in meters). The numbers are negative because it represents being below the shoreline.

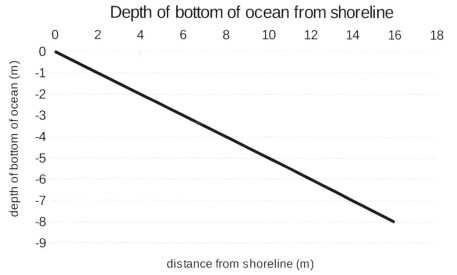

a) What is the depth of the bottom of the ocean when you are 10 meters away from the shoreline?
b) What is the depth of the bottom of the ocean when you are 6 meters away from the shoreline?
c) If the depth of the bottom of the ocean is 8 meters, how far are you from the shoreline?

5. The depth of the bottom of the ocean from the shoreline on the beach chart is shown below. The horizontal axis represents the distance (in meters) from the shoreline on a beach while the vertical axis represents the depth of the bottom of the ocean (in meters). The numbers are negative because it represents being below the shoreline.

270

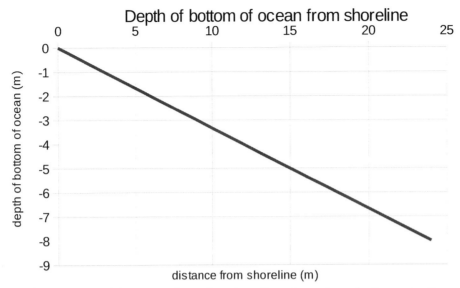

Depth of bottom of ocean from shoreline

distance from shoreline (m)

a) When you are 15 meters from the shoreline, how deep is the water?
b) When you are 6 meters from the shoreline, how deep is the water?
c) When the water is 6 meters deep, how far are you from the shoreline?
d) When the water is 2 meters deep, how far are you from the shoreline?

6. Alison runs at a constant speed of 5 miles per hour. The distance she has traveled in the first two hours is shown in the graph below.

Distance Alison Ran in First 2 Hours

time (hours)

Use this graph to answer the following questions:
   a) Approximately how far did Alison run after 45 minutes (0.75 hours)?
   b) Approximately how far did Alison run after 1 hour and 15 minutes (1.25 hours)?
   c) Approximately how long did it take Alison to run 10 miles?
   d) Approximately how long did it take Alison to run 3 miles?

7. Alfonso has a car that gets 30 mpg (miles per gallon). His car has a 13 gallon gas tank. Assume

the following graph shows how much gas is left after a certain amount of distance is traveled.

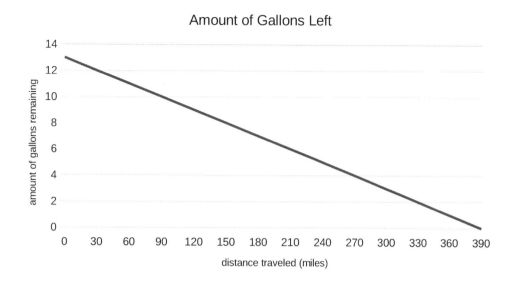

a) Approximately how many gallons are left after Alfonso has driven 180 miles?
b) Approximately how many gallons are left after Alfonso has driven 255 miles?
c) Approximately how far has Alfonso driven when there are 10 gallons remaining?
d) Approximately how far has Alfonso driven when there are 2 gallons remaining?
e) Can Alfonso drive 400 miles without filling up his car again?

Chapter 8 Practice Test

1. Identify the coordinates of the points on the graph

A =   B =   I =

C =   D =

E =   F =

G =   H =

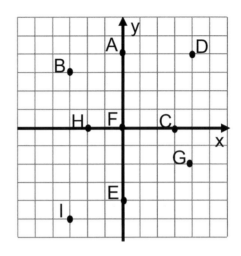

2. Construct a grid and plot the following points:

A = $(-2,3)$ B = $(4,0)$ C = $(-3,-4)$ D = $(0,-1)$ E = $(2, 5)$

F = $(5,-1)$ G = $(0,0)$ H = $(0,2)$ I = $(-5,-3)$

Find the slope of the line through the given points, and state what it means.

3. $(1,3)$, and $(2,5)$ 4. $(10,4)$, and $(1,7)$ 5. $(-1,-3)$, and $(2,5)$

6. Determine which, if any, of the following points are solutions to the equation: $y=2x-7$

a) (2, –3)   b) (–1, –8)   c) (–2, –11)   d) (0, 7)

Find the slopes of the following linear equations:

7. $y=8x+7$ 8. $y=14$ 9. $2y-4=x$ 10. $3y=-\dfrac{6}{5}x-9$

11. Graph the equation: $y=2x+1$, by plotting points

12. Graph the equation: $y=-x+5$, by plotting points

13. Graph the equation: $y=-4x+7$, by plotting points

14. Graph the line by finding the slope and y-intercept: $y=x-6$

15. Graph the line by finding the slope and y-intercept: $y=-3x+2$

Answer the following. If an exact answer is not possible, then round all answers to the hundredths place, unless stated otherwise.

16. Find the "best-fit" line using a ruler or a straight edge, and then compute the slope of the line.

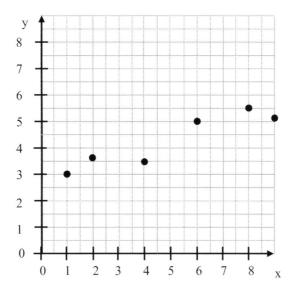

17. Plot the data points, construct a "best-fit" line with a ruler or straight edge, and then approximate the slope, and x– and y–intercepts.

x	y
1.0	1.5
2.5	3.5
4.0	4.0
5.5	6.5
6.5	8.0
7.0	9.0

18. During a lab experiment a student made measurements of the position of an object at different times. The results of the experiment are provided in the table below.

Position in Feet	Time in seconds
0	0
3	5
6.5	10
9	15
12	20
15.5	25

Plot this data with the time on the horizontal axis, and the position on the vertical axis and then estimate the slope of this line using a straight edge or a ruler. The slope is rate of change, or how the position of the object changes with time (feet per second). What is this a measure of?

19. Beer's Law describes the linear relationship between absorbance and concentration of a particular solution. A graphical example illustrating the law is shown below.

Beer's Law Calibration Graph

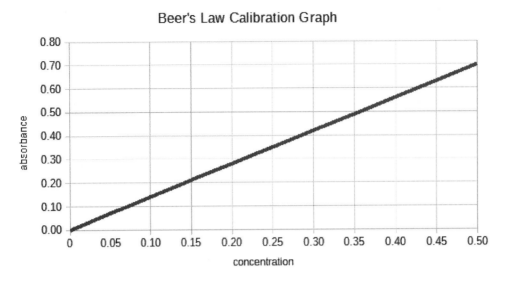

Use this graph to answer the following questions:
a) Approximately what is the concentration of the solution when the absorbance is 0.1?
b) Approximately what is the concentration of the solution when the absorbance is 0.6?
c) Approximately what is the absorbance when the concentration of the solution is 0.15?
d) Approximately what is the absorbance when the concentration of the solution is 0.5?

CHAPTER 9

Systems of Equations

We sometimes come across problems that require two or more equations to solve. For example, consider the problem that Danise has:

Danise is responsible for buying a week's supply of food and medication for the dogs and cats at a local shelter. The food and medication for the dogs, costs three times as much as those supplies for the cats. She needs to feed 125 cats and 40 dogs. Her budget is $3,675. How much can she spend on each dog and cat, for food and medication?

This problem can be expressed mathematically as a solution to two coupled linear equations of the form:

$$40x + 125y = 3675$$
$$x - 3y = 0$$

Where x represents the money needed for each cat, and y is the money for each dog.

Also, consider Maria who is a chemist and she needs 15ml of a 30% acidic solution for a lab experiment, but she only has a 20% and a 50% solution available. How does she mix the two solutions to obtain 15ml of a 30% acidic solution? This problem can also be expressed mathematically as a solution to two coupled linear equations of the form:

$$0.5x + 0.2y = 4.5$$
$$x + y = 15$$

Both of these problems come down to solving something we call a system of equations, or a system of two equations in two variables (unknowns). To solve the system means we need to find the values of the two variables, that when substituted into each equation separately will yield a true statement. We say that the solution satisfies the equations simultaneously.

In general, any system of two equations in two unknowns (x and y) can be written with unspecified numerical coefficients, a, b, c, d, e, and f, as

$$ax + by = c$$
$$dx + ey = f$$

To solve this system, means we are trying to determine if these two equations have the same ordered-pair, which solves both equations at the same time. (x, y)

Since any linear equation has a line as its graph, we can also study the problem graphically. We are looking for any points, which the two graphs have in common, i.e. where they intersect. The points in common between the two graphs are solutions to the system.

Extending the graphical picture approach to solving systems further, we see that when solving a system of two equations in two unknowns, we have three possible outcomes. The first is that the graphs intersect at one unique point. In this case we say the solution is consistent and has a unique solution as shown below:

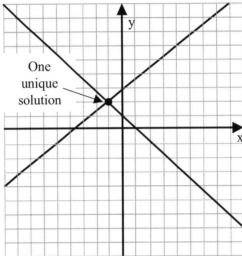

The second possibility, is that the two lines are parallel, but do not overlap. In this case, the system is inconsistent and we have no solution, e.g.

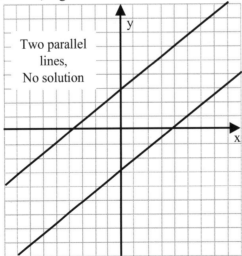

The third possibility is that the lines are coincident (the same line.) In this case we say the system is consistent, but dependent. Furthermore, we say that this system has infinitely many solutions, and these solutions are shown by using either equation to express the solution set.

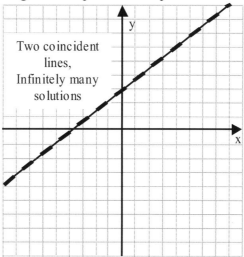

Two coincident lines, Infinitely many solutions

The focus of this chapter is to show how to find these solutions. We will present and demonstrate three different techniques for solving these systems – graphing, elimination, and substitution. The last of these techniques, substitution, can be omitted without causing problems later on in the book. It is, however, a required topic in the next level algebra course.

9.1 SOLVING SYSTEMS BY GRAPHING

As the title implies, to solve a system of equations by graphing, means we have to graph the two equations. We then determine where the two lines intersect, and this is the solution. This technique works nicely provided that the solution is an integer value. Otherwise, we can only approximate the solution by estimating the location of the point-of-intersection on the graph.

In the problems that follow, the solutions will be integer values, which will show up clearly on the graph. We mainly introduce this technique for illustration purposes, and to highlight a conceptual idea that will be used for more complicated equations and concepts that will be presented in future courses.

EXAMPLE 1: Solve the system of equations by graphing:

$$y = 2x - 3$$
$$y = -x + 6$$

These equations are written in the slope-intercept form, so we will use that graphing technique. The top equation, or what we'll call equation (1), has a slope of 2 and a y-intercept of -3. The second equation, (2), has a slope of -1 and a y-intercept of 6.

Graphing both of these equations on the same grid shows us that they intersect at the point (3, 3).

See the graph below:

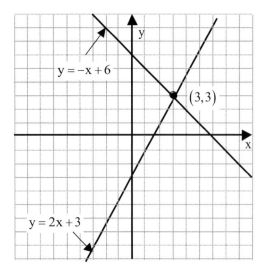

We could verify our findings by checking the solution. To do so, we just substitute for x and y the ordered pair (3, 3) into the original equations to make sure they yield true statements. We'll leave that for the reader to verify.

EXAMPLE 2: Solve the system of equations by graphing:

$$-4x - 2y = 4$$
$$-2x + y = 6$$

These equations are written in the standard form, but the x-, and y-coefficients divide evenly into the constant on the right-hand-side of each equation, so for this system we'll graph the equations using intercepts.

The x- and y-intercepts of the first equation are $x = -1$, and $y = -2$.

The x- and y-intercepts of the second equation are $x = -3$, and $y = 6$.

Plotting these points on the same graph and drawing each line, we see they both intersect at the point (–2, 2), as shown in the graph below:

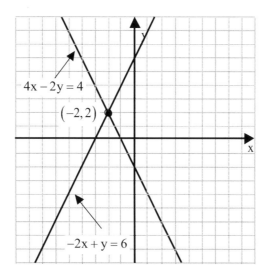

EXERCISES 9.1

Solving Systems Using Graphing Techniques

Solve the systems of equations by graphing

1.

$y = x - 1$

$y = -x + 3$

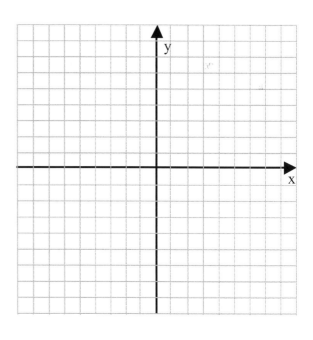

280
2.

$y = 2x + 6$

$y = -2x + 6$

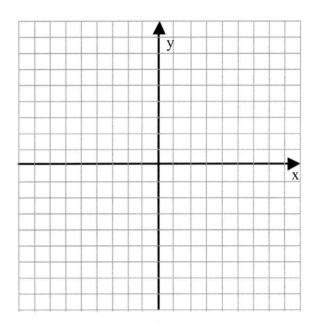

3.

$y = \dfrac{1}{3}x - 5$

$y = -3x + 5$

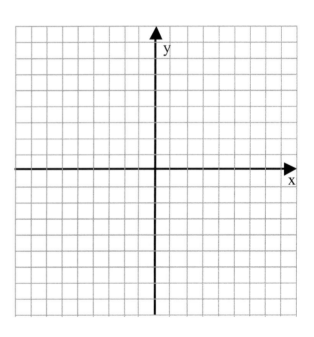

4.

$y = 3x$

$y = -4x$

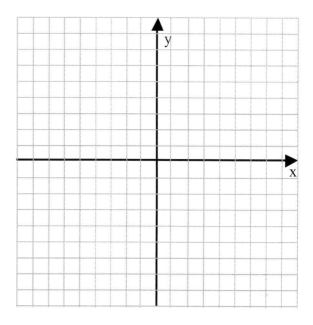

5.

$y = -2x + 5$

$y = \dfrac{2}{5}x - 7$

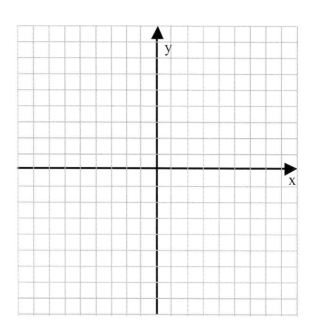

282

6.

$$y = \frac{1}{4}x - 1$$

$$y = -x + 4$$

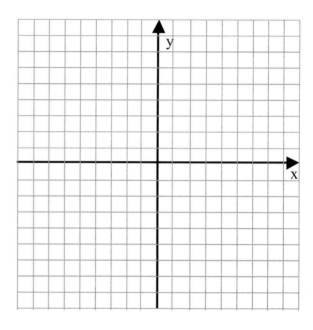

7.

$$y = -5x - 1$$

$$y = x - 7$$

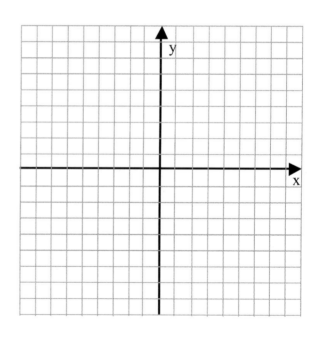

8.

$y = -2x - 5$

$y = \dfrac{1}{2}x + 5$

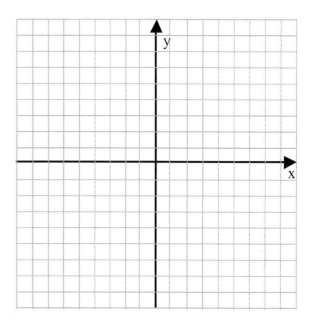

9.

$y = \dfrac{2}{3}x - 1$

$y = -\dfrac{2}{3}x - 5$

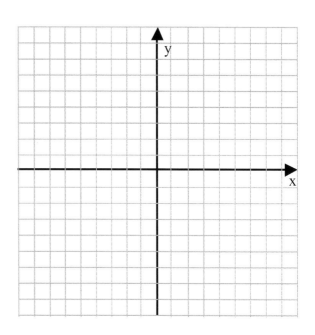

284

10.

$$y = -\frac{5}{2}x - 3$$

$$y = -\frac{1}{2}x + 5$$

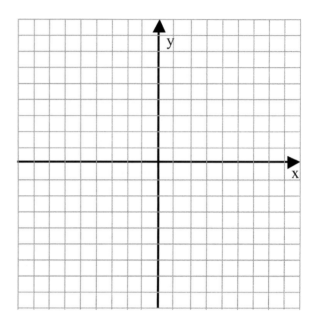

11.

$$y = x + 5$$

$$y = -4x + 5$$

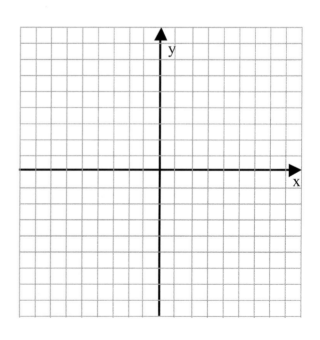

12.
$3x + y = -3$
$x + 3y = 15$

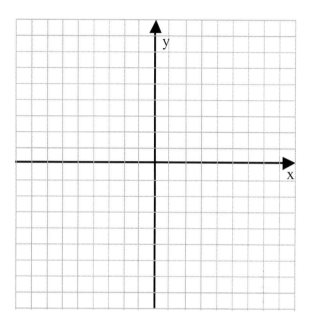

13.
$y = x + 2$
$y = -3x - 6$

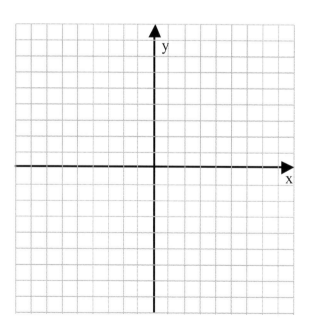

14.

$$y = \frac{1}{2}x - 2$$

$$y = -\frac{1}{6}x + 2$$

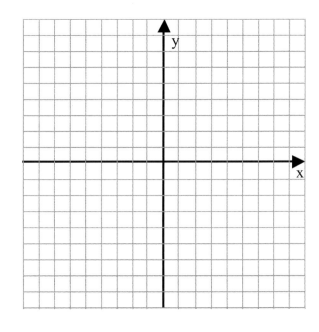

15.

$$y = -x + 5$$

$$y = \frac{2}{7}x - 4$$

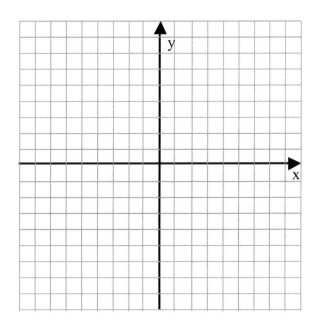

16.

$$y = \frac{4}{3}x - 3$$
$$y = -5x - 3$$

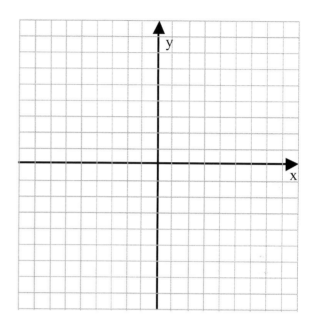

17.

$$-2x + 3y = -6$$
$$x + y = 8$$

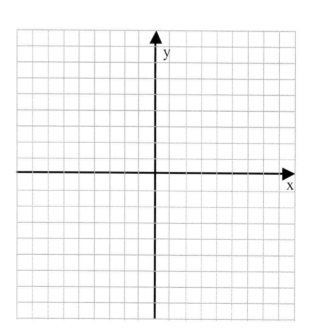

18.

$$x + y = 2$$
$$2x + 2y = 4$$

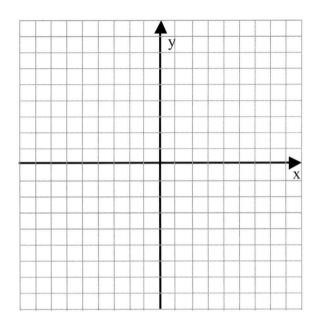

19.

$$x - y = 4$$
$$-2x + 2y = -8$$

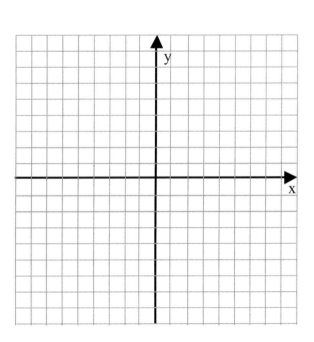

20.
$x + y = 8$
$x - y = -4$

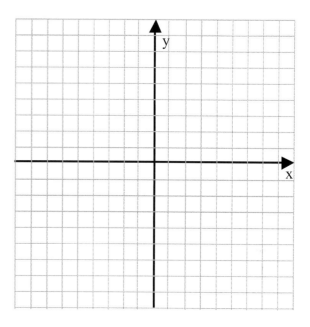

21.
$-4x + 3y = 12$
$8x - 6y = 24$

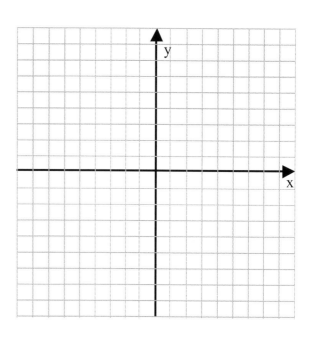

22.
$$y = 2x + 3$$
$$-6x + 3y = 9$$

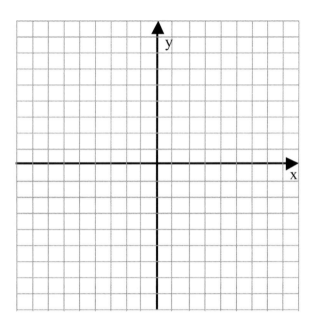

23.
$$-x - 2y = 4$$
$$5x + 10y = 30$$

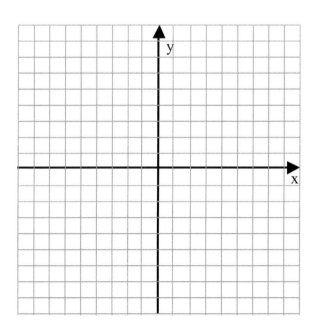

9.2 SOLVING SYSTEMS BY ELIMINATION

The elimination technique is based upon the concept that, we can add, subtract, multiply, or divide any equation by any number, other than zero, and not change the solution to that equation. This is a basic property of the equality of numbers and multiplication. Let's illustrate how we use this property to solve a system of equations with an example.

Given the system of equations:

$$x + y = 5 \qquad (1)$$
$$-x + 3y = 3 \qquad (2)$$

We note that we can add equation (1) and equation (2) to each other and not change the solution to the equation, since we are adding equal amounts on both side of the equations. If we do this, we still have equal amounts on either side of the equal-sign. If we do this, notice what happens:

$$x + y = 5 \qquad (1)$$
$$\underline{-x + 3y = 3} \qquad (2)$$
$$x - x + y + 3y = 5 + 3 \qquad (1)+(2)$$
$$4y = 8$$
$$y = 2$$

By adding the two equations together the x variable was eliminated from the problem. However, we did not change the solution, since we added like amounts to both sides of the equation. The resulting equation, (1) + (2), is now an equation in only y, so we can solve this for y and obtain $y = 2$, as shown above.

Now that we know the value of y, we can use it in either equation (1) or (2) to find the value that x must be. We'll use equation (1) and obtain:

$$x + y = 5$$
$$x + (2) = 5$$
$$x = 5 - 2$$
$$x = 3$$

Thus, the solution to this system is $x = 3$ and $y = 2$.

We can also check this result by putting the solution into both of the original equations, and show that it satisfies both equations:

$$x + y = 5$$
$$(3) + (2) = 5$$
$$5 = 5$$

$$-x + 3y = 3$$
$$-(3) + 3(2) = 3$$
$$-6 + 6 = 3$$
$$3 = 3$$

This example illustrates what we are trying to accomplish. We would like to be able to add the 2 equations together to eliminate one of the variables. However, not all systems are written in the form where we can do this right from the start. Instead, we may have to multiply one or both of the equations by a constant before we add the equations together. We illustrate this procedure in the following examples.

EXAMPLE 1: Solve the system by elimination

$$2x + 5y = 3 \quad (1)$$
$$x - 3y = -4 \quad (2)$$

We begin by choosing to eliminate the x variable first. The reason we chose x was because we noticed that we only need to multiply the equation (2) by a -2, and then we would have the same magnitude coefficients, but with opposite signs in front of the x terms. Thus, when we add the equations together, the x term will vanish.

We demonstrate this as follows:

$$2x + 5y = 3 \quad (1)$$
$$x - 3y = -4 \quad (2)$$

Multiply equation (2) by -2:
$$2x + 5y = 3$$
$$-2(x - 3y) = (-2)(-4)$$

Rewrite the system:
$$2x + 5y = 3$$
$$-2x + 6y = 8$$

In the above steps we have simply rewritten equation (2) with an equivalent equation.

We can now solve the resulting equivalent system.

$$2x + 5y = 3 \quad (1)$$
$$-2x + 6y = 8 \quad (2)$$

We should note that we are trying to transform a problem we don't know how to solve into one we already know how to solve. This happens frequently in mathematics. The new system is very similar to the motivating example from above.

Now if we add these two equations together, we can eliminate one of the variables (x again), and proceed as before.

$$2x + 5y = 3 \qquad (1)$$
$$\underline{-2x + 6y = 8} \qquad (2)$$
$$2x - 2x + 5y + 6y = 3 + 8 \qquad (1)+(2)$$
$$11y = 11$$
$$y = 1$$

We now know that $y = 1$, so we can use this in the original equation (2) to obtain:

$$x - 3y = -4$$
$$x - 3(1) = -4$$
$$x - 3 = -4$$
$$x = -1$$

Thus, the final solution is x = −1, and y = 1 . Again, we could check to verify, but we'll leave that to the reader. You, however, should make it a habit to always check the results.

EXAMPLE 2: Solve the system by elimination

$$5x - 3y = 1 \qquad (1)$$
$$-3x + 2y = -4 \qquad (2)$$

We begin by choosing to eliminate the y variable this time. We could have chosen x or y to eliminate, but by eliminating y, the numbers we need to multiply by will be slightly smaller. This is just an arbitrary choice, though.

What is now different is that we cannot multiply a single equation by a single number and eliminate a variable by adding the equations together. We now have to multiply both equations and then add to eliminate the variable. For this problem we see that if we multiply equation (1) by 2, two, and equation (2) by 3, three, the coefficients in front of the y terms will be −6 and 6 for equation (1) and (2) respectively. Thus, when we add the equations together the y term will be eliminated.

We demonstrate this as follows:

$$5x - 3y = 1 \qquad (1)$$
$$-3x + 2y = -4 \qquad (2)$$

Multiply equation (1) by 2: $\qquad 2(5x - 3y) = 1(2)$

Multiply equation (2) by 3: $\qquad 3(-3x + 2y) = -4(3)$

Rewrite both equations:

$$10x - 6y = 2$$
$$-9x + 6y = -12$$

We can now solve the equivalent system:

$$\begin{aligned} 10x - 6y &= 2 \qquad (1) \\ -9x + 6y &= -12 \qquad (2) \end{aligned}$$

Add the two equations together to obtain:

$$\begin{aligned} 10x - 6y &= 2 \qquad (1) \\ \underline{-9x + 6y} &\underline{= -12} \qquad (2) \end{aligned}$$

$$10x - 9x - \cancel{6y} + \cancel{6y} = 2 - 12 \quad (1)+(2)$$
$$x = -10$$

We now know that $x = -10$, so we can use this in the original equation (1) to obtain:

$$\begin{aligned} 5x - 3y &= 1 \\ 5(-10) - 3y &= 1 \\ -50 - 3y &= 1 \\ -3y &= 51 \\ \frac{-3y}{-3} &= \frac{51}{-3} \\ y &= -\frac{51}{3} \\ y &= -17 \end{aligned}$$

Thus, the final solution is $x = -10$, and $y = -17$. Again, we could check to verify, but we'll leave that to the reader. Again, you should make it a habit to always check the results.

EXAMPLE 3: Solve the system by elimination

$$\begin{aligned} 5x - 3y &= 1 \qquad (1) \\ -10x + 6y &= -2 \qquad (2) \end{aligned}$$

We see that we can multiply equation (1) by 2 and add it to equation (2). Doing this, however, we end up with the final result:

$$5x - 3y = 1 \qquad (1)$$
$$-10x + 6y = -2 \qquad (2)$$
$$2(5x - 3y) + (-10x + 6y) = 2(1) + (-2) \quad 2(1) + (2)$$
$$10x - 10x - 6y + 6y = 2 - 2$$
$$0 = 0$$

Both equations vanished and we are left with a true (consistent) statement. This means that the two lines were the same. Thus, when we subtracted like things from each other we got zero.

Since they are the same line, the solutions are given using either line, e.g.

$$5x - 3y = 1$$

is the solution, or if we solve this for y

$$y = \frac{5}{3}x - \frac{1}{3}$$

is the solution.

There are an infinite number of solutions. We can substitute any value for x into the above equation and find the y-value associated with it. We could also graph the line to show the solution set.

EXAMPLE 4: Solve the system by elimination

$$3x - y = 6 \qquad (1)$$
$$9x - 3y = 2 \qquad (2)$$

Multiplying equation (1) by -3 and adding it to equation (2) we find

$$3x - y = 6 \qquad (1)$$
$$9x - 3y = 2 \qquad (2)$$
$$-3(3x - y) + (9x - 3y) = -3(6) + (2) \qquad -3(1) + (2)$$
$$-9x + 9x + 3y - 3y = -18 + 2$$
$$0 = -16$$

The result is inconsistent (an untrue statement). This means that the two lines are parallel and that they do not intersect. This means there is no solution. The method we used tries to force the two lines to intersect, but because they cannot, it gives us an inconsistent result $0 = -16$. Whenever this occurs we simply state this system has no solutions.

EXERCISES 9.2

Solving Systems Using Elimination

Solve the following systems by the elimination method:

<u>Basic Problems</u>

1. $\begin{array}{l} -2x - y = -6 \\ 2x + 4y = -12 \end{array}$

3. $\begin{array}{l} -7x + 2y = 6 \\ 7x - y = 3 \end{array}$

5. $\begin{array}{l} 7x + 2y = 24 \\ 8x + 2y = 30 \end{array}$

7. $\begin{array}{l} -6x + 5y = 2 \\ 6x - 6y = -4 \end{array}$

2. $\begin{array}{l} 3x - y = 7 \\ 4x + y = 14 \end{array}$

4. $\begin{array}{l} x + 3y = -2 \\ x + 4y = 4 \end{array}$

6. $\begin{array}{l} -2x - 7y = -17 \\ -4x - 7y = -15 \end{array}$

8. $\begin{array}{l} 5x + y = 3 \\ 3x - y = 13 \end{array}$

<u>Intermediate to Advanced Problems</u>

9. $\begin{array}{l} -3x + 2y = 4 \\ 6x - 4y = 4 \end{array}$

13. $\begin{array}{l} -4x + 13y = -22 \\ -x + 2y = 2 \end{array}$

17. $\begin{array}{l} 3x - 2y = -4 \\ 5x - 5y = 10 \end{array}$

21. $\begin{array}{l} 5x - 3y = 3 \\ -15x + 9y = -1 \end{array}$

10. $\begin{array}{l} -7x + y = -5 \\ 2x - 3y = -23 \end{array}$

14. $\begin{array}{l} 5x + 14y = 8 \\ -3x - 7y = -3 \end{array}$

18. $\begin{array}{l} 2x - 3y = 5 \\ -4x + 6y = -10 \end{array}$

22. $\begin{array}{l} -2x - y = -7 \\ -10x + 7y = 1 \end{array}$

11. $\begin{array}{l} -4x + 6y = 6 \\ x - 3y = 3 \end{array}$

15. $\begin{array}{l} 2x - 3y = -7 \\ -3x - 7y = -1 \end{array}$

19. $\begin{array}{l} x - 6y = 8 \\ -4x + 24y = -32 \end{array}$

23. $\begin{array}{l} 14x - 13y = 1 \\ 7x - 3y = -3 \end{array}$

12. $\begin{array}{l} 16x - 9y = 7 \\ -8x + 6y = 4 \end{array}$

16. $\begin{array}{l} 4x - 3y = -4 \\ 5x + 4y = -5 \end{array}$

20. $\begin{array}{l} x - 3y = 12 \\ 2x - 6y = -2 \end{array}$

24. $\begin{array}{l} 5x + 4y = -12 \\ 3x - 4y = -12 \end{array}$

9.3 SOLVING SYSTEMS BY SUBSTITUTION

The third technique to solve systems is very similar to the elimination method. Instead of eliminating a variable by adding the two equations together, we eliminate a variable by substituting the value of the variable written in terms of the other variable from one equation into the other. This is best illustrated through an example.

Consider the following system:

$$y = 3x - 1 \quad (1)$$
$$x + y = 3 \quad (2)$$

Since equation (1) is already solved for y, we will substitute what y is equal to, in terms of x, into equation (2), i.e.

$$y = 3x - 1 \quad (1)$$
$$x + y = 3 \quad (2)$$

$$x + (3x - 1) = 3 \quad \text{Solve for x}$$
$$x + 3x - 1 = 3$$
$$4x - 1 = 3$$
$$4x = 4$$
$$x = 1$$

We now know that $x = 1$, and to find y we simply substitute $x = 1$ into either equation (1) or (2) above to find y. We choose equation (1), because it is already solved for y, i.e.

$$y = 3x - 1$$
$$y = 3(1) - 1$$
$$y = 3 - 1$$
$$y = 2$$

The solution to this system then is $x = 1$, and $y = 2$.

EXAMPLE 1: Solve the system by substitution

$$x = 2y - 6 \qquad (1)$$
$$2x + y = 3 \qquad (2)$$

Since equation (1) is already solved for x, we substitute this equation into equation (2) for x to obtain:

$$2x + y = 3$$
$$2(2y - 6) + y = 3$$
$$4y - 12 + y = 3$$
$$5y - 12 = 3$$
$$5y = 15$$
$$y = 3$$

We now substitute $y = 3$ into equation (1) and obtain:

$$x = 2y - 6$$
$$x = 2(3) - 6$$
$$x = 6 - 6$$
$$x = 0$$

The solution is $x = 0$, $y = 3$.

EXAMPLE 2: Solve the system by substitution

$$5x - y = 4 \qquad (1)$$
$$2x + 3y = 5 \qquad (2)$$

For this system, neither equation is solved for a variable. Therefore, we have to do this ourselves. We can take equation (1) and solve this for y and obtain:

$$y = 5x - 4$$

Substituting this for y in equation (2) we obtain:

$$2x + 3y = 5$$
$$2x + 3(5x - 4) = 5$$
$$2x + 15x - 12 = 5$$
$$17x - 12 = 5$$
$$17x = 17$$
$$x = 1$$

We substitute for $x = 1$ in either of the original equations to find y:

$$5x - y = 4$$
$$5(1) - y = 4$$
$$5 - y = 4$$
$$-y = -1$$
$$y = 1$$

The solution is $x = 1$, $y = 1$.

EXAMPLE 3: Solve the system by substitution

$$2x - 3y = 6 \qquad (1)$$
$$y = 2x + 1 \qquad (2)$$

Equation (2) is already solved for y in terms of x, so we can use this to replace the y-term in equation (1), i.e.

$$2x - 3y = 6 \qquad (1)$$
$$y = 2x + 1 \qquad (2)$$

Replace y in equation (1) with $2x + 1$, and combine like terms and solve for x:

$$2x - 3(2x + 1) = 6$$
$$2x - 6x - 3 = 6$$
$$-4x - 3 = 6$$

$$-4x = 9$$
$$\frac{-4x}{-4} = \frac{9}{-4}$$
$$x = -\frac{9}{4}$$

Now that we have found the value of x, we use this in either equation to find y. However, since equation (2) is already solved for y we'll use this equation.

$$y = 2x + 1$$
$$= 2\left(-\frac{9}{4}\right) + 1$$
$$= -\frac{9}{2} + 1$$
$$= -\frac{9}{2} + \frac{2}{2}$$
$$= -\frac{7}{2}$$

The solution then is $x = -\dfrac{9}{4}$, and $y = -\dfrac{7}{2}$.

EXERCISES 9.3

Solving Systems by Substitution

Solve the following systems by substitution.

<u>Basic Problems</u>

1. $3x - y = 2$
$y = 1$

2. $-3x + 2y = 5$
$y = -2$

3. $x = -5$
$2x - 4y = -2$

4. $3x - y = -2$
$y = x - 6$

5. $x = 14$
$3x - 6y = 12$

6. $x + y = 4$
$y = 2x - 8$

7. $3x - 4y = 6$
$y = -x + 2$

8. $-2x - 3y = 5$
$y = 3x - 9$

9. $x = 2y - 3$
$y + x = 3$

10. $x = 4y - 3$
$2x - 3y = 4$

11. $x = -2y - 3$
$-2x + y = -9$

12. $x = 13 - y$
$-3x + 4y = 10$

13. $3x - 2y = 8$
$y = -3x + 5$

14. $4x - 13y = -24$
$y = x - 3$

15. $12x - y = 9$
$y = 13x - 3$

16. $2x + 7y = 19$
$y = -2x + 1$

17. $-x - 8y = -115$
$y = -2x + 5$

18. $6x + 6y = 12$
$y = -4x + 8$

19. $x - y = 3$
$y = x$

20. $5x - y = -9$
$y = 5x - 4$

21. $-2x + y = -9$
$y = 2x - 9$

22. $6x - 2y = 12$
$y = 3x - 6$

<u>Intermediate to Advanced Problems</u>

23. $4x - y = 4$
$2y = 6x - 2$

24. $-2x - y = -9$
$4y = -12x + 4$

25. $-x + y = -9$
$2x + y = -3$

26. $-2x + y = 4$
$x - y = 3$

27. $3x + y = 4$
$x + y = -2$

28. $x + 4y = 4$
$x - y = 3$

29. $2x + y = -3$
$x + 2y = 3$

9.4 APPLICATIONS

Let's consider the two application problems presented at the beginning of the chapter.

EXAMPLE 1:

Danise is responsible for buying a week's supply of food and medication for the dogs and cats at a local shelter. The food and medication for the dogs, costs three times as much as those supplies for the cats. She needs to feed 125 cats and 40 dogs. Her budget is $3,675. How much can she spend on each dog and cat, for food and medication?

When we first presented this problem we gave the associated system of equations. Now, we show how we came up with this system. First we have to decide on the variables. This is the – how much, question.

"How much can she spend on each dog and cat, for food and medication?"
Let's choose x to represent how much we can spend on a dog, and y the amount we can spend on a cat.

Then we look at the other statements.

"The food and medication for the dogs, costs three times as much as those supplies for the cats."

This can be written mathematically as: $x = 3y$.

"She needs to feed 125 cats and 40 dogs. Her budget is $3,675."

This can be written mathematically as: $40x + 125y = 3675$.
Written all together we have:

$$40x + 125y = 3675$$
$$x - 3y = 0$$

We now solve this system using one of the above techniques. We shall use elimination.

$$40x + 125y = 3675$$
$$-40(x - 3y) = 0(-40)$$

$$40x + 125y = 3675$$
$$-40x + 120y = 0$$

$$245y = 3675$$
$$\frac{\cancel{245}}{\cancel{245}} y = \frac{3675}{245} = 15$$

Thus, y, which is the amount Danise can spend on a cat, is $15, and since x, the amount Danise can spend on a dog is three times that, x = $45.

EXAMPLE 2:

Maria is a chemist and she needs 15ml of a 30% acidic solution for a lab experiment, but she only has a 20% and a 50% solution available. How does she mix the two solutions to obtain 15ml of a 30% acidic solution?

First, we must choose the variables. Since we are trying to find out how much to mix each of the two solutions to obtain another solution, we choose x to represent the amount of the 50% solution and y to represent the amount of the 20% solution. These are the two unknowns we are trying to solve for.

The first equation is simple. Since we want a total of 15ml then: $x + y = 15$.

Now we need to formulate the second equation. This is a bit more complicated. Anytime we are working with percentages, we will need to focus on the actual amount of what we are working with. In this case it is alcohol. Thus, a 50% solution with have a total $0.5x$ alcohol in it, and a 20% solution will have a total of $0.2y$ alcohol in it.

We can equate the total alcohol needed, which is 30% of 15ml or 4.5 ml, with the two mixtures to obtain the following equation: $0.5x + 0.2y = (0.3)(15) = 4.5$.

The two equations together form the system:

$$0.5x + 0.2y = 4.5$$
$$x + y = 15$$

We now solve this system using the elimination method.

$$0.5x + 0.2y = 4.5$$
$$-0.5(x + y) = (-0.5)(15) \quad \text{Multiply both sides by} -0.5$$

$$0.5x + 0.2y = 4.5$$
$$-0.5x - 0.5y = -7.5$$

$$-0.3y = -3 \qquad \text{Add the two equations together}$$
$$\frac{-0.3y}{-0.3} = \frac{-3}{-0.3} \qquad \text{Solve for y}$$
$$y = 10$$

With y = 10, we can obtain x by replacing it in either of the original equations. We chose the second equation and solve for x as follows:

$$x + y = 15$$
$$x + 10 = 15$$
$$x = 5$$

Thus, Maria must combine 5ml of the 50% solution with 10ml of the 20% solution to obtain 15ml of a 30% solution.

EXAMPLE 3:

A dietician needs a patient to have a meal that has 70 grams (g) of protein and 120g of carbohydrates. The hospital food service tells the dietician that the dinner is chicken with corn. Each serving of chicken has 30g of protein and 35g of carbohydrates. Each serving of corn has 3g of protein and 15g of carbohydrates. How many servings of chicken and corn should the patient have?

First we must identify the variables. Since we are trying to find the amount of chicken and corn to serve, the variables are x, the number of servings of chicken, and y, the number of servings of corn.

We'll use one equation to balance the amount of protein in the meal and the other to balance the amount of carbohydrates in the meal.

Protein: $\qquad$ $30x + 3y = 70$
Carbohydrates: $\qquad$ $35x + 15y = 120$

We solve this system using the elimination method:

$$30x + 3y = 70$$
$$35x + 15y = 120$$

$$-5(30x + 3y) = (70)(-5)$$
$$35x + 15y = 120$$

$$-150x - 15y = -350$$
$$35x + 15y = 120$$

$$-115x = -230$$
$$\frac{\cancel{-115}x}{\cancel{-115}} = \frac{-230}{-115}$$
$$x = 2$$

This tells us that the meal must have two (2) servings of chicken. We use $x = 2$ to find y, the servings of corn, by using either original equation to obtain the y amount, i.e.

$$30x + 3y = 70$$
$$30(2) + 3y = 70 \quad \text{Replace x with 2}$$
$$60 + 3y = 70 \quad \text{Solve for y}$$
$$3y = 10$$
$$\frac{\cancel{3}y}{\cancel{3}} = \frac{10}{3}$$

Thus, the patient needs to receive two (2) servings of chicken and $10/3$, or 3 and $1/3$ servings of corn.

EXERCISES 9.4

1. A chemist needs 20ml of a 35% acidic solution for a lab experiment, but she only has a 10% and a 50% solution available. How does she mix the two solutions to obtain 20ml of a 35% acidic solution?

2. A dietician needs a patient to have a meal that has 90 grams (g) of protein and 150g of carbohydrates. The hospital food service tells the dietician that the dinner is chicken with corn. Each serving of chicken has 30g of protein and 35g of carbohydrates. Each serving of corn has 3g of protein and 15g of carbohydrates. How many servings of chicken and corn should the patient have?

3. The manager at pet store is responsible for buying a week's supply of food and medication for the dogs and cats at a local shelter. The food and medication for the dogs, costs twice as much as those supplies for the cats. She needs to feed 250 cats and 75 dogs. Her budget is $4,800. How much can she spend on each dog and cat, for food and medication?

4. A food manufacturer needs produce a 25 pound mixture of peanuts and cashew nuts. The peanuts cost $3.50 per pound and the cashews are $6.75 per pound. The manufactures wants a mixture that costs $5.00 per pound. How much of the peanut and cashew nuts must the manufacturer use?

5. A dieter wishes to have a meal of spaghetti with meatballs and a salad that contains no more than 600 calories and a maximum of 62 grams of carbohydrates. A one-cup serving of spaghetti and meatballs is 275 calories with 30 g of carbohydrates. One cup of salad is 25 calories and 2 grams of carbohydrates. How many servings of each should the dieter eat?

6. A coffee manufacturer wishes to create a blend of coffee worth $8.00 per pound. She needs to mix Columbian coffee at $6.50 per pound with a French Roast at $9.50 per pound. How much should she mix to obtain a total of 100 pounds?

Chapter 9: Practice Test

Solve the systems of equations by graphing

1.

$y = x + 1$

$y = -x + 7$

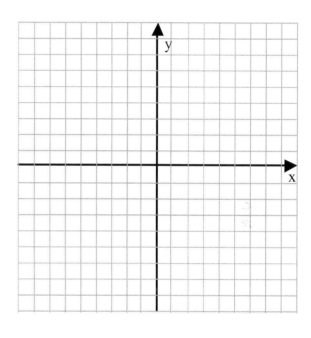

2.

$y = 2x + 1$

$y = -2x - 7$

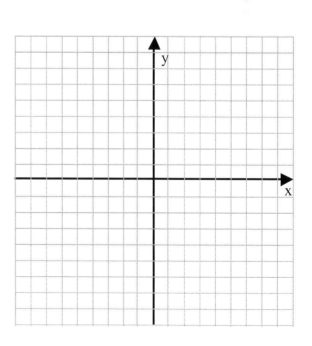

Solve the following systems by the elimination method:

3. $\begin{aligned} -2x - y &= -6 \\ 2x + 4y &= -12 \end{aligned}$

6. $\begin{aligned} 14x - 13y &= -24 \\ 7x + 3y &= 26 \end{aligned}$

4. $\begin{aligned} 3x - y &= 7 \\ 4x + y &= 14 \end{aligned}$

7. $\begin{aligned} 5x + 4y &= -12 \\ 3x - 4y &= -12 \end{aligned}$

5. $\begin{aligned} -7x + 2y &= 6 \\ 7x - y &= 3 \end{aligned}$

Solve the following systems by substitution.

8. $\begin{aligned} 3x - y &= 2 \\ y &= 1 \end{aligned}$

11. $\begin{aligned} x &= 14 \\ 3x - 6y &= 12 \end{aligned}$

9. $\begin{aligned} -3x + 2y &= 5 \\ y &= -2 \end{aligned}$

12. $\begin{aligned} x + y &= 4 \\ y &= 2x - 8 \end{aligned}$

10. $\begin{aligned} x &= -5 \\ 2x - 4y &= -2 \end{aligned}$

13. A chemist needs 30ml of a 40% acidic solution for a lab experiment, but she only has a 20% and a 50% solution available. How does she mix the two solutions to obtain 30ml of a 40% acidic solution?

14. The manager at a pet store is responsible for buying a month's worth of food for cats and birds at his store. The food for the cats costs four times as much as for the birds. She needs to feed 25 cats and 100 birds. Her budget is $800. How much can she spend on each cat and bird for food?

15. A food manufacturer needs produce a 100 pound mixture of peanuts blended coffee. They will use a mixture of Columbian coffee that costs $9.50 per lb and French Roast coffee that cost them$12 per lb. The manufacture wants a mixture that costs $11.00 per pound. How much of the Columbian and French Roast must the manufacturer use?

CHAPTER 10

Basic Math for Lab Science Courses

In this final chapter we focus on a series of miscellaneous topics that are often needed in the different laboratory science classes you need to take. These are techniques and concepts that deal with measured data, and how to begin to organize and analyze it, as well as the formulas that are used to describe the data.

10.1 SCIENTIFIC NOTATION

In performing calculations we may encounter numbers that are very large, such as

$$602,210,000,000,000,000,000,000$$

which is Avogadro's number. Or numbers that are very nearly equal to zero, such as

$$0.000000000000000000000000166$$

which equals one atomic mass unit measured in grams.

When we encounter numbers of this type, it is usually convenient, and often necessary, to represent them in an alternative form, called Scientific Notation. Scientific notation is a method, which condenses the writing of very small (close to zero) and very large numbers. It removes the excess zeros from the number using the exponent notation we learned earlier in the book.

In scientific notation, numbers are expressed as the product of a decimal number, whose whole number part is a number from 1 to 9, and a base, 10, raised to an integer power (exponent).

The exponent indicates how many times a number must be multiplied by itself (positive exponent). We noticed that it was always one more than the number of times it was multiplied:

$$10^1 = 10$$
$$10^2 = 10 \times 10 = 100$$
$$10^3 = 10 \times 10 \times 10 = 1000$$

Division by factors of 10 is indicated using negative exponents, as shown below:

$$10^{-1} = \frac{1}{10} = 0.1$$

$$10^{-2} = \frac{1}{10 \times 10} = \frac{1}{100} = 0.01$$

$$10^{-3} = \frac{1}{10 \times 10 \times 10} = \frac{1}{1000} = 0.001$$

Examples of numbers and their equivalent representation in scientific notation:

$$30 = 3 \times 10^1 \qquad\qquad 150 = 1.5 \times 10^2 \qquad\qquad 60,367 = 6.0367 \times 10^4$$

$$0.3 = 3 \times 10^{-1} \qquad\qquad 0.046 = 4.6 \times 10^{-2} \qquad\qquad 0.000002 = 2 \times 10^{-6}$$

Writing Numbers in Scientific Notation Form

Writing a number in scientific notation involves rewriting a number as a decimal part, times 10 raised to an appropriate power.

If the original number is greater than 1 in magnitude (magnitude means we don't consider the sign of the number, only the number without the sign), then we can treat the two types of numbers, negative and positive, in a similar way. Thus, $3,200,000,000$ and $-3,200,000,000$ both have the same magnitude, which is greater than 1. This means the power is non-negative, indicating that we are multiplying by factors of 10. If the number is less than 1 in magnitude, then the exponent is negative, indicating that we are dividing by factors of 10. We'll illustrate the process with some examples.

In scientific notation we move the decimal point from the left or right (depending upon the number), so that it is located with just one non-zero digit (1-9) to its left. We count the number of place-values we moved to relocate the decimal point. If we moved the decimal point to the left we count the exponent as positive. This is because we would have to multiply our number by this many factors of ten, i.e. 10 raised to the positive exponent, to obtain the decimal representation of the number.

For example, consider the number: 3,500

The scientific notation form would be: 3.5×10^3

This is because 10^3 is equivalent to $10 \times 10 \times 10 = 1,000$ and 3.5 times 1,000 is equal to 3,500, our original number. The two representations are equivalent forms of the same number. This required us moving the decimal point three places to the left.

If we had to move the decimal point to the right, the exponent will be negative. This is due to the fact that we are dividing by this many factors of 10, and we recall that exponentiating to a negative power is equivalent to division.

For example, consider the decimal number, 0.00057 (57 hundred-thousandth's)

The scientific-notation form would be 5.7×10^{-4}

This is because 10^{-4} is equivalent to $\dfrac{1}{10^4} = \dfrac{1}{10 \times 10 \times 10 \times 10} = \dfrac{1}{10,000}$ and multiplying 5.7 times this number yields our original number 0.00057. Alternatively we moved the decimal point four places to the right. Again, the two representations are equivalent.

EXAMPLE 1: Write 325 in scientific notation

The first step is to rewrite the number with only one digit to the left of the decimal point, i.e. 3.25.

Next we have to determine what power to write for the exponent for the base of 10. Since we need the product to be three hundred and twenty five, we have to multiply 3.25 times one hundred to achieve this. Thus, the exponent term is $100 = 10^2$.

The final form is: 3.25×10^2.

We should notice that since 325 is greater than 1, we moved the decimal place, two places to the left. Thus, to accommodate this we have to have a positive exponent of 2 for the base ten. Moving the decimal place to the left gives a positive exponent.

EXAMPLE 2: Write 0.093 in scientific notation,

The first step is to rewrite the number with only one digit to the left of the decimal point, i.e. 9.3. Next we have to determine what power to write for the exponent for the base of 10. Since we need the product to be ninety three thousandths, we have to divide 9.3 by one hundred to achieve this. Thus, the exponent term is:

$$\frac{1}{100} = \frac{1}{10^2} = 10^{-2}$$

The final form is: 9.3×10^{-2}.

We should notice that since 0.093 is less than 1, we moved the decimal place, two places to the right. Thus, to accommodate this we have to have a negative exponent of -2 for the base ten. Moving the decimal place to the right will always give a negative exponent.

We should point out that the sign of the number does not affect the sign of the exponent!

EXAMPLE 3: Write -0.000456 in scientific notation,

The first step is to rewrite the number with only one digit to the left of the decimal point, i.e. -4.56.

Next since we moved the decimal point 4 places to the right the exponent is -4.

Thus, the final form is: -4.56×10^{-4}.

EXAMPLES: Write each number in scientific notation.

$$1,230,000,000 = 1.23 \times 10^9$$

(Since we moved the decimal point 9 places to the left. Moving a place to the left accounts for multiplying by a factor of ten.)

$$0.000035465 = 3.5465 \times 10^{-5}$$

(Since we moved the decimal point 5 places to the right. Moving a place to the right accounts for dividing by a factor of ten.)

$$245,000 = 2.45 \times 10^5$$
$$0.00628 = 6.28 \times 10^{-3}$$
$$2,209,000 = 2.209 \times 10^6$$
$$0.0005507 = 5.507 \times 10^{-4}$$

To write a number that is originally in scientific form as a fixed decimal number, we just reverse the process from above. If the exponent is positive, we move the decimal point that (the exponent value) many places to the right. If the exponent is negative, we move the decimal point that (the positive exponent value) many places to the left.

EXAMPLES: Convert each number which is expressed in scientific notation into decimal (base-10) notation:

1. $1.88465 \times 10^7 = 18,846,500$ Moved seven places to the right
2. $3.95 \times 10^{-4} = 0.000395$ Moved four places to the left
3. $6.19 \times 10^{-7} = 0.000000619$ Moved seven places to the left
4. $3.42 \times 10^6 = 3,420,000$ Moved six places to the right
5. $4.9712 \times 10^7 = 49,712,000$ Moved seven places to the right
6. $-2.906 \times 10^{-8} = -0.00000002906$ Moved eight places to the left
7. $-5.03 \times 10^7 = -50,300,000$ Moved seven places to the right

Computations Involving Scientific Notation

Computations involving numbers written in scientific notation are easy to perform. To multiply numbers written in scientific notation, multiply the number parts and add the powers (exponents). To divide numbers written in scientific notation, divide the number parts and subtract the power (exponent) of the denominator from the power of the numerator. We illustrate this with the following examples:

EXAMPLE 1: Multiply $(3 \times 10^4) \times (5 \times 10^3)$

$$(3 \times 10^4) \times (5 \times 10^3) = (3 \times 5) \times (10^4 \times 10^3)$$

$$= 15 \times 10^7 \qquad \text{but } 15 = 1.5 \times 10^1$$

$$= 1.5 \times 10^1 \times 10^7$$

$$= 1.5 \times 10^{1+7}$$

$$= 1.5 \times 10^8$$

EXAMPLE 2: Multiply $(5 \times 10^2) \times (2 \times 10^3) \times (1.5 \times 10^{-4})$

$$(5 \times 10^2) \times (2 \times 10^3) \times (1.5 \times 10^{-4}) = (5 \times 2 \times 1.5) \times (10^2 \times 10^3 \times 10^{-4})$$

$$= 15 \times 10^{2+3-4}$$

$$= 15 \times 10^1 \qquad \text{but } 15 = 1.5 \times 10^1$$

$$= 1.5 \times 10^2 \qquad \text{rewriting in scientific form we have}$$

EXAMPLE 3: Divide $\dfrac{6 \times 10^8}{3 \times 10^3}$

$$\frac{6 \times 10^8}{3 \times 10^3} = \left(\frac{6}{3}\right) \times \left(\frac{10^8}{10^3}\right)$$

$$= 2 \times 10^{8-3}$$

$$= 2 \times 10^5$$

EXAMPLE 4: Divide $\dfrac{6 \times 10^{-2}}{2 \times 10^{-4}}$

$$\frac{6 \times 10^{-2}}{2 \times 10^{-4}} = \left(\frac{6}{2}\right) \times \left(\frac{10^{-2}}{10^{-4}}\right)$$

$$= 3 \times 10^{-2-(-4)}$$

$$= 3 \times 10^{-2+4}$$

$$= 3 \times 10^2$$

Scientific Notation on a Calculator

In this course, as well as your laboratory science courses, you will use the TI-30X IIS calculator. With that in mind, we show you how to properly use it. One important skill, is to know how to enter scientific notation numbers. The figure below shows to to enter Avogadro's number, 6.0221×10^{23}.

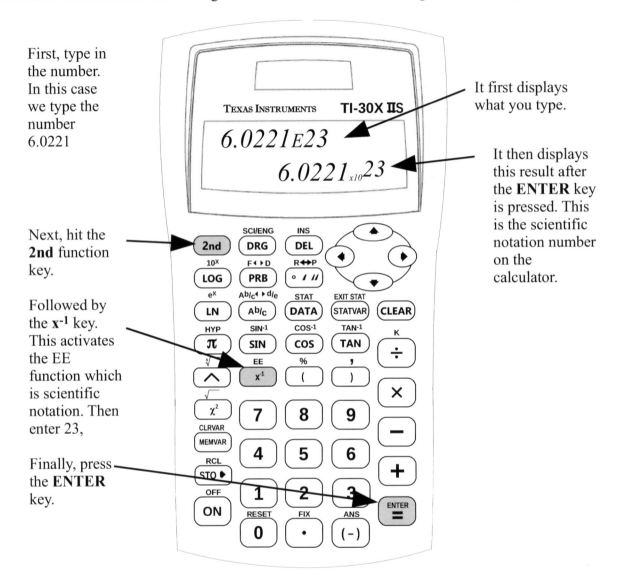

First, type in the number. In this case we type the number 6.0221

Next, hit the **2nd** function key.

Followed by the **x⁻¹** key. This activates the EE function which is scientific notation. Then enter 23,

Finally, press the **ENTER** key.

It first displays what you type.

It then displays this result after the **ENTER** key is pressed. This is the scientific notation number on the calculator.

This calculator also has a key to change from a fixed floating point number (this is how we typically write our numbers) to scientific notation, and then back again. First we show how to enter a decimal number, and using your calculator translate that number into scientific form. Let's use the number 0.00000801. We begin by typing this into the calculator, and then follow the keystrokes in the figures below.

First, type in the number. In this case we type the number 0.00000801

It first displays what number you type.

It then displays this window. The underlined feature is FLO meaning it is set for a floating point number.

Next, hit the **2nd** function key.

Followed by the **DRG** key. This activates the SCI/ENG function which allows you to change between floating point and scientific numbers.

Now use the arrow key to move the underline to SCI, for scientific notation.

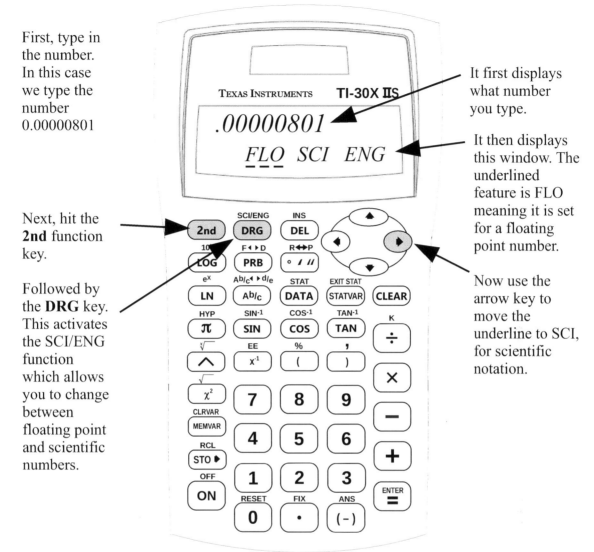

Continue on with the following keystrokes.

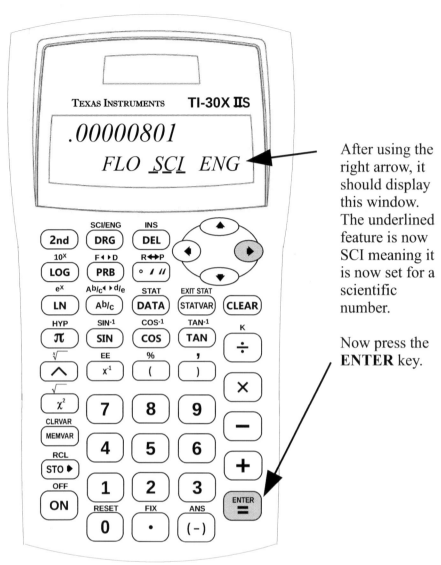

After using the right arrow, it should display this window. The underlined feature is now SCI meaning it is now set for a scientific number.

Now press the **ENTER** key.

The final result should look something like the figure below.

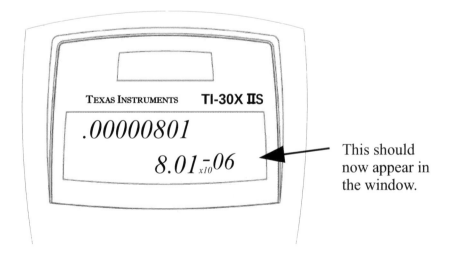

This should now appear in the window.

The result shows that 0.00000801 has been translated into 8.01×10^{-6}. See if you can use this approach to change the answer back into a floating point number. Note: the other ENG option on the screen has not been used. This is the engineering notation, and will not be used in this course, so you simply ignore this option.

We end with an example to show you how to divide two scientific numbers correctly. In this example we are going to perform the following division on the calculator.

$$7.1 \times 10^6 \div 3.8 \times 10^{-4}$$

We begin by entering each number as we did for our first example, but this time we separate each by the division symbol, /. Follow along with the figures below.

First, type in the number. In this case we type the number 7.1

This should be on your display.

Next, hit the **2nd** function key.

Followed by the **x⁻¹** key. This activates the EE function which is scientific notation. Then type in 6.

Hit the **divide** key and repeat with the new number, 3.8, **2nd** function key, **x⁻¹** key, then the **negative sign** key, followed by 4

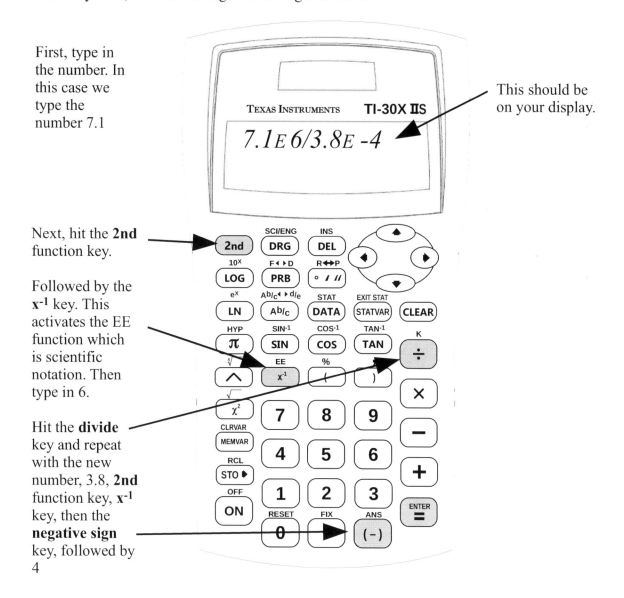

316

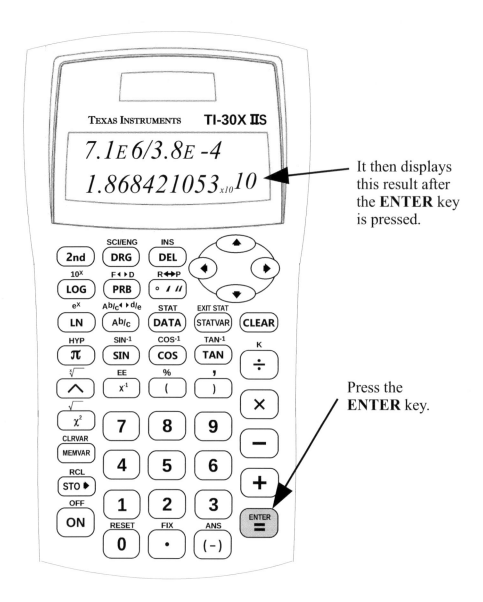

It then displays this result after the **ENTER** key is pressed.

Press the **ENTER** key.

EXERCISES 10.1

Write each number in scientific notation form, by hand and on your calculator:

1. 5,280,450,000

2. 0.0000967

3. 109,230,000,000

4. 55,000

5. 0.0000000534

6. 100,000,000,000

7. −0.0000007950

8. −2,300,500,000,000

9. 24

10. −0.1

11. −99,999,000,000

12. −0.000333

13. 0.0005

14. 333,020,000,000,000,000

15. −459

16. −23,490,000,000

Write each number in decimal form, by hand and on your calculator:

17. 3.0×10^3

18. -1.9×10^{-7}

19. 4.2×10^{-8}

20. 7.2×10^{12}

21. -2.0×10^{-4}

22. 4.5×10^{-13}

23. -8.79×10^{10}

24. 3.89×10^7

25. 3.345×10^8

26. -5.34×10^{-10}

27. 3.256×10^{-7}

28. 2.5×10^{-9}

29. -3.1×10^6

30. 7.7×10^{18}

31. 1.8532×10^3

32. 3.71×10^{-1}

33. -2.31×10^{-1}

34. 5.5×10^6

35. -5.3×10^0

36. 2.0×10^0

Perform the following operations by hand and on your calculator, write the final answer in scientific notation form. Round your answer to one digit to the right of the decimal point.

37. $\left(2.3 \times 10^4\right)\left(3.0 \times 10^3\right)$

38. $\left(5.0 \times 10^{-3}\right)\left(-1.9 \times 10^{-7}\right)$

39. $\left(4.2 \times 10^{-8}\right)\left(5.1 \times 10^9\right)$

40. $\left(7.2 \times 10^{12}\right)\left(4.5 \times 10^{10}\right)$

41. $\left(-2 \times 10^{-4}\right)\left(3 \times 10^4\right)$

42. $\left(8.9 \times 10^{-15}\right)\left(4.5 \times 10^{-13}\right)$

43. $\left(-4.51 \times 10^{-6}\right)\left(-8.79 \times 10^{10}\right)$

44. $\left(3.89 \times 10^7\right)\left(-4 \times 10^9\right)$

45. $\dfrac{\left(3.3 \times 10^8\right)}{\left(3 \times 10^2\right)}$

46. $\dfrac{\left(1.3 \times 10^{-6}\right)}{\left(-5 \times 10^{-10}\right)}$

47. $\dfrac{\left(9.6 \times 10^{12}\right)}{\left(3.2 \times 10^{-7}\right)}$

48. $\dfrac{\left(2.5 \times 10^{-9}\right)}{\left(7.5 \times 10^{-9}\right)}$

49. $\dfrac{\left(-6.2 \times 10^{-6}\right)}{\left(-3.1 \times 10^6\right)}$

50. $\dfrac{\left(7.7 \times 10^{18}\right)}{\left(1.1 \times 10^{-2}\right)}$

51. $\dfrac{\left(1.85 \times 10^3\right)}{\left(9.5 \times 10^{13}\right)}$

52. $\dfrac{\left(2.2 \times 10^5\right)\left(5.0 \times 10^3\right)}{\left(5.0 \times 10^{-5}\right)}$

53. $\dfrac{\left(8.9 \times 10^5\right)\left(-2.8 \times 10^{-9}\right)}{\left(1.4 \times 10^{10}\right)}$

54. $\dfrac{\left(-3.8 \times 10^{-4}\right)\left(3.3 \times 10^{-7}\right)}{\left(-1.9 \times 10^3\right)}$

Perform the following operations on your calculator and write the final answer in scientific notation form: Round your answer to one digit to the right of the decimal point.

55. $\left(3 \times 10^2\right)^4$

56. $\left(-1.5 \times 10^{-8}\right)^3$

57. $\left(5.0 \times 10^8\right)^{-1}$

58. $\left(4.3 \times 10^5\right)^{-2}$

59. $\left(5.31 \times 10^{12}\right)^{-3}$

60. $\left(2.5 \times 10^{-5}\right)^{-3}$

61. $\dfrac{\left(999.0 \times 10^5\right)}{\left(111.555 \times 10^3\right)}$

62. $\left(20.5 \times 10^4\right) \times \left(2.5 \times 10^{-4}\right)$

63. $\left(0.012 \times 10^2\right)^3$

64. $\dfrac{\left(6.66 \times 10^{-3/4}\right)}{\left(6.66 \times 10^{3/4}\right)}$

10.2 APPLICATIONS OF SCIENTIFIC NOTATION

In this section we show how to work application problems with scientific notation and units.

EXAMPLE 1: The average American consumes about 63.3 pounds of beef each year. There are approximately 3.0×10^8 people living in the US. What is the total yearly US beef consumption in pounds? What is it in kg?

To find the total you must multiply the average individual consumption by the total number of people. We first rewrite 63.3 pounds in scientific notation as:

$$6.33 \times 10^1 \frac{\text{pounds}}{\text{person}}$$

Now find the product:

$$
\begin{aligned}
6.33 \times 10^1 \frac{\text{pounds}}{\text{people}} \cdot 3.0 \times 10^8 \ \text{people} &= \left(6.33 \times 10^1\right)\left(3.0 \times 10^8\right)\left(\frac{\text{pounds}}{\text{people}}\right)(\text{people}) \\
&= (6.33 \times 3.0)\left(10^1 \times 10^8\right) \text{pounds} \\
&= 18.99 \times 10^9 \ \text{pounds} \\
&= 1.899 \times 10^{10} \ \text{pounds}
\end{aligned}
$$

Thus, the US consumes 1.899×10^{10} pounds of beef, or 18.99 billion pounds per year. Since 1 pound = 2.2043 kg we multiply the two to obtain:

$$\left(1.899 \times 10^{10} \ \text{pounds}\right)\left(\frac{\text{kg}}{2.2043 \ \text{pound}}\right) = 0.861498 \times 10^{10} \ \text{kg} = 8.61498 \times 10^9 \ \text{kg}$$

or 8.6 billion kg each year.

EXAMPLE 2: In 2007 there were 2.7 billion emails sent in the US per day. There were approximately 3.0×10^8 people living in the US at that time. What is the average number of emails per person in the US sent each day in 2007?

To find the average we need to divide the total number of emails by the total number of people, or

$$\frac{2.7 \text{ billion emails}}{3.0 \times 10^8 \text{ people}} = \frac{2.7 \times 10^9 \text{ emails}}{3.0 \times 10^8 \text{ people}}$$

$$= 0.9 \times 10^{9-8} \text{ emails/person}$$

$$= 0.9 \times 10^1 \text{ emails/person}$$

$$= 9.0 \times 10^0 \text{ emails/person}$$

$$= 9 \text{ emails/person}$$

Thus, on average there were 9 emails sent per person each day in the US in 2007.

EXAMPLE 3: The mass of the earth is given by the formula

$$M_E = \frac{g \, R_E^2}{G}$$

Where the gravitational constant is:

$$G = 6.673 \times 10^{-11} \frac{m^3}{kg \, s^2}.$$

The radius of the earth is, $R_E = 6.38 \times 10^6 \, m$, and the gravitational attraction is:

$$g = 9.8 \frac{m}{s^2}.$$

Find the mass of the earth.

Substitute into the formula all the values given:

$$M_E = \frac{g \, R_E^2}{G} = \frac{\left(9.8 \frac{m}{s^2}\right)\left(6.38 \times 10^6 \, m\right)^2}{6.673 \times 10^{-11} \frac{m^3}{kg \, s^2}}$$

Now reorganize the numerical parts and the unit parts.

$$M_E = \frac{g \, R_E^2}{G} = \frac{(9.8)\left(6.38 \times 10^6\right)^2}{6.673 \times 10^{-11}} \frac{\left(\frac{m}{s^2}\right)(m)^2}{\frac{m^3}{kg \, s^2}}$$

Simplify the numerical and the unit parts separately

Numerical:

$$= \frac{(9.8)(6.38)^2 \times (10^6)^2}{6.673 \times 10^{-11}}$$

$$= \frac{(9.8)(40.7044) \times 10^{12}}{6.673 \times 10^{-11}}$$

$$= \frac{398.9 \times 10^{12}}{6.673 \times 10^{-11}}$$

$$= \left(\frac{398.9}{6.673}\right) \times \left(\frac{10^{12}}{10^{-11}}\right)$$

$$= 59.8 \times 10^{12-(-11)}$$

$$= 59.8 \times 10^{23}$$

$$= 5.98 \times 10^{24}$$

Units:

$$\frac{\left(\dfrac{m}{s^2}\right)(m)^2}{\dfrac{m^3}{kg\ s^2}} = \frac{\dfrac{m \cdot m^2}{s^2}}{\dfrac{m^3}{kg\ s^2}} = \frac{\dfrac{m^3}{s^2}}{\dfrac{m^3}{kg\ s^2}} = \frac{m^3}{s^2} \div \frac{m^3}{kg\ s^2} = \frac{m^3}{s^2} \cdot \frac{kg\ s^2}{m^3} = kg$$

Putting both the numerical and the unit parts together we have the total mass of the earth equal to $5.98 \times 10^{24}\,kg$ ·

EXERCISES 10.2

Answer each of the following using scientific notation and round your answer to one digit to the right of the decimal point, unless indicaated otherwise.

1. The mass of the Earth's crust is $2.8 \times 10^{22}\,kg$, volume of the crust is $1 \times 10^{19}\,m^3$. What is the density of the crust, if density is the ratio of the mass divided by its volume?

2. A computer can process 800,000 bits of data in 0.00000625 seconds. How many bits per second can it process?

3. An adult human being has about 5,200,000 red blood cells (RBCs) per μL of blood. If Sally has 4.75 L of blood, how many RBCs does she have?

4. The frequency of an electromagnetic wave (how often it goes through a complete peak and valley cycle) is given by the formula, $\text{Freq} = \dfrac{c}{\lambda}$, where c is the speed of light and λ is the length of the wave. With $c = 3.0 \times 10^8 \dfrac{m}{s}$, and the wavelength of an x-ray at $\lambda = 6.0 \times 10^{-10} m$ for x-rays, what is the frequency of these x-rays? Also, given the following two frequencies: $1.8 \times 10^{-17} Hz$, $5.0 \times 10^{16} Hz$, find their associated wavelengths.

5. Debris is constantly hitting the earth from space. Each day it is estimated that this change in mass is equivalent to, $\Delta M = 6.0 \times 10^5 \, kg/day$. The age of the earth has been estimated to be, $A_E = 4.5 \times 10^9 \, years$. How much has the mass of the earth increased from its beginning, due to this added space debris?

6. The total energy received from the sun each second is 1.7×10^{17} calories. The surface area of the earth exposed to the sun is $2.55 \times 10^{14} m^2$. The amount of energy received per square meter every second (the solar constant) is

$$\frac{1.7 \times 10^{17} \ calories/s}{2.55 \times 10^{14} m^2}$$

Simplify this expression. What are the final units?

7. Population density is the ratio of the population divided by its area. Compute and compare the population densities of the following countries.
    a. Russia: Population 1.43×10^8 people, Land area 6.59×10^6 square miles
    b. Canada: Population 3.3×10^7 people, Land area 3.85×10^6 square miles
    c. US: Population 3.0×10^8 people, Land area 3.7×10^6 square miles
    d. China: Population 1.3×10^9 people, Land area 3.7×10^6 square miles
    e. Japan: Population 1.27×10^8 people, Land area 1.46×10^5 square miles
    f. Monaco: Population 3.3×10^4 people, Land area 7.7×10^{-1} square miles

8. The volume of the Milky Way can be approximated using the shape of a disk. The volume of a disk is given by the formula as: $V = \pi r^2 h$, where r is the radius of the disk and h is its height or thickness. With a thickness of 1000 light years and a radius of 50,000 light years. Compute the volume of the Milky Way in light years. If 1 light year = 9.5 x 10^{17} centimeters, what is the volume in cubic centimeters?

9. There are approximately 300 billion stars in the Milky Way, each with an average mass of approximately 2 x 10^{33} grams. Compute the total approximate mass of the Milky Way in grams and kg. What are we ignoring is this calculation?

10. Find the total US consumption of soft drinks in gallons for each year. How many liters is this? The following are statistics on soft drink consumption in the US since 1970. Round your answers to two digits to the right of the decimal point.
    1970: 22.7 gallons consumed per person with a total US population of 2.03×10^8 people
    1980: 34.1 gallons consumed per person with a total US population of 2.27×10^8 people

1990: 47 gallons consumed per person with a total US population of 2.49×10^8 people

2000: 60 gallons consumed per person with a total US population of 2.81×10^8 people

11. If you were to take all of the stars and gas in the Milky Way and spread them out throughout the entire volume of the Milky Way, about what would be the density of the Milky Way in:
    a. grams/cm³   b. kilograms/m³

    Note: you will need the results from exercises 7 and 8 above and the density formula: $D = \dfrac{M}{V}$

12. Nuclear fission can be used as an energy source. Each gram of uranium-235 provides 2.0×10^{10} calories of energy. If 1 calorie of energy provides 1.163×10^{-3} kWh (kilo Watt hours) of energy, how much energy in kWh is provided by 1 kg of uranium-235?

13. The velocity of light can be measured by dividing the distance from the earth to the sun, which is approximately 1.5×10^{11} meters, by the time it takes the light to travel this distance of about 500 seconds. Find the velocity of light in meters per second by taking the total distance traveled and dividing it by the total time.

14. Astronomers often measure distances in light-years. The distance that light travels in a year is 9.46×10^{22} kilometers. The closet star to our solar system is Proxima Centauri, which is 4.22 light-years away. How many kilometers are we from this star? How many miles is this? Round your answer to two digits to the right of the decimal point.

15. The average American consumes 4.2 kg of coffee every year. Since there are 3.0×10^8 people in the US, what is our total coffee consumption in kg? What is it in pounds?

16. As mentioned in Section 9.1, a number you will encounter in chemistry class, is Avogadro's number, 6.0221×10^{23} or 6.022×10^{23} , which represents the number of molecules or atoms a substance contains. Use Avogadro's number to answer the following questions. Round your answer to two digits to the right of the decimal point.
    a. How many molecules are present in 0.500 moles of water? 1 mol of water = 6.022×10^{23} molecules of water
    b. How many molecules are present in 2.76 moles of nitrous oxide? 1 mol of nitrous oxide = 6.022×10^{23} molecules of nitrous oxide
    c. How many molecules are present in 32.7 g of carbon tetrachloride? 1 mol of carbon tetrachloride = 6.022×10^{23} molecules of carbon tetrachloride. The molar mass of carbon tetrachloride is 153.81 g (i.e. 1 mol of carbon tetrachloride = 153.81 g).

17. The force between two masses, m_1 and m_2 is calculated from the formula:
$$F = G \frac{m_1 \, m_2}{r^2},$$
where $G = 6.67 \times 10^{-11} \dfrac{m^3}{kg \, s^2}$, and r is the distance between the two masses. Use this formula to determine the force acting between the Earth and the Moon given the $m_E = 5.97 \times 10^{24} kg$, $m_M = 7{,}35 \times 10^{22} kg$ and the distance between them is $r = 3.848 \times 10^8$ m

10.3 BASIC STATISTICS

Statistics is all about data. What is data? Data is information about something you are interested in. It could be the heights of a group of people, or the color of people's hair, or the income of families, or the gender of doctors and lawyers, or the population density of countries, or the size of apples in an orchard, etc.

When it was first used, data meant information about a political state, and it was the rulers of these states that wanted to collect this data. They wanted to keep track of their people, their land, and the taxes that needed to be collected, among other things. This is probably why the word statistics started out as a derivative of the Latin word status. It was concerned with the status of the state. As time evolved, so did the word, as well as its meaning.

Statistics is a branch of mathematics that is comprised of a series of tools that enable us to analyze data to uncover patterns in what we observe.

By its very definition, statistics, only leads to possible conclusions or predictions. However, it is extremely important because most things cannot be determined deductively.

Types of Statistics

Statistics is the collection, display, and analysis of data. We can further classify statistics in one of two ways; Descriptive Statistics and Inferential Statistics.

Descriptive Statistics: Is involved with a process of organizing and describing data. It provides us with a snapshot of the status of what we are investigating. Organizing the data can be visual or numerical.

Inferential Statistics: Is involved with a process of using data to make some conjectures or assertions. It provides us with predictive capabilities when combined with the concept of probability which we will not talk about in this book.

In this book we focus our attention exclusively on descriptive statistics, and leave inferential statistics for another course. We begin our introduction into statistics by first defining the various terms associated with it.

Types of Data

Key to the definitions above, is what is referred to as data. Data comes in many different forms, however, there are two general classifications or types of data. One is qualitative data, and the other is quantitative data.

Qualitative Data: Data that describes general characteristics of an object, such as color, gender, shape, etc. This is also called **categorical data**.

Quantitative Data: Data that provides a numerical description of the object or event, such as the height in inches, the weight in pounds of kilograms, the time in seconds, the velocity in miles per hour, the frequency in times per day, or shots per game. This is also called **numerical data**.

Some data seem to fall in between these two categories. Data such as size classifications; small, medium, or large, or height classifications; high or low, or weight classifications; heavy, very heavy, light, and very light. If we can't attach a number to the data, then we will still classify it as qualitative data. To be quantitative, we need a number. However, not all numerical data is quantitative. For example, zip codes and social security numbers are numbers, but we do not treat them as quantitative data, since many of the analyses we do with quantitative data do not apply. To understand this better, we provide more details below. For now, quantitative data must be numerical, but if it is numerical, it may, or may not be quantitative.

Quantitative data can be divided further, as either continuous or discrete data.

Discrete data can only come in fixed increments, such as whole numbers for the number of basketball shots in a game, or pitches in a baseball game, or fixed fractions such as change when we are dealing with fractions of a dollar. They can be 1/2 for a 50-cent piece, 1/4 for a quarter, 1/10 for a dime, 1/20 for a nickel, and 1/100 for a penny. The smallest increment we can have in real money is 1/100th of a dollar. Discrete data has no "in–between" values.

Data on the other hand could also be continuous.

Continuous data is data that can be measured to any degree of accuracy you are capable of, such as time an event takes, or the distance traveled, of the length of something. It can take on any value in a "continuous" way. Continuous data always has another value in–between two data points, no matter how close they are to each other.

EXAMPLES: Identify the type of data in each.

1. A survey of the satisfaction level of customers; qualitative data
2. The time it takes to run a marathon; quantitative, and continuous data
3. The number of points scored in a baseball game; quantitative, and discrete
4. Which shampoos consumers buy on a regular basis; qualitative data
5. The yearly household salary; quantitative, and also considered continuous, but could be discrete if you had categories down to the penny. Typically if the number of data categories is very large (depends upon the number, but usually greater than 50) then we consider it to be continuous data, although it is often a judgment call.
6. The brand of shoes women at a social event are wearing; qualitative

In what follows our focus will be on quantitative (numerical) data. We leave the other aspects to another course and textbook.

Data Collection and Organization

As the definitions above indicate, collection of data is an integral part of statistics. In fact, it is so important that you can take a whole course on how to effectively and properly collect data. Obviously, this is only a basic introduction to the concepts of statistics, so here in our treatment of the subject we will not go to the level of detail of data collection that a professional would need to know. For those with an interest beyond a basic presentation of the subject, we suggest that you look for reference material on "Design of Experiments" to better understand the data collection process.

After we record our data, we then look to organize it. Let's illustrate the basic process with a few examples. Consider the first example below, which involves quantitative or numerical data.

Numerical Data

EXAMPLE 1:

Consider the survey below. This could be a survey of the age of individuals in a group of students in a class, similar to the one you are taking.

25	18	19	17	21	23	59	33
32	20	21	27	30	19	22	19
24	19	24	19	18	18	24	22
23	20	19	26	22	20	18	19

It is hard to see any patterns here. This is where proper organization can help. First we see the youngest age is 17, and the oldest is 59. However, some ages are missing and some ages are repeated. If we organize the data in the following way, it becomes more apparent what we are looking at.

Age	Tally Marks	Frequency
17	I	1
18	IIII	4
19	IIII II	7
20	III	3
21	II	2
22	III	3
23	II	2
24	III	3
25	I	1
26	I	1

Age	Tally Marks	Frequency
27	I	1
28		0
29		0
30	I	1
31		0
32	I	1
33	I	1
...		
59	I	1
Total		**32**

This table is called a frequency distribution table. It tells us how many people in this group are a specific age.

A **frequency distribution table** is a list of data values along with their corresponding counts or frequency.

An organized approach to obtaining this table is as follows. Every time we see a specific age in the original list of data, we simply cross off each occurrence of that age and add a tally mark in our table. We do this until all the data values are crossed off, and then sum the tally marks and write the equivalent whole number in the frequency column for each data value. As an added check, we typically count the total tally marks, and the total frequency number, and compare these numbers to the original data number, to make sure we that they all add to the same number and that we did this process correctly. In the example above, they must all equal the value of 32.

Sometimes, however, it is not convenient having only a single value for counting. Especially when all the data values are unique, that often occurs with data that comes from a continuous data variable. In these cases, we need to devise an alternative representation technique as shown in the example below.

EXAMPLE 2:
Consider the table of measurements made in a laboratory.

1.25	2.46	0.95	2.98	0.54	0.23	1.45	1.19
1.05	2.81	1.93	1.11	0.36	1.09	2.15	1.75
1.21	1.99	2.31	2.12	1.22	1.29	0.15	2.63
0.83	0.97	1.34	1.17	1.24	2.01	1.85	1.22

If we use all the individual measurements and try and tally the results we may only have one tally mark per measurement. Thus, we will need to divide the data in classes or bins of data. For this introduction, you can assume that this will be done for you. For this data we chose to create the following classes or bins to place our data in to construct the frequency distribution table.

Value (classes)	Tally Marks	Frequency
0.00-0.49	|||	3
0.50-0.99	||||	4
1.00-1.49	++++ ++++ |||	13
1.50-1.99	||||	4
2.00-2.49	++++	5
2.50-2.99	|||	3
Total	**32**	**32**

Data Visualization

The next step in the process is to take this table and create a visualization of this data. When we actually see a picture of the data, it becomes easier to see a pattern in it. One of the tools used to visualize these data is through something called a histogram.

Histogram

A histogram is a chart that uses adjacent rectangles to indicate how many data values are contained in a particular class. It provides a quick and easy way to "see" our data. We illustrate this chart using an example below.

EXAMPLE 3: Using the frequency distribution data from Example 1 above, but now broken down into the indicated classes, as was done in Example 2, construct a histogram of the data.

Age (Classes)	Tally Marks	Frequency
17-19	++++ ++++ ||	12
20-22	++++ |||	8
23-25	++++		6
26-28	||	2
29-31			1
32-34	||	2
35-37		0

Age (Classes)	Tally Marks	Frequency
38-40		0
41-43		0
44-46		0
47-49		0
50-52		0
53-55		0
56-58		0
59-61	|	1
Total	**32**	**32**

We start by plotting the data as rectangular bars that touch each other. The touching rectangles represents the fact that the data is related in a quantitative way, and we are showing the full and complete range of data values.

The histogram, just described, takes the following form:

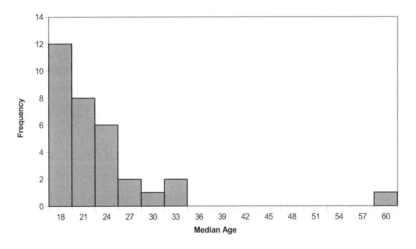

Notice that to create numbers along the horizontal axis of our chart, we picked the value in the middle of our bins. Thus, for an age range of 17 through 19, we chose to write the middle value of 18. This is called the **class mid-point**, but we don't need to focus on that too much here, as we will not be constructing histograms to a sophisticated level in this book, but rather we will mostly focus our attention on interpreting already constructed histograms and trying to ascertain what they tell us.

You can now see a pattern begin to emerge in the data! Most of the data falls to the left in our chart, with a steep drop off as we move to the right, or to the older students.

Similarly, the histogram for the frequency table in Example 2 from above would look like:

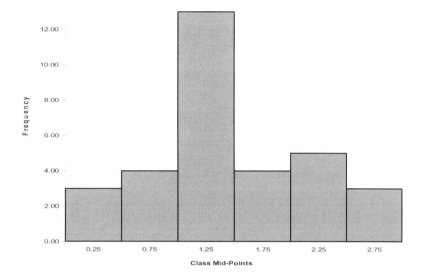

Also with appropriately chosen class mid-points. In this histogram, we see the data values are more populous in the middle of the chart, and drop off as we move to either the left or right.

In both of these examples we see a unique visual pattern in how the data are related. This starts to show us the potential power of statistics.

Data Quantification

In the previous section we showed how to visualize data. We will now show how to turn raw data into some useful numbers. If the data is representative of some pattern, we may be able to see or understand this pattern if we can quantify it. The visual representations of the data give us a hint of this. Consider the following histograms of these four different data sets shown below. Looking at the shapes of the histograms you can see a pattern. Some of the data sets seem to peak at a certain value, and drop off as we get further from that value. They seem to be tending to take on some average or central value. Furthermore, we can see that how rapidly the histograms decline or spread out from this central value also varies.

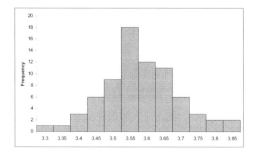 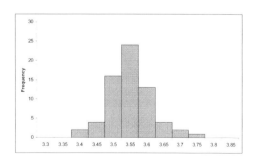

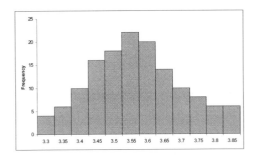 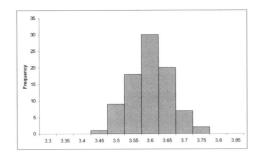

These are features we would like to make more precise. In particular, we would like to quantify this average or central tendency value, as well as the degree of spread or variation of the data. We begin with measures of central tendency.

Measures of Central Tendency

In statistics we have three ways to measure the central tendency of a data set. They are the mean, the median, and the mode.

> The **mean** is the arithmetic average of the data values. The average is obtained by adding all the data values together and then dividing by the total number of data values in the set.

or

$$\text{Mean} = \frac{\text{Sum of all data values}}{\text{Total number of data values}}$$

In mathematics we like to use symbols to abbreviate a process. The above formula is no exception. The first is to use a letter to represent a data value. For us we will use the standard variable, x. To signify that we are talking about different data values we will use a subscript of 1, 2, 3, 4, ... if we are talking about a specific data value or we shall use an "i" meaning any generic or non-specific data value. Thus a generic data value could be written as x_i (in the example below we'll illustrate this more precisely).

Next we would like to use a symbol that indicates we want to sum a list of numbers. For this we use the capital Greek letter Sigma (starts with s for sum), Σ. The number of data values will be indicated by the letter N, and finally, the mean will be indicted by the lower-case Greek letter mu, μ, the letter m in Greek for mean value. Using this convention the above formula can be rewritten as:

$$\mu = \frac{\sum x_i}{N}$$

This formula simply means that we are adding all the data values together (given the generic data value x_i), and then dividing by the total number of values. It is the average value of all the data.

EXAMPLE: Again, consider the data set of measurements used in a previous example:

1.25	2.46	0.95	2.98	0.54	0.23	1.45	1.19
1.05	2.81	1.93	1.11	0.36	1.09	2.15	1.75
1.21	1.99	2.31	2.12	1.22	1.29	0.15	2.63
0.83	0.97	1.34	1.17	1.24	2.01	1.85	1.22

First there are 32 data values, so N = 32. Next the x_i are the individual data values themselves, or x_1 = 1.25, x_2 = 2.46, x_3 = 0.95, etc.

Using the formula from above, we have:

$$\mu = \frac{\sum x_i}{N} = \frac{1.25+2.46+0.95+2.98+0.54+0.23+...+1.17+1.24+2.01+1.85+1.22}{32} \approx 1.46$$

Thus, we would say the mean of this data set, μ = 1.46.

Another measure of the central tendency value is called the median.

> The **median** is the middle value of the data set. If the set contains an odd number of data points, we will always have one data point in the middle. If, however, it contains an even number, then there are two values in the middle. Thus, we simply average these two values to get our median value for the data set.

In the example below we have 32 data points, thus we will have to take the average of the two "middle" values.

1.25	2.46	0.95	2.98	0.54	0.23	1.45	1.19
1.05	2.81	1.93	1.11	0.36	1.09	2.15	1.75
1.21	1.99	2.31	2.12	1.22	1.29	0.15	2.63
0.83	0.97	1.34	1.17	1.24	2.01	1.85	1.22

Before we can find the middle value we need to sort the data set from lowest to highest or highest to lowest.

0.15	0.23	0.36	0.54	0.83	0.95	0.97	1.05
1.09	1.11	1.17	1.19	1.21	1.22	1.22	1.24
1.25	1.29	1.34	1.45	1.75	1.85	1.93	1.99
2.01	2.12	2.15	2.31	2.46	2.63	2.81	2.98

In this sorted list of the above data we have highlighted the two middle values, so the median is simply their average or:

$$\text{Median} = \frac{1.24 + 1.25}{2} = 1.245$$

Which we could round up to 1.25.

The final measure of central tendency is called the mode.

> The **mode** is the data value that occurs most frequently in the data set. If two values occur the same number of times and they are the largest number of occurrences, then there is no mode. If all data values are unique, then there also is no mode for the data set.

In our data set above, we see that the value 1.22 occurs twice, and no other data values occurs more than once, so this is our mode. We should note that the above definition is for data that is called single–mode data. In a more advanced presentation than we are giving here, we could extend this definition to include something called multi-modal data, but for now that is beyond the scope of what we want to discuss. You should, however, be aware that this definition may change in a follow–up course on statistics.

EXAMPLES: Find the mean, median and mode of the following data sets.

  1.

3.2	2.9	4.7	3.6	3.7	3.7	3.4	4.1
3.7	3.9	3.0	3.6	3.7	4.4	3.8	4.0

<u>Mean</u>
There are 16 values in our data set, so to find the mean we simply add these values together and divide by 16.

$$\mu = \frac{\sum x_i}{N} = \frac{3.2 + 2.9 + 4.7 + 3.6 + 3.7 + 3.7 + \ldots + 3.6 + 3.7 + 4.4 + 3.8 + 4.0}{16}$$

$$= \frac{59.4}{16} = 3.7125$$

Which we can round to $\mu = 3.7$

<u>Median</u>
Order the data and identify the middle value. Since there is an even number of data points, we take the average of the two middle numbers, underlined below:

2.9, 3.0, 3.2, 3.4, 3.6, 3.6, 3.7, <u>3.7</u>, <u>3.7</u>, 3.7, 3.8, 3.9, 4.0, 4.1, 4.4, 4.7

$$\text{Median} = \frac{3.7 + 3.7}{2} = 3.7$$

Mode

Again, using the ordered data, we can easily see that 3.7 occurs 4 times, which is more than any other data value, so

Mode = 3.7

2. {17, 11, 15, 13, 14, 12, 13, 11, 16, 13, 10}

Mean

There are 11 values in our data set, so to find the mean we simply add these values together and divide by 11.

$$\mu = \frac{\sum x_i}{N} = \frac{17+11+15+13+14+12+13+11+16+13+10}{11}$$

$$= \frac{145}{11} = 13.\overline{18}$$

Which we can round to $\mu = 13$

Median

Order the data and identify the middle value underlined below:

10, 11, 11, 12, 13, <u>13</u>, 13, 14, 15, 16, 17

Median = 13

Mode

Again, using the ordered data, we can easily see that 13 occurs 3 times, which is more than any other data value, so

Mode = 13

3.

8.3	7.5	7.0	6.9	6.9	8.3
7.4	7.6	7.7	7.8	7.1	

Mean

There are 11 values in our data set, so to find the mean we simply add these values together and divide by 11.

$$\mu = \frac{\sum x_i}{N} = \frac{8.3+7.5+7.0+6.9+6.9+8.3+7.4+7.6+7.7+7.8+7.1}{11}$$

$$= \frac{82.5}{11} = 7.5$$

<u>Median</u>
Order the data and identify the middle value underlined below:

6.9, 6.9, 7.0, 7.1, 7.4, <u>7.5</u>, 7.6, 7.7, 7.8, 8.3, 8.3

Median = 7.5

<u>Mode</u>
Again, using the ordered data, we see that both 6.9 and 8.3 occur twice, so there is NO Mode.

<u>Measures of Spread (Dispersion)</u>

Just as we have three ways to measure the central value of a data set, we also have three ways to measure the spread of a data set. Two of the measures are sort of intuitive, while the third is not, but is a very useful quantity throughout other aspects of statistics. The three measures of spread or dispersion are the range, the variance, and the standard deviation.

We will introduce and define these measures through an example. Consider the data set we first introduced above:

1.25	2.46	0.95	2.98	0.54	0.23	1.45	1.19
1.05	2.81	1.93	1.11	0.36	1.09	2.15	1.75
1.21	1.99	2.31	2.12	1.22	1.29	0.15	2.63
0.83	0.97	1.34	1.17	1.24	2.01	1.85	1.22

We will now define and calculate three different measures of the spread of a data set – the range, the variance, and the standard deviation.

> The **range** of the data set is the difference between the largest and smallest values in the data set.

$$\text{Range } = \text{ max data value - min data value}$$

EXAMPLE: The minimum value of the data set above is 0.15, and the maximum value is 2.98. Thus, the
$$\text{Range} = 2.98 - 0.15 = 2.83$$

The variance, on the other hand, is quite a bit more involved. It defines the spread in terms of how far a data point is away from the mean, as an average of the square of each of these distances together, to get an average distance the entire data set differs from the mean, i.e.

> The **variance** measures the average of the square of the distance away from the mean of the data set of all the values in the data set. NOTE: The reason we use the distance squared is so that distances on different sides of the mean do not cancel out (all the values become positive numbers). It is identified with the Greek letter lower-case sigma squared, σ^2. The square is to indicate the square of the distance.

In mathematical notation this becomes:

$$\sigma^2 = \frac{\sum (x_i - \mu)^2}{N}$$

At first glance this formula seems very complicated, however, it is just a short-hand notation for the process illustrated in the example below.

EXAMPLE: In this example we will consider only part of the data set above (this will reduce the number of calculations, and make the example a little easier to follow.)

1.25	2.46	0.95	2.98	0.54	0.23	1.45	1.19
1.05	2.81	1.93	1.11	0.36	1.09	2.15	1.75

We first need to find the mean. For this data set the mean, $\mu \approx 1.46$, and N = 16. We now construct a table where we perform the calculations in the formula:

x_i	$x_i - \mu$	$(x_i - \mu)^2$
1.25	$1.25 - 1.46 = -0.21$	$(-0.21)^2 = 0.0441$
2.46	$2.46 - 1.46 = 1.00$	$(1.00)^2 = 1$
0.95	$0.95 - 1.46 = -0.51$	$(-0.51)^2 = 0.2601$
2.98	$2.98 - 1.46 = 1.52$	$(1.52)^2 = 2.3104$
0.54	$0.54 - 1.46 = -0.92$	$(-0.92)^2 = 0.8464$
0.23	$0.23 - 1.46 = -1.23$	$(-1.23)^2 = 1.5129$
1.45	$1.45 - 1.46 = -0.01$	$(-0.01)^2 = 0.0001$
1.19	$1.19 - 1.46 = -0.27$	$(-0.27)^2 = 0.0729$
1.05	$1.05 - 1.46 = -0.41$	$(-0.41)^2 = 0.1681$
2.81	$2.81 - 1.46 = 1.35$	$(1.35)^2 = 1.8225$
1.93	$1.93 - 1.46 = 0.47$	$(0.47)^2 = 0.2209$
1.11	$1.11 - 1.46 = -0.35$	$(-0.35)^2 = 0.1225$
0.36	$0.36 - 1.46 = -1.10$	$(-1.10)^2 = 1.21$

x_i	$x_i - \mu$	$(x_i - \mu)^2$
1.09	$1.09 - 1.46 = -0.37$	$(-0.37)^2 = 0.1369$
2.15	$2.15 - 1.46 = 0.69$	$(0.69)^2 = 0.4761$
1.75	$1.75 - 1.46 = 0.29$	$(0.29)^2 = 0.0841$
		$\sum (x_i - \mu)^2 = 10.288$
		$\dfrac{\sum (x_i - \mu)^2}{N} = \dfrac{10.288}{16} = 0.643$

The first column contains the 16 data values. In the second column we subtract the mean equal to 1.46 from each of these values. In the last column we square the values in column two. Finally, at the end of column three we first add all the squared terms in column three together to get 10.288, and then we divide this number by 16, our total number of data values, to obtain:

$$\sigma^2 = 0.643$$

This number is called the variance.

The last measure of spread is the standard deviation. The standard deviation tends to be more useful and intuitive than the variance. It is simply the square root of the variance. Why is it more intuitive? The variance is the average of the distances the data values are away from the mean *squared*, but for the standard deviation it is more like the average distance from the mean of the data values in the data set.

> The **standard deviation** is the average distance from the mean of the data values in the data set, and is obtained by taking the square root of the variance.

In mathematical notation this becomes:

$$\sigma = \sqrt{\frac{\sum (x_i - \mu)^2}{N}}$$

The standard deviation is measured in the same units as the data set, whereas the variance is measured in units squared. This is another reason why it is a more intuitive measure of spread than the variance.

EXAMPLE: The standard deviation of the example from above is just the square root of the variance, i.e.

$$\sigma = \sqrt{0.643} \approx 0.802$$

EXAMPLES: Find the range, variance and standard deviation of the following data sets.

1.

3.2	2.9	4.7	3.6	3.7	3.7	3.4	4.1
3.7	3.9	3.0	3.6	3.7	4.4	3.8	4.0

<u>Range</u>
Order the data set, and find the largest and smallest value, underline below

<u>2.9</u>, 3.0, 3.2, 3.4, 3.6, 3.6, 3.7, 3.7, 3.7, 3.7, 3.8, 3.9, 4.0, 4.1, 4.4, <u>4.7</u>

$$\text{Range} = 4.7 - 2.9 = 1.8$$

<u>Variance</u>
First find the mean, which we did earlier in EXAMPLE 1 in the previous section as
$$\mu = 3.7$$
We should note that the exact value is 3.7125, and we should use this value whenever possible, however, as we'll see the answer is the same rounded to two places after the decimal. We will use the value of 3.7125 in our example, but you can use the rounded value 3.7

We now compute the variance, $\sigma^2 = \dfrac{\sum (x_i - \mu)^2}{N}$

Using a table of values we have,

x_i	$x_i - \mu$	$(x_i - \mu)^2$
3.2	$3.2 - 3.7125 = -0.5125$	$(-0.5125)^2 = 0.2627$
2.9	$2.9 - 3.7125 = -0.8125$	$(-0.8125)^2 = 0.6602$
4.7	$4.7 - 3.7125 = 0.9875$	$(0.9875)^2 = 0.9752$
3.6	$3.6 - 3.7125 = -0.1125$	$(-0.1125)^2 = 0.0127$
3.7	$3.7 - 3.7125 = -0.0125$	$(-0.0125)^2 = 0.0002$
3.7	$3.7 - 3.7125 = -0.0125$	$(-0.0125)^2 = 0.0002$
3.4	$3.4 - 3.7125 = -0.3125$	$(-0.3125)^2 = 0.0977$
4.1	$4.1 - 3.7125 = 0.3875$	$(0.3875)^2 = 0.1502$
3.7	$3.7 - 3.7125 = -0.0125$	$(-0.0125)^2 = 0.0002$
3.9	$3.9 - 3.7125 = -0.1875$	$(-0.1875)^2 = 0.0352$

x_i	$x_i - \mu$	$(x_i - \mu)^2$
3.0	$3.0 - 3.7125 = -0.7125$	$(-0.7125)^2 = 0.5077$
3.6	$3.6 - 3.7125 = -0.1125$	$(-0.1125)^2 = 0.0127$
3.7	$3.7 - 3.7125 = -0.0125$	$(-0.0125)^2 = 0.0002$
4.4	$4.4 - 3.7125 = 0.6875$	$(0.6875)^2 = 0.4727$
3.8	$3.8 - 3.7125 = 0.0875$	$(0.0875)^2 = 0.0077$
4.0	$4.0 - 3.7125 = 0.2875$	$(0.2875)^2 = 0.0827$
		$\sum (x_i - \mu)^2 = 3.2782$
		$\dfrac{\sum (x_i - \mu)^2}{N} = \dfrac{3.2782}{16} \approx 0.205$

The variance rounded to two decimal places is $\sigma^2 \approx 0.21$

Standard deviation

To find the standard deviation we simply take the square root of the variance.

$$\sigma = \sqrt{0.21} \approx 0.46$$

2. $\{17, 11, 15, 13, 14, 12, 13, 11, 16, 13, 10\}$

Range

Order the data set, and find the largest and smallest value, underlined below

10, 11, 11, 12, 13, 13, 13, 14, 15, 16, 17

$$\text{Range} = 17 - 10 = 7$$

Variance

First find the mean, which we did earlier in EXAMPLE 2 in the previous section as
$$\mu = 13.1818 \ldots$$

We will round this to 13.18

Now compute the variance $\sigma^2 = \dfrac{\sum (x_i - \mu)^2}{N}$

Using a table of values we have,

x_i	$x_i - \mu$	$(x_i - \mu)^2$
17	$17 - 13.18 = 3.82$	$(3.82)^2 = 14.59$
11	$11 - 13.18 = -2.18$	$(-2.18)^2 = 4.75$
15	$15 - 13.18 = 1.82$	$(1.82)^2 = 3.31$

x_i	$x_i - \mu$	$(x_i - \mu)^2$
13	$13 - 13.18 = -0.18$	$(-0.18)^2 = 0.03$
14	$14 - 13.18 = 0.82$	$(0.82)^2 = 0.67$
12	$12 - 13.18 = -1.18$	$(-1.18)^2 = 1.39$
13	$13 - 13.18 = -0.18$	$(-0.18)^2 = 0.03$
11	$11 - 13.18 = -2.18$	$(-2.18)^2 = 4.75$
16	$16 - 13.18 = 2.82$	$(2.82)^2 = 7.95$
13	$13 - 13.18 = -0.18$	$(-0.18)^2 = 0.03$
10	$10 - 13.18 = -3.18$	$(-3.18)^2 = 10.11$
		$\sum (x_i - \mu)^2 = 47.61$
		$\dfrac{\sum (x_i - \mu)^2}{N} = \dfrac{47.61}{11} \approx 4.33$

The variance $\sigma^2 \approx 4.33$

Standard deviation
To find the standard deviation we take the square root of the variance.

$\sigma = \sqrt{4.33} \approx 2.1$

3.

8.3	7.5	7.0	6.9	6.9	8.3
7.4	7.6	7.7	7.8	7.1	

Range
Order the data set, and find the largest and smallest value, underline below

6.9, 6.9, 7.0, 7.1, 7.4, 7.5, 7.6, 7.7, 7.8, 8.3, 8.3

$$\text{Range} = 8.3 - 6.9 = 1.4$$

Variance
First, find the mean, which we did earlier in EXAMPLE 3 in the previous section as
$$\mu = 7.5$$

Now compute the variance $\sigma^2 = \dfrac{\sum (x_i - \mu)^2}{N}$

Using a table of values we have,

x_i	$x_i - \mu$	$(x_i - \mu)^2$
8.3	$8.3 - 7.5 = 0.8$	$(0.8)^2 = 0.64$
7.5	$7.5 - 7.5 = 0$	$(0)^2 = 0$
7.0	$7.0 - 7.5 = -0.5$	$(-0.5)^2 = 0.25$
6.9	$6.9 - 7.5 = -0.6$	$(-0.6)^2 = 0.36$
6.9	$6.9 - 7.5 = -0.6$	$(-0.6)^2 = 0.36$
8.3	$8.3 - 7.5 = 0.8$	$(0.8)^2 = 0.64$
7.4	$7.4 - 7.5 = -0.1$	$(-0.1)^2 = 0.01$
7.6	$7.6 - 7.5 = 0.1$	$(0.1)^2 = 0.01$
7.7	$7.7 - 7.5 = 0.2$	$(0.2)^2 = 0.04$
7.8	$7.8 - 7.5 = 0.3$	$(0.3)^2 = 0.09$
7.1	$7.1 - 7.5 = -0.4$	$(-0.4)^2 = 0.16$
		$\sum (x_i - \mu)^2 = 2.56$
		$\dfrac{\sum (x_i - \mu)^2}{N} = \dfrac{2.56}{11} \approx 0.23$

The variance $\sigma^2 \approx 0.23$

Standard deviation
To find the standard deviation we take the square root of the variance.

$\sigma = \sqrt{0.23} \approx 0.48$

EXERCISES 10.3

Determine which data sets are qualitative and which are quantitative. If quantitative tell whether the data is discrete or continuous.

1. The win and loss record of a sports team.
2. The arrival time of airplanes at an airport.
3. Types of cars for sale at an automotive dealership.
4. The average daily temperature in a particular city.
5. The cost for a gallon of gas at different cities throughout the country.

6. The political party affiliation of the members of a social club.

7. The number of points scored by a team in a football game.

8. Types of hits in a baseball game.

9. Types of pitches a pitcher can throw in a baseball game.

10. The winnings and losses of a gambler at a casino.

11. The list of new movies for the summer.

12. The box office earnings of films.

13. The number of spectators at a golf tournament.

14. The possible choices of hair color in a beauty salon.

Draw a Histogram of the data sets

15. The number of hits a baseball player had for the last 12 games, is as follows (use 0, 1, 2, 3 as the categories along the bottom of the histogram.)

1	0	3	0	1	1
2	1	0	0	3	1

16. The frequency distribution table of measurements in a laboratory.

Value (classes)	Frequency
0.00-0.49	2
0.50-0.99	7
1.00-1.49	10
1.50-1.99	17
2.00-2.49	9
2.50-2.99	1
Total	46

Find the Mean, Median, Mode, Range, Variance, and Standard Deviation of the following data sets:

17. {21, 18, 20, 20, 19, 24, 18, 17, 20, 21, 20, 16, 23}

18. {2.5, 2.9, 3.2, 2.4, 2.6, 2.7, 2.6, 3.0}

19. {113, 229, 186, 187, 192, 156, 203, 186, 187, 175}

20. {61, 64, 58, 57, 57, 72, 84, 65, 65, 87, 59, 65, 72}

21. The average August rainfall in inches, at a city in the state of Texas was recorded for the last 28 years as follows:

1.7	2.2	4.9	3.1	1.8	4.6	2.7
3.4	1.5	1.3	3.0	2.7	1.9	1.8
2.9	0.7	1.3	1.1	2.8	1.3	1.7
2.0	1.9	0.3	0.2	0.4	0.3	0.0

22. A quality control worker at a factory has recorded the following list of monthly defects for the year:

325	768	543	186	449	486
298	463	389	416	587	201

23. The number of hits a baseball player had for the last 12 games, is as follows:

1	0	3	0	1	1
2	1	0	0	3	1

10.4 SCIENTIFIC FORMULAS

The real utility of Algebra is in its application to mathematical models that can be used to calculate and predict specific outcomes. These models can come in the form of formulas. Formulas are equations in two or more variables that allow us to compute the relationships between these variables. Formulas can be classified in three different ways.

The first group contains formulas used to transform from one unit of measure to another, such as from Fahrenheit to Celsius temperature scales, or from English to Metric scales. Another type of formula is used to compute exact values of quantities to be used, such as the area and volume formulas in mathematics, as well as the average of a set of measurements, or the voltage or power given a current and a resistance. The final classification of formulas, are those that are derived from approximate laws of nature, which may change in the future based upon new discoveries. These are the models of the phenomena under consideration. These latter type of formulas are what we strive for, but are beyond the scope of this book. They are, however, the real goal that we will pursue in later mathematics courses. However, before we attempt to work with these formulas we need to get some basic skills under our belts.

In this chapter we will learn how to work with basic formulas. In particular, how to compute specific values produced by these formulas, or how to rewrite these formulas to determine other values.

EVALUATING FORMULAS

To evaluate a formula you need to do two important things. You must first be able to read and distinguish what values are called out in the problem, and what result you need to calculate. The next step is to be able to substitute these specified values into the formula and perform the mathematical calculations necessary to obtain the value of the desired variable.

The best way to learn this technique is to see lots of different formulas being evaluated. We learn by seeing and then committing the various solution approaches to memory and recalling them for when we see them again. We will start with some relatively simple examples, and then get progressively more complicated, and also introduce some new mathematical notations along the way.

EXAMPLE 1: Given the formula for the Celsius to Fahrenheit conversion:

$$^0F = 1.8\,^0C + 32 \text{, or}$$

$$^0F = \frac{9}{5}\,^0C + 32 \text{,}$$

you are asked to find the Fahrenheit temperature associated with the following Celsius temperatures: $22\,^0C$, $-30\,^0C$, $0\,^0C$ and $100\,^0C$. Since there are only two variables and the value of one is given, it is relatively easy. We simply replace the ^0C temperature with the desired values and calculate the ^0F temperature, e.g.:

For $^0C = 22$

$$^0F = \frac{9}{5}(22) + 32 = \frac{198}{5} + 32 = 39.6 + 32 = 71.6\,^0F$$

For $^0C = -30$

$$^0F = \frac{9}{5}(-30) + 32 = -54 + 32 = -22\,^0F$$

For $^0C = 0$

$$^0F = \frac{9}{5}(0) + 32 = 32\,^0F \text{, the freezing point of water at sea level.}$$

For $^0C = 100$

$$^0F = \frac{9}{5}(100) + 32 = 180 + 32 = 212\,^0F \text{, the boiling point of water at sea level.}$$

EXAMPLE 2: Given the formula for the volume of a cylinder in a car engine:

$$V = \pi r^2 h \text{,}$$

where r is the radius of the cylinder and h is its height, find the volume of a cylinder with a radius of 1.50 inches and a height of 3.50 inches.

$$V = \pi r^2 h$$
$$= \pi (1.50 \text{ in})^2 (3.50 \text{ in})$$
$$= \pi (2.25 \text{ in}^2)(3.50 \text{ in})$$
$$= \pi (7.875)(\text{in}^3)$$
$$= 7.875\pi \text{ in}^3$$
$$\approx 7.875(3.141592)\text{in}^3$$
$$\approx 24.74 \text{ in}^3$$

EXAMPLE 3: Given the formula for computing the final amount on an investment, invested using compounded interest:

$$A = A_0 \left(1 + \frac{r}{n}\right)^{nt},$$

Here A is the final amount the investment is worth, A_0 is the initial value of the investment, r is the interest rate written in decimal form, n is the number of times the investment is **compounded** per year, and t is the number of years (time in years) the investment is invested for.

Find the value of the investment after an initial investment of $25,000 is invested at 8.5%, compounded monthly for 15 years.

First we must determine all the values of the variables we need to evaluate the formula with. From what is provided above we see that:

$A_0 = \$25,000$, r = 8.75% = 0.0875, n = 12 (12 months per year means 12 compoundings per year) and t = 15 (for 15 years.) Substituting for these variables into the formula we obtain:

$$A = A_0 \left(1 + \frac{r}{n}\right)^{nt}$$
$$= \$25,000 \left(1 + \frac{0.0875}{12}\right)^{(12)(15)}$$
$$= \$25,000 (1 + 0.007291\overline{6})^{180}$$
$$= \$25,000 (1.007291\overline{6})^{180}$$
$$= \$25,000 (3.697799742 \ldots)$$
$$= \$92,444.99$$

EXAMPLE 4: In statistics there is a formula for computing the average of a set of data called the mean. The formula is quite complicated looking, but once you understand the language it is written in, it is not that difficult. Here is the formula:

$$\overline{x} = \frac{\sum_i x_i}{n}$$

First, you read this formula as x-bar is equal to the sum of the n distinct x sub i data values, then you divide by the total number of data values, n. The capital Greek letter Σ stands for summation, and simply means that we will add a series of numbers together. In what follows we show how to interpret and work with this abstract notation.

Here is the data: $n = 5$, $x_1 = 2.5$, $x_2 = 2.0$, $x_3 = 2.75$, $x_4 = 2.95$, $x_5 = 3.2$, and we evaluate the formula as follows:

$$\begin{aligned}
\overline{x} &= \frac{\sum_i x_i}{n} \\
&= \frac{\left(x_1 + x_2 + x_3 + x_4 + x_5\right)}{n} \\
&= \frac{\left(2.5 + 2.0 + 2.75 + 2.95 + 3.2\right)}{5} \\
&= \frac{\left(13.4\right)}{5} \\
&= 2.68
\end{aligned}$$

EXAMPLE 5: Another complicated formula from statistics calculates the measure of how a data set varies. This is called the standard deviation and is defined with the lower case Greek letter sigma, σ, and is given by the formula:

$$\sigma = \sqrt{\frac{\Sigma \left(x_i - \mu\right)^2}{N}}$$

If we use the same data from Example 4 above, except in this case we'll substitute $\overline{x}$ for μ:

$N=5$, $x_1=2.5$, $x_2=2.0$. $x_3=2.75$, $x_4=2.95$, $x_5=3.2$, as well as the result, $\mu=2.68$. Then we evaluate this formula as follows:

$$\sigma = \sqrt{\frac{\Sigma(x_i - \mu)^2}{N}}$$

$$= \sqrt{\frac{\left[(x_1 - \mu)^2 + (x_2 - \mu)^2 + (x_3 - \mu)^2 + (x_4 - \mu)^2 + (x_5 - \mu)^2\right]}{N}}$$

$$= \sqrt{\frac{\left[(2.5 - 2.68)^2 + (2.0 - 2.68)^2 + (2.75 - 2.68)^2 + (2.95 - 2.68)^2 + (3.2 - 2.68)^2\right]}{5}}$$

$$= \sqrt{\frac{\left[(-0.18)^2 + (-0.68)^2 + (0.07)^2 + (0.27)^2 + (0.52)^2\right]}{5}}$$

$$= \sqrt{\frac{\left[0.0324 + 0.4624 + 0.0049 + 0.0729 + 0.2704\right]}{5}}$$

$$= \sqrt{\frac{0.843}{5}}$$

$$= \sqrt{0.1686}$$

$$= 0.41$$

EXAMPLE 6: This example we show is also from statistics. Again, it is very complex looking, but we will try and break it down and simplify it as best we can. It is a formula used to compute something called a probability. Probability is related to the chances that some event will occur. At this time we don't need to understand the mathematics that defines this formula, we just want to learn how to evaluate it.

$$P = \frac{n!}{(n-x)!x!} p^x q^{n-x}$$

In this formula we are introduced to another new mathematical object called a factorial, the number followed by an exclamation point, !, which we'll describe in further detail below.

We will evaluate this formula for $n = 6$, $x = 4$, $p = 0.6$, $q = 0.4$. Substituting into the formula yields:

$$P = \frac{n!}{(n-x)!x!} p^x q^{n-x}$$

$$= \frac{6!}{(6-4)!4!} (0.6)^4 [0.4]^{6-4}$$

$$= \frac{6!}{2!4!} (0.6)^4 [0.4]^2$$

We are now faced with a new mathematical entity. What is 6! (six factorial), or 2! (two factorial), or 4! (four factorial). The answer is quite simple. This is just an abbreviation for taking the number in front of the exclamation point and multiplying by that number, times a number one less than that number, times another number one less than the previous number, and continuing to subtract and multiply until we reach the number one. For example:

$$6! = (6)(5)(4)(3)(2)(1)$$
$$= 720$$

$$4! = (4)(3)(2)(1)$$
$$= 24$$

$$2! = (2)(1)$$
$$= 2$$

We can now use these values to complete the calculation started above:

$$P = \frac{6!}{2!4!}(0.6)^4[0.4]^2$$
$$= \frac{720}{(2)(24)}(0.6)^4[0.4]^2$$
$$= 0.31104$$

As we have shown, formulas can take on many forms, and are frequently necessary every time we wish to obtain a numerical description of phenomena of interest. The more we practice and become competent at this, the easier it becomes.

Other important exponent notations that you should be familiar with (common in Astronomy).

1. $r^{-1} = \dfrac{1}{r}$ 3. $r^{-2} = \dfrac{1}{r^2}$ 4. $D = \dfrac{1}{p} = p^{-1}$ 6. $M^{5/2} = M^{2.5} = M^2\sqrt{M}$

2. $s^{-1} = \dfrac{1}{s}$ 5. $M^{1/2} = M^{0.5} = \sqrt{M}$

EXAMPLES: These examples highlight some of the notations above.

1. The distance a star is away from us is given by the formula, $D = \dfrac{1}{p}$, where D is the distance to the star measured in parsecs (one parsec is about 3.26 light years), and p is the parallax angle measured in arc-seconds (1 arc-second ≈ 0.00028 degrees). Using the above notation this

formula can also be written as, $D = p^{-1} = \dfrac{1}{p}$. Find D for $\rho = 0.015$ arc-seconds

Substitute into the formula to obtain:

$$D = (0.015 \text{ arc}-\text{seconds})^{-1} = \frac{1}{0.015 \text{ arc}-\text{seconds}} = 66.\overline{6} \text{ parsecs} \approx 67 \text{ parsecs}$$

2. In Astronomy the time a star spends burning hydrogen, versus any other element, is called its Main Sequence (MS) time. For our sun its MS time is about 10 billion years. For other stars, their MS time is found using the formula:

$$\tau \approx 1 \times 10^{10} \frac{1}{M_{\odot}^{2.5}} \text{ years}$$, where τ is the MS time measured in years, and $M_{\odot}$ is the mass of the star relative to our sun's mass.

Find τ for $M_{\odot} = 0.9$

Substitute into the formula to obtain:

$$\tau \approx 1 \times 10^{10} \frac{1}{M_{\odot}^{2.5}} \text{ years}$$

$$\approx 1 \times 10^{10} \frac{1}{0.9^{2.5}} \text{ years}$$

$$\approx 1.3 \times 10^{10} \text{ years}$$

Or roughly 13 billion years.

3. The radius of a star relative to our sun's radius of 695,700 km, is given by the formula:
$$R \approx R_{\odot} \times [M_{\odot}]^{0.5} \text{ km},$$
where R is the radius measured as a proportion of the solar radius, $R_{\odot} = 675,700 \text{ km}$, and $M_{\odot}$ is the proportion of how many solar masses the star is relative to our sun's mass.

Find the radius, R, of a star with a mass 5 times our solar mass.

Substitute into the formula to obtain:
$$R \approx 675,700 \times [M_{\odot}]^{0.5} \text{ km}$$

$$\approx 675,700 \times [5]^{0.5} \text{ km}$$

$$\approx 675,700 \times [2.2360...] \text{ km}$$

$$\approx 1,510,911 \text{ km}$$

Or roughly 1.5 million km.

SOLVING FOR OTHER VARIABLES

In all the problems we evaluated above, the formulas were solved for the variable whose value we were trying to determine. This, however, will not always be the case. Frequently we will be called upon to rewrite a formula so that a different variable is solved for. The approach is very similar to what was done in Chapter 8 when we were asked to rewrite the equation for a line from the standard form into the slope-intercept form, so we had to solve for the y variable.

Again, let's look at a few examples to illustrate this process.

EXAMPLE 1: Consider the Celsius to Fahrenheit conversion formula:

$$^0\text{F} = \frac{9}{5}\,^0\text{C} + 32$$

Suppose we want to now solve this formula for ^0C instead of ^0F.

We do this as follows:

$$^0\text{F} = \frac{9}{5}\,^0\text{C} + 32 \qquad \text{subtract 32 from both sides of the equation}$$

$$^0\text{F} - 32 = \frac{9}{5}\,^0\text{C} + 32 - 32$$

$$^0\text{F} - 32 = \frac{9}{5}\,^0\text{C} \qquad \text{now multiply by 5 and divide by 9 (or multiply by } \frac{5}{9}\text{)}$$

$$\frac{5}{9}\left(^0\text{F} - 32\right) = \frac{\cancel{5}}{\cancel{9}}\frac{\cancel{9}}{\cancel{5}}\,^0\text{C}$$

$$\frac{5}{9}\left(^0\text{F} - 32\right) = \,^0\text{C} \qquad \text{or rewrite as}$$

$$^0\text{C} = \frac{5}{9}\left(^0\text{F} - 32\right)$$

Thus, our new formula is: $^0\text{C} = \frac{5}{9}\left(^0\text{F} - 32\right)$

EXAMPLE 2: Consider the area of a triangle formula: $A_T = \frac{1}{2}\,b\cdot h$

Suppose we want to now solve this formula for h instead of A_T.

We do this as follows:

$$A_T = \frac{1}{2}\, b \cdot h \qquad \text{Multiply by 2}$$

$$(2)\, A_T = (2)\frac{1}{2}\, b \cdot h$$

$$2\, A_T = b \cdot h \qquad \text{Divide by b}$$

$$\frac{2A_T}{b} = \frac{b \cdot h}{b}$$

$$\frac{2A_T}{b} = h$$

or

$$h = \frac{2A_T}{b}$$

EXAMPLE 3: Consider the Ideal Gas formula: $PV = nRT$

Suppose we want to now solve this formula for T instead of PV.

We do this as follows:

$$PV = nRT \qquad \text{divide by nR}$$

$$\frac{PV}{nR} = \frac{nRT}{nR}$$

$$\frac{PV}{nR} = T \qquad \text{or}$$

$$T = \frac{PV}{nR}$$

This is probably one of the more difficult procedures to master for students at this level. It requires a high level of abstraction, and the only way to be successful is to work numerous problems to train your brain how to react to all the different possibilities.

EXERCISES 10.4

Evaluating Formulas
Evaluate the formulas for the given values of the variable/variables

1. $°F = 1.8\, °C + 32$, $°C = -5$

2. $A = A_0\left(1 + \dfrac{r}{n}\right)^{nt}$, $A_0 = \$10{,}000$, $r = 0.075$, $n = 4$ compoundings / year, $t = 25$ yrs

3. $P = \dfrac{nRT}{V}$, $V = 2.5 \text{ m}^3$, n = 3 mol, R = 8.31 J/(K. Mol), T = 293 K

4. $I = \dfrac{E}{R}$, E = 112 Volts, R = 10 Ohms

5. $W = I^2 R$, I = 11 Amps, R = 50 Ohms

6. $D = \dfrac{M}{V}$, M = 125 g, V = 5 cm^3

7. $z = \dfrac{x - \mu}{\sigma}$, x = 85 m, μ = 79 m. σ = 2 m

8. %Weight Change $= \dfrac{W_f - W_i}{W_i} \times 100$, W_f = 1.35 lbs, W_i = 1.25 lbs

9. $M = \dfrac{Pr/12}{\left[1 - (1 + r/12)^{-12t}\right]}$, P = \$300,000, r = 0.05, t = 25 yrs

10. $R_T = \dfrac{R_1 R_2}{R_1 + R_2}$, R_1 = 15 Ohms, R_2 = 10 Ohms

11. $V_S = \dfrac{4}{3} \pi r^3$, r = 3 in

12. $A = 4\pi r^2$, r = 1.5 m

13. $K = {}^\circ C + 273$, ${}^\circ C = 21$

14. $V_{cyl} = \pi r^2 \cdot h$, r = 2.5 cm, h = 5 cm

15. $p = 2l + 2w$, l = 9 m, w = 15 m

16. $A = A_0 (1 + rt)$, A_0 = \$15,000, r = 0.085, t = 15 yrs

17. Total Magnification = Power of Ocular $\times$ Power of Objective ,

    Power of Ocular = 20 x, Power of Objective = 50 x

18. Image size = Total Magnification × Objective Size ,

   Total Magnification = 500 x, Objective size = 25 μm

19. $C = 2\pi r$, $r = 3.5$ in

20. $A_T = \frac{1}{2} b \cdot h$, $b = 5$ m, $h = 8$ m

21. $A = \pi r^2$, $r = 4\,cm$

22. $\bar{x} = \frac{\Sigma x}{n}$, $n = 3$, $x_1 = 21$ g, $x_2 = 18$ g, $x_3 = 24$ g

23. $\sigma = \sqrt{\frac{\Sigma (x - \mu)^2}{N}}$, $N = 2$, $x_1 = 12$ hrs, $x_2 = 15$ hrs, $\mu = 13.5$ hrs

24. $P = \frac{n!}{(n-x)!x!} p^x q^{n-x}$, $n = 3$, $x = 2$, $p = 0.25$, $q = 0.75$

25. $Q = mc\Delta T$, $m = 50$ kg, $c = 0.386$ J/(kg K), $\Delta T = 80$ K

26. $z = \frac{\bar{x} - \mu}{\sigma / \sqrt{n}}$, $\bar{x} = 20.2\,m, \mu = 20\,m, \sigma = 1\,m, n = 49$

27. $V_2 = \frac{P_1 V_1}{P_2}$, $V_1 = 2.5\,m^3, P_1 = 760\,Pa, P_2 = 630\,Pa$

28. $\tau \approx 1 \times 10^{10} \frac{1}{M_\odot^{2.5}}$ years , for $M_\odot = 1.5$

29. $R \approx 675{,}700 \times [M_\odot]^{0.5}$ km , for $M_\odot = 10$

30. The formula for density is $\frac{mass(g)}{volume(cm^3)}$. Given an object's mass is 80 grams and its volume is 400 cm^3, find this object's density.

31. The density formula is a formula you will encounter in chemistry: $d = \frac{m}{v}$, where d = density (g/mL), m = mass (g), and v = volume (mL). Use this formula to answer the following questions. Note that sometimes it may be helpful to rearrange the formula so that the variable you are looking for is isolated first.

a) What is the density of a sample of 65.0 mL of automobile oil having a mass of 59.82 g?
b) What is the density of a mineral if 421 g of the mineral occupy a volume of 35.0 mL?
c) Linseed oil has a density of 0.929 g/mL. How many mL are in 25 g of the oil?
d) Glycerol has a density of 1.20 g/mL. How many mL are in 65 g of glycerol?
e) The density of gold is 19.3 g/mL. What is the mass of 30.0 mL of gold?
f) The density of iron is 7.9 g/mL. What is the mass of 2.0 mL of iron?

32. The distance a star is away from us is given by the formula, $D = \dfrac{1}{p}$, where D is the distance to the star measured in parsecs, and p is the parallax angle measured in arc-seconds. If the parallax is measured as 0.02 arc-seconds, how many parsecs is it away?

33. Clouds form when warm, moist air rises vertically. As a mass of air rises it expands in the reducing air pressure, causing the air mass to cool and greatly reducing the air's ability to hold moisture. At some point, the moisture in the air exceeds the value which that air mass can hold, forcing the water vapor to condense, forming clouds. The point at which air is cooled to the temperature that it can no longer hold its moisture is called the dew point. The level where this condensation first occurs is the altitude of the cloudbase, and is given by the formula:
$$H = 1000 \times \frac{(T - T_D)}{4.5} \text{ ft},$$
where H is the height of the cloudbase in feet, T is the air temperature in °F, and T_D is the dewpoint temperature in °F. Find the height of the cloudbase if T = 92 °F and T_D = 71 °F.

34. The power of wind can be both destructive and beneficial. Wind energy can be converted to mechanical or electrical energy. The power of wind blowing perpendicular to a surface is given by the formula: $P = \dfrac{1}{2} DV^2 (\text{watts}/\text{m}^2)$, where V = velocity (m/sec) and D = density (kg/m³).

Find the power for the following:
a) D = 1.18, V = 10   Summer sea breeze
b) D = 1.14, V = 70   Hurricane Andrew
c) D = 1.19, V = 135   F5 tornado
d) D = 1.53, V = 15   Mid-latitude winter storm
e) D = 1.19, V = 30   Mesoscale Convective Complex

35. The acceleration is given by the formula, $a = \dfrac{F}{m}$ (the units for acceleration are ft/s²), where F is force and m is mass. A car with a 3000 pound mass experiences a 750 pound force (the units for force are lbs ft/s²) in the forward direction. What is the acceleration?

36. Velocity is given by the formula, $V = \dfrac{D}{T}$, where D is distance, and T is time. A car travels a distance of 10 miles in 15 minutes. What is its speed in miles per hour? What is its speed in feet per second?

37. Work is given by the formula, $W = F \times D$, where F is a force and D is a distance traveled. A car exerts a traction force of 150 pounds over a distance of 880 ft. How much work does it do?

38. The Schwarzschild radius (sometimes historically referred to as the gravitational radius) is the radius of a sphere such that, if all the mass of an object were to be compressed within that

sphere, the escape velocity from the surface of the sphere would equal the speed of light, and is given by the formula: $R_{Sch} \approx 3\,M_\odot$, where R_{Sch} is measured in km, and $M_\odot$ is the proportion of how many solar masses the star is, relative to our sun. Thus, our sun has a Schwarzschild radius of 3km. Use this formula to find the following:

a) $M_\odot$ = 10 solar masses

b) $M_\odot$ = 100 solar masses

c) $M_\odot$ = 0.1 solar masses

d) $M_\odot$ = 0.5 solar masses

39. Density (D) is calculated as mass (M) / volume (V), $D = M/V$. You are measuring the density of a 292 ml sample of seawater. If the mass of the sample is 300.0 g, what is the density of the seawater in g/cm^3?

40. Density (D) is calculated as mass (M) / volume (V), $D = M/V$. Suppose you have already determined that the density of a 150 ml sample of seawater is 1.03 g/cm^3. What is the mass of the sample?

41. Wave depth (d) is calculated as wavelength (λ) / 2, $d = \lambda/2$. If a wave is 1.5 m long, how deep is it?

42. Wave depth (d) is calculated as wavelength (λ) / 2, $d = \lambda/2$. If a wave is 1.35 m deep, how long is it?

43. Wave frequency (f) is calculated as 1 / period (T), $f = 1/T$. If the wave period is 5 seconds, what is the wave frequency?

44. Wave frequency (f) is calculated as 1 / period (T), $f = 1/T$. If the wave frequency is 0.125 seconds^{-1}, what is the period?

45. Wave amplitude (A) is calculated as wave height (h) / 2, $A = h/2$. If a wave is 0.5 m high, what is its amplitude?

46. Wave amplitude (A) is calculated as wave height (h) / 2, $A = h/2$. If the wave amplitude is 3.65 m, what is the wave height?

47. Mean arterial pressure (MAP) is calculated as follows:

$$MAP = diastolic\ pressure + \frac{pulse\ pressure}{3}$$

Joe's diastolic pressure is 80 mmHg (mmHg represents millimeters of mercury, a measurement of pressure). If his pulse pressure is 33 mmHg, compute his MAP.

48. Mean arterial pressure (MAP) is calculated as follows:

$$MAP = diastolic\ pressure + \frac{pulse\ pressure}{3}$$

Tania's diastolic pressure is 78 mmHg (mmHg represents millimeters of mercury, a measurement of pressure). If her MAP is 90 mmHg, what is her pulse pressure?

49. To determine the % of a specific gas relating to the given atmospheric pressure, we

multiply %N_2 by the barometric pressure at a certain elevation. That is,

$$\text{Partial Pressure} = \%N_2 \times \text{barometric pressure}$$

at a certain elevation. Suppose the atmospheric air contains 78% N_2, and the barometric pressure at 20,000 feet is 553mm Hg (millimeter in mercury), what is the partial pressure of N_2 at that elevation in the inspired air? Round your answer to the nearest whole number.

50. Evaluate C_2 using the formula, $C_1V_1 = C_2V_2$, if $C_1 = 1$ mg/ml, $V_1 = 10$ml, and $V_2 = 11$ml.

51. The formula for density is $\dfrac{mass(g)}{volume(cm^3)}$. Given an object's mass is 2 kilograms and its volume is 1.2 m^3, find this object's density.

Solving Formulas for Other Variables

Solve the formulas for the indicated variable:

52. $^\circ F = 1.8 \,^\circ C + 32$, Solve for $^\circ C$

53. $PV = nRT$, Solve for T

54. $I = \dfrac{E}{R}$, Solve for E

55. $W = I^2R$, Solve for R

56. $D = \dfrac{M}{V}$, Solve for V

57. $z = \dfrac{x - \mu}{\sigma}$, Solve for μ

58. $A_T = \dfrac{1}{2}b \cdot h$, Solve for b

59. $A = \pi r^2$, Solve for r

60. $K = ^\circ C + 273$, Solve for $^\circ C$

61. $V_{cyl} = \pi r^2 \cdot h$, Solve for h

62. $p = 2l + 2w$, Solve for w

63. $A = A_0(1 + rt)$, Solve for t

64. $C = 2\pi r$, Solve for r

65. $Q = mc\,\Delta t$, Solve for m

66. $z = \dfrac{\bar{x} - \mu}{\sigma/\sqrt{n}}$, Solve for $\bar{x}$

67. $P_1V_1 = P_2V_2$, Solve for P_2

68. In Biology, if a person has a vital capacity (VC) of 6000ml, an expiratory reserve volume (ERV) of 2000 ml and an inspiratory reserve volume (IRV) of 3000 ml, what is the tidal volume (TV)? Formula required: VC = TV + IRV + ERV

Chapter 10 Practice Test

Write each number in Scientific Notation

1. 5,280,450,000   2. 0.0000967   3. 109,230,000,000   4. 55,000   5. 0.000000053

Write each number in decimal form.

6. 3.0×10^3 7. -1.9×10^{-7} 8. 4.2×10^{-8} 9. 7.2×10^{12}

Perform the operations with Scientific Notation. Round to digits to the right of the decimal point.

10. $(2.3 \times 10^4)(3.0 \times 10^3)$ 12. $(4.2 \times 10^{-8})(5.1 \times 10^9)$

11. $(5.0 \times 10^{-3})(-1.9 \times 10^{-7})$

13. $\dfrac{(-3.8 \times 10^{-4})(3.3 \times 10^{-7})}{(-1.9 \times 10^3)}$

14. The mass of the Earth's crust is 2.8×10^{22} kg, volume of the crust is 1×10^{19} m^3. What is the density of the crust, if density is the ratio of the mass divided by its volume?

Determine which data sets are qualitative and which are quantitative. If quantitative tell whether the data is discrete or continuous.

15. College student's major   17. Amount of summer rainfall
16. Number of people at a restaurant on a Saturday night   18. New York State cell phone carriers

Find the Mean, Median, Mode, Range, Variance, and Standard Deviation of the following data sets. Round your answers to two digits to the right of the decimal point.

19. Atomic Mass of Alkaline Earth Metals: {9.012, 24.31, 40.08, 87.62, 137.3, 226.0}
20. Number of credits enrolled: {15, 9, 12, 6, 3, 4, 15}

Use a calculator to calculate the following. Round your answers to two digits to the right of the decimal point.

21. There are about 3.15576×10^7 seconds in a year. After 24 years, for how long has a person lived?
22. According to the Center for Disease Control (CDC), about 3.00×10^6 cells are regenerated per minute in the average human body. Use this to determine how many cells your body regenerates every day.

Evaluate the formulas for the given values of the variable/variables

23. $°F = 1.8\,°C + 32$, $°C = 10$ 26. $W = I^2 R$, $I = 25$, $R = 5$

24. $P = \dfrac{nRT}{V}$, $V = 1000$, $n = 5$, $R = 8.31$, $T = 321$ 27. $z = \dfrac{x - \mu}{\sigma}$, $x = 100$, $\mu = 91$, $\sigma = 3$

25. $I = \dfrac{E}{R}$, $E = 108$, $R = 36$

Solve the formulas for the indicated variable:

28. $°F = 1.8\,°C + 32$, Solve for $°C$

29. $PV = nRT$, Solve for n

30. $I = \dfrac{E}{R}$, Solve for R

31. $z = \dfrac{x - \mu}{\sigma}$, Solve for σ

32. $p = 2l + 2w$, Solve for l

33. $C = \pi D$, Solve for D

APPENDIX A

Fractions – A Conceptual Approach

As we saw in Chapter 1, dividing integers can lead to a new type of number. This new number represents parts of a whole and can allow us to identify values in–between the integers on the number line. These new numbers are known as fractions.

In this section we will show that fractions (parts of a whole) are numbers, just like the natural counting numbers, 1, 2, 3, … Numbers are essentially used to quantify things or phenomena we see in the world. The first numbers that came about were the natural or counting numbers. These numbers allowed us to answer questions like; How many? and Which sets are larger than other sets? When we count whole objects there are no in-between values. For example it does not make sense to talk about five and one-third people. There are either five or six people, there is no in-between value. However, as we stated early in Chapter 1, numbers need to be thought of more generally, as they can be used to quantify things that do have in-between values, such as, distances, time, weight, etc.. It is this more general idea of numbers that we explore further here.

A useful tool to understand the number concept, is the number line:

$$\ldots\ ^-3\quad ^-2\quad ^-1\quad 0\quad 1\quad 2\quad 3\ \ldots$$

On this number line we have placed the integer as points on the line. We have placed the numbers in order, with increasing values to the right. In this way we can think of the numbers as representing a distance from zero on the line. Thus, the numbers are not simply counting whole objects, but are now used to measure (count) distance. In this way we have extended the concept of number to quantify things that have values in-between the integers.

To explore this further we focus our attention on a portion of the number line (a line segment) from zero to one:

$$0 \qquad\qquad\qquad\qquad\qquad\qquad\qquad 1$$

From this unit line segment we will begin to develop the concept of fractions and its use as a number, just like the counting numbers. We should point out that this is a fairly abstract concept. The Ancient Egyptians in 3,000 BCE were some of the earliest users of fractions, but did not consider them numbers. To the early Egyptians and most ancient civilizations, the only numbers were the counting numbers, and fractions such as:

$$\frac{1}{4},\quad \frac{1}{2},\quad \frac{1}{7}\quad \frac{3}{5},\ \text{etc.}$$

were simply two numbers used to represent parts of a whole, but were not thought to be a single numerical quantity themselves. One early civilization, the Babylonians (3,000 BCE), however, did view fractions as numbers.

A.1 UNIT FRACTIONS

We now begin to develop the concept of the two numbers as written above as actually representing a single numerical quantity, something the Ancient Egyptians could not grasp.

We start with our unit line segment:

We now look to break this segment into equal parts, or sub-segments. The first thing we notice is that we can break this segment into many different equal parts. We could break it into two equal parts, three equal parts, four equal parts, etc. If we break our line segment into two equal parts we call each of the parts a half of the whole and represent this using the symbol that places the number 1 representing the whole line segment over the number 2 representing the number of equal parts were are breaking (or dividing) our unit line segment into, with a horizontal line between them:

$$\frac{1}{2}$$

This is the unit fraction called one half.

Using our line segment we can illustrate the two equal parts and their size represented by one-half:

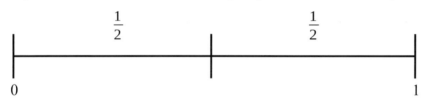

Breaking the line segment into three equal parts we have:

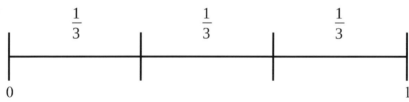

With each equal part called a third.

Breaking into four equal parts we have:

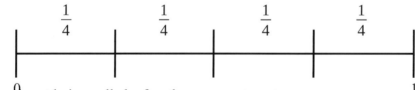

With each equal part being called a fourth, or a quarter, etc.

Each of the equal parts: $\frac{1}{2}$, $\frac{1}{3}$, $\frac{1}{4}$, etc. are called unit fractions, since we are breaking a single part (the unit), into two or more equal parts.

Another useful way to understand fractions is by using an area approach in addition to the number line. Thus, along with the number line we can consider an area bar to represent a fraction. Sometimes we will refer to this as a "fraction bar," or "tape diagram."

Below are examples of representing one-half, one-third, and one-fourth, using a fraction bar:

$\dfrac{1}{2}$	$\dfrac{1}{2}$

$\dfrac{1}{3}$	$\dfrac{1}{3}$	$\dfrac{1}{3}$

$\dfrac{1}{4}$	$\dfrac{1}{4}$	$\dfrac{1}{4}$	$\dfrac{1}{4}$

At this point the fractions are not quite numbers, but instead illustrate a process where we take a whole object (the unit), and divide it into "n" equal parts to get the fraction,

$$\frac{1}{n}$$

To begin to develop the number concept for a fraction we see that any whole object can be decomposed (broken down into) into a sum of unit fractions. For example:

$$1 = \frac{1}{2} + \frac{1}{2}$$

We call this an addition sentence.

As another example, consider breaking one into thirds. We see that we can add it in two different ways. We could add all the thirds together in our addition sentence to get three thirds or one, or we could first add the two thirds together and then the one third and get the same result.

$$1 = \frac{1}{3} + \frac{1}{3} + \frac{1}{3}$$

$$= \frac{2}{3} + \frac{1}{3}$$

$$= \frac{3}{3}$$

We can also see this using a tape diagram below.

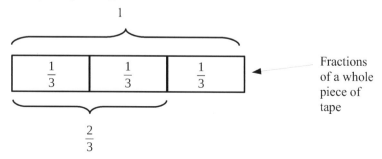

360

Furthermore, we see that two thirds can be thought of as adding two one thirds together or equivalently as multiplying 2 times a third, since multiplication as was defined earlier is simply repeated addition.

Written out in mathematical language, we call the first line an addition sentence and the second line a multiplication sentence.

$$\frac{2}{3} = \frac{1}{3} + \frac{1}{3} \qquad \text{Addition Sentence}$$

$$\frac{2}{3} = 2 \times \frac{1}{3} \qquad \text{Multiplication Sentence}$$

We can extend this process to consider fractions where the top number (the numerator) is greater than the bottom number (the denominator.) For example let's break up the fraction four thirds as shown below:

$$\frac{4}{3} = \frac{3}{3} + \frac{1}{3}$$

$$= 1 + \frac{1}{3}$$

$$= 1\,\frac{1}{3}$$

$$= 4 \times \frac{1}{3}$$

As we see fractions are beginning to look more like numbers than we first thought. We can add them and we can multiply them by whole numbers.

EXAMPLE 1: Decompose the fraction $\frac{7}{5}$ in at least three different ways, and also write using a multiplication sentence.

Using the ideas from above we can represent $\frac{7}{5}$ in a few different ways, for example:

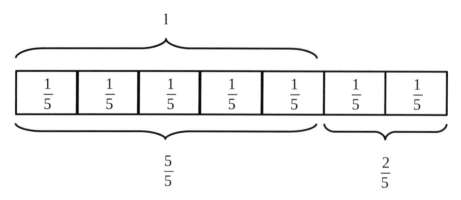

$$\frac{7}{5} = \frac{1}{5} + \frac{1}{5} + \frac{1}{5} + \frac{1}{5} + \frac{1}{5} + \frac{1}{5} + \frac{1}{5}$$

$$= \frac{6}{5} + \frac{1}{5}$$

$$= \frac{5}{5} + \frac{2}{5}$$

$$= 1 + \frac{2}{5}$$

$$= 1\frac{2}{5}$$

$$= 7 \times \frac{1}{5}$$

EXAMPLE 2: Given the fraction $\frac{5}{6}$ use an addition sentence to represent it using unit fractions. Then show an alternative way to write it. Finally represent its value using multiplication.

$$\frac{5}{6} = \frac{1}{6} + \frac{1}{6} + \frac{1}{6} + \frac{1}{6} + \frac{1}{6}$$

$$= \frac{4}{6} + \frac{1}{6}$$

$$= 5 \times \frac{1}{6}$$

Decompose Fractions Into Smaller Units

We just learned how to take a whole unit and decompose it into smaller unit fractions. We now extend the idea of decomposition to fractions themselves. We now ask the question; Can we decompose a unit fraction into smaller unit fractions?

EXAMPLE 1: Write $\frac{1}{3}$ as smaller unit fractions.

We could break $\frac{1}{3}$ into two smaller but equal pieces. As we see from the tape diagram, half of its size is equal to $\frac{1}{6}$, and it takes two $\frac{1}{6}$'s to equal $\frac{1}{3}$.

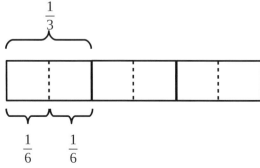

We can also write with number sentences, both addition and multiplication:

$$\frac{1}{3} = \frac{1}{6} + \frac{1}{6} = \frac{2}{6}$$

$$\frac{1}{3} = 2 \times \frac{1}{6} = \frac{2}{6}$$

Using the tape diagram, addition and multiplication sentences, we see that $\frac{1}{3}$ is equivalent to $\frac{2}{6}$. We can take this even further.

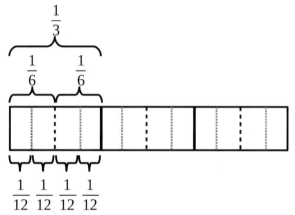

What does the figure above tell us about the relationships between $\frac{1}{3}$, $\frac{1}{6}$, and $\frac{1}{12}$?

We can write addition and multiplication sentences that describe this relationship.

$$\frac{1}{3} = \frac{1}{12} + \frac{1}{12} + \frac{1}{12} + \frac{1}{12} = \frac{4}{12}$$

$$\frac{1}{3} = 4 \times \frac{1}{12} = \frac{4}{12}$$

$$\frac{1}{3} = \frac{1}{6} + \frac{1}{6} = \frac{2}{6}$$

$$\frac{1}{3} = 2 \times \frac{1}{6} = \frac{2}{6}$$

EXAMPLE 2: Break $\frac{1}{5}$ into 3 equal sized pieces. Write addition and multiplication sentences that describe this relationship.

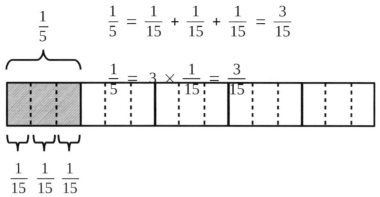

$$\frac{1}{5} = \frac{1}{15} + \frac{1}{15} + \frac{1}{15} = \frac{3}{15}$$

$$\frac{1}{5} = 3 \times \frac{1}{15} = \frac{3}{15}$$

EXAMPLE 3: Break $\frac{2}{6}$ into 4 equal sized pieces.

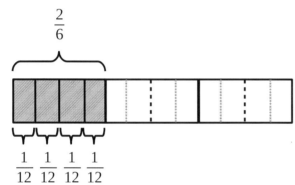

$$\frac{2}{6} = \left(\frac{1}{12} + \frac{1}{12}\right) + \left(\frac{1}{12} + \frac{1}{12}\right) = \frac{4}{12}$$

$$\frac{2}{6} = \left(2 \times \frac{1}{12}\right) + \left(2 \times \frac{1}{12}\right) = \frac{4}{12}$$

$$\frac{2}{6} = 4 \times \frac{1}{12} = \frac{4}{12}$$

This shows the relationship between $\frac{2}{6}$ and $\frac{4}{12}$

From this we see that $\frac{2}{6}$ and $\frac{4}{12}$ are equivalent. They have the same value. We call these two **Equivalent Fractions**. In the next section we look at many different ways to understand and find equivalent fractions.

EXERCISES A.1

Unit Fractions

Represent the unit fractions using a number line or tape diagram.

1. $\frac{1}{5}$ 2. $\frac{1}{9}$ 3. $\frac{1}{6}$ 4. $\frac{1}{10}$

Represent the fractions using a number line or tape diagram. Then decompose the fractions into unit fractions in at least 2 different ways. Then write a suitable addition and multiplication sentences for each.

5. $\frac{4}{5}$ 6. $\frac{5}{6}$ 7. $\frac{7}{9}$ 8. $\frac{4}{7}$

9. $\dfrac{7}{4}$ 10. $\dfrac{9}{5}$ 11. $\dfrac{3}{4}$ 12. $\dfrac{5}{8}$

Decompose Fractions Into Smaller Units

Break the given fractions into the given number of equal sized pieces. Show with a tape diagram. Then write addition and multiplication sentences that describe the relationship.

13. $\dfrac{1}{4}$ into 2 pieces   17. $\dfrac{4}{5}$ into 4 pieces   21. $\dfrac{5}{8}$ into 10 pieces.

14. $\dfrac{1}{2}$ into 5 pieces   18. $\dfrac{7}{9}$ into 7 pieces   22. $\dfrac{2}{3}$ into 4 pieces.

15. $\dfrac{1}{6}$ into 3 pieces   19. $\dfrac{2}{4}$ into 8 pieces   23. $\dfrac{2}{5}$ into 6 pieces.

16. $\dfrac{1}{8}$ into 2 pieces   20. $\dfrac{4}{6}$ into 12 pieces   24. $\dfrac{3}{10}$ into 9 pieces.

A.2 FRACTIONS AS NUMBERS AND EQUIVALENT FRACTIONS

We have shown that it is possible to count parts, just like we count wholes. For example, we can have one half, two halves, three halves, etc. Furthermore, in the same way we can count multiples of wholes, simply by multiplying by our counting numbers, such as:

$$1 = 1 \times 1,$$
$$2 = 1 + 1 = 2 \times 1,$$
$$3 = 1 + 1 + 1 = 3 \times 1,$$
etc.

we can do the same for unit fractions, and write the results as follows:

$$\frac{1}{2} = 1 \times \frac{1}{2},$$
$$\frac{2}{2} = \frac{1}{2} + \frac{1}{2} = 2 \times \frac{1}{2},$$
$$\frac{3}{2} = \frac{1}{2} + \frac{1}{2} + \frac{1}{2} = 3 \times \frac{1}{2},$$
etc.

Where the number on top (the numerator) tells us how many copies we have of the unit fraction, and the number on the bottom (the denominator) indicates the size; halves, thirds, fourths, etc., of what we are counting. We can now associate a numerical value with these multiples of parts of a whole.

Using our number line and starting with two to divide the distance between the whole numbers into halves we have:

The number line shows:

0 at $\frac{0}{2}$, 1 at $\frac{2}{2}$, 2 at $\frac{4}{2}$, with $\frac{1}{2}$ and $\frac{3}{2}$ in between.

The first thing we notice is that we now have two equivalent ways of representing the same numerical value, or location, on the number line.

For example, we can either write 0, or we can write $\frac{0}{2}$. meaning if we take zero and divide it in half we still have zero.

Next, we can either write 1 or we can write $\frac{2}{2}$. meaning if we take two objects and divide that in half we have one object.

Continuing, we can either write 2 or we can write $\frac{4}{2}$. Again, two is equivalent to taking four objects and dividing that in half.

We don't have to stop here.

EXAMPLE 1: Write equivalent fractions for 3, 4, 5, 6, using halves.

$$3 = \frac{6}{2}, \quad 4 = \frac{8}{2}, \quad 5 = \frac{10}{2}, \quad \text{and} \quad 6 = \frac{12}{2}$$

Another way to view an equivalent fraction is a different fraction that puts you at the same location on the number line.

The next thing we notice is that we can now think of the symbol for representing parts of a whole as a number that can identify a location on a number line.

$\frac{1}{2}$ represents the location halfway between 0 and 1 on the number line,

$\frac{3}{2}$ represents the location halfway between 1 and 2 on the number line.

Continuing in this way, $\frac{5}{2}$ would represent the location halfway between 2 and 3 on the number line, etc.

EXAMPLE 2: What are the fractions representing numbers halfway between 7 and 8, between 9 and 10?

$$\frac{15}{2} \quad \text{and} \quad \frac{19}{2} \text{ respectively.}$$

We also see that there is an order to these fractions, just like there is an order to the counting numbers. Fractions can be less than or greater than other fractions and their position can be precisely identified on a number line. They really are numbers in their own right.

Now let's consider thirds with a number line.

$$0 \qquad\qquad 1 \qquad\qquad 2 \quad \cdots$$
$$\frac{0}{3} \quad \frac{1}{3} \quad \frac{2}{3} \quad \frac{3}{3} \quad \frac{4}{3} \quad \frac{5}{3} \quad \frac{6}{3} \quad \cdots$$

One thing to notice is that we now have more equivalent ways to write 0, 1, and 2.

For example, we can either write 0, or we can write $\frac{0}{3}$. We can either write 1, or we can write $\frac{3}{3}$.

We can either write 2, or we can write $\frac{6}{3}$.

Since they refer to the same location on the number line, they have an equivalent numerical value.

Furthermore, we can say that $\frac{0}{2}$ is equivalent to $\frac{0}{3}$, and that $\frac{2}{2}$ is equivalent to $\frac{3}{3}$ and $\frac{4}{2}$ is equivalent to $\frac{6}{3}$

Continuing in this way we establish a pattern of equivalent fractions related to our counting numbers:

$$0 = \frac{0}{2} = \frac{0}{3} = \frac{0}{4} = \frac{0}{5}, \; \cdots \qquad\qquad 2 = \frac{4}{2} = \frac{6}{3} = \frac{8}{4} = \frac{10}{5}, \; \cdots$$

$$1 = \frac{2}{2} = \frac{3}{3} = \frac{4}{4} = \frac{5}{5}, \; \cdots \qquad\qquad 3 = \frac{6}{2} = \frac{9}{3} = \frac{12}{4} = \frac{15}{5}, \; \cdots$$

EXAMPLE 3: Write equivalent fractions for 4, 5, 6, and 7 using thirds.

$$4 = \frac{12}{3} \qquad\qquad 5 = \frac{15}{3} \qquad\qquad 6 = \frac{18}{3} \qquad\qquad 7 = \frac{21}{3}$$

Each fraction above represents the exact same place on the number line as the other equal fractions and whole numbers.

Now let's consider fourths with a number line.

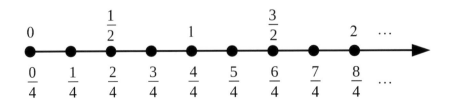

One thing we notice is that we now have more equivalent ways to write $\frac{1}{2}$ and $\frac{3}{2}$

For example, we can either write $\frac{1}{2}$ or we can write $\frac{2}{4}$

Similarly, we can either write $\frac{3}{2}$ or we can $\frac{6}{4}$ write

Since they refer to the same location on the number line, they have an equivalent numerical value.

Continuing in this way we establish a pattern of equivalent fractions for $\frac{1}{2}, \frac{3}{2}, \frac{5}{2}, \frac{7}{2}$, etc.

$$\frac{1}{2} = \frac{2}{4} = \frac{3}{6} = \frac{4}{8}, \; ... \qquad\qquad \frac{5}{2} = \frac{10}{4} = \frac{15}{6} = \frac{20}{8}, \; ...$$

$$\frac{3}{2} = \frac{6}{4} = \frac{9}{6} = \frac{12}{8}, \; ... \qquad\qquad \frac{7}{2} = \frac{14}{4} = \frac{21}{6} = \frac{28}{8}, \; ...$$

EXAMPLE 4: Write equivalent fractions for 4, 5, 6, and 7 using fourths.

$$4 = \frac{16}{4} \qquad 5 = \frac{20}{4} \qquad 6 = \frac{24}{4} \qquad 7 = \frac{28}{4}$$

Again, each fraction above represents the exact same place on the number line as the other equivalent fractions.

Using Multiplications and Division to Find Equivalent Fractions

To learn more about equivalent fractions we use the fraction bar (tape diagram), along with our number line. In the example below, we start by taking a whole and breaking it into three equal parts. We then break two of the three parts each into four equal parts.

From the picture below we can see that $\frac{2}{3}$ is equivalent to $\frac{8}{12}$, or stated more directly $\frac{2}{3} = \frac{8}{12}$

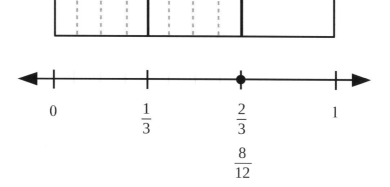

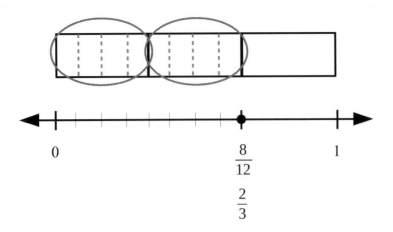

$$\frac{2}{3} = \frac{2 \times 4}{3 \times 4} = \frac{8}{12}$$

We can also show this using multiplication. We simply start with $\frac{2}{3}$ and multiply the top number 2 (called the numerator) by 4, and the bottom number 3 (called the denominator) also by 4, and get the equivalent fraction $\frac{8}{12}$.

In a similar way we can start with $\frac{8}{12}$, and take the numerator and divide by 4, and do the same to the denominator and get the equivalent fraction $\frac{2}{3}$.

$$\frac{8}{12} = \frac{8 \div 4}{12 \div 4} = \frac{2}{3}$$

The number line and the fraction bar illustrate the concept of what we are doing, while the multiplication and division approach is the "shorthand" algorithm we will use to find equivalent fractions more quickly.

We show this in the figures below.

$$\frac{2}{3} = \frac{2 \times 4}{3 \times 4} = \frac{8}{12} \qquad\qquad \frac{8}{12} = \frac{8 \div 4}{12 \div 4} = \frac{2}{3}$$

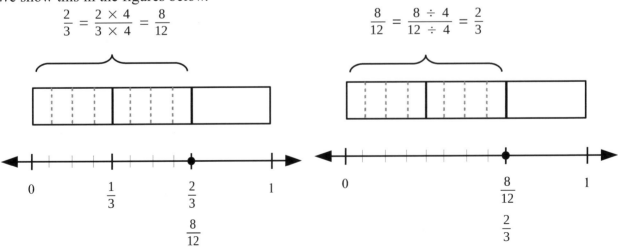

Area Model of Fractions

To help us understand this even better we develop an area model to show the relationship between multiplication and an equivalent fraction. For example, let's show $\frac{1}{2}$ using an area model. For this we simply draw a box and shade in half the box to represent our fraction, as shown below.

Now if we want to rewrite $\frac{1}{2}$ using four times as many units, we can simply divide up the one-half area into four equal sub-areas as shown.

Then we can see the equivalence that $\frac{1}{2} = \frac{4}{8}$. Which we could also have obtained through multiplication: $\frac{1}{2} = \frac{1 \times 4}{2 \times 4} = \frac{4}{8}$

In the same way the following area chart,

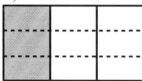

shows that

$$\frac{1}{3} = \frac{1 \times 3}{3 \times 3} = \frac{3}{9}$$

Another example:

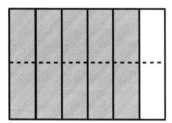

$$\frac{5}{6} = \frac{5 \times 2}{6 \times 2} = \frac{10}{12}$$

One final example:

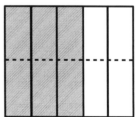

$$\frac{3}{5} = \frac{3 \times 2}{5 \times 2} = \frac{6}{10}$$

EXERCISES A.2

Fractions as Numbers and Equivalent Fractions

1. Write equivalent fractions for 3, 4, 5, 6, using thirds.

2. Write equivalent fractions for 3, 4, 5, 6, using fourths.

3. Write equivalent fractions for 3, 4, 5, 6 using fifths.

4. Write equivalent fractions for 3, 4, 5, 6 using sixths.

5. What are the fractions representing numbers halfway between 6 and 7, between 11 and 12?

6. What are the fractions representing numbers halfway between 8 and 9, between 13 and 14?

7. Show using a number line that $\frac{2}{5}$ is equivalent to $\frac{4}{10}$. Then show through either multiplication

   or division that $\frac{2}{5} = \frac{4}{10}$

8. Show using a number line that $\frac{4}{5}$ is equivalent to $\frac{12}{15}$. Then show through either multiplication

   or division that $\frac{4}{5} = \frac{12}{15}$

9. Show using a number line that $\frac{9}{18}$ is equivalent to $\frac{1}{2}$. Then show through either multiplication

   or division that $\frac{9}{18} = \frac{1}{2}$

10. Show using a number line that $\frac{9}{15}$ is equivalent to $\frac{3}{5}$. Then show through either multiplication

    or division that $\frac{9}{15} = \frac{3}{5}$

11. Use an area chart to show that $\frac{2}{3} = \frac{6}{9}$

12. Use an area chart to show that $\frac{5}{6} = \frac{20}{24}$

13. Use an area chart to show that $\frac{4}{7} = \frac{8}{14}$

14. Use an area chart to show that $\frac{3}{4} = \frac{12}{16}$

15. Use the area chart to show that $\frac{2}{3} = \frac{12}{18}$

16. Use the area chart to show that $\frac{4}{5} = \frac{12}{15}$

A.3 ADDING AND SUBTRACTING FRACTIONS

Now that we see fractions as really numbers, we want to be able to perform arithmetic with them, just like we did with our integers. In this section we introduce the process of how we add and subtract fractions. Ordinarily this can be quite confusing, especially since fractions look and behave quite different than our integers. If, however, we keep the number line idea in mind, we can more easily explain and grasp the arithmetic of fractions.

Adding Fractions

Adding wholes is much easier, since we can simply count how many we have. Fractions, however, are like being given a bag of different length sticks and then asked to determine how long it would be if they were laid end to end. The only way to be able to measure how long the entire bag is in this way, is to have to measure each piece with a common measuring device, and then add all these different lengths of measure together. In fractions, this common device is called the equivalent unit fraction. We have to transform each fraction into a sum of equivalent unit fractions, and then we can "count" how many equivalent unit fractions we have. Thus, we try and turn adding fractions into a process similar to adding whole numbers, but instead of counting wholes, we count unit fractions. Let's illustrate this with a few examples.

In our first example we start with a problem that is already set up for us, with each fraction based upon a common unit fraction. This means that they have the same denominator.

EXAMPLE 1: Add the following fractions: $\frac{3}{7} + \frac{2}{7}$

Since each fraction is based upon the common unit fraction of $\frac{1}{7}$, we can add all the seventh's together. First let's show this visually using the number line.

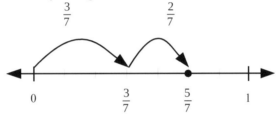

As you can see we start at zero then move $\frac{3}{7}$ unit to the right and then move another $\frac{2}{7}$ units to the right again, placing us at $\frac{5}{7}$. We can also see this using our tape diagrams.

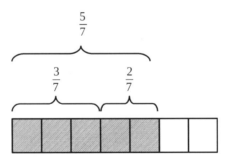

Finally, we show this with the unit fraction notation.

$$\frac{3}{7} = \frac{1}{7} + \frac{1}{7} + \frac{1}{7} \text{ and } \frac{2}{7} = \frac{1}{7} + \frac{1}{7}, \text{ so } \frac{3}{7} + \frac{2}{7} = \left(\frac{1}{7} + \frac{1}{7} + \frac{1}{7}\right) + \left(\frac{1}{7} + \frac{1}{7}\right) = \frac{5}{7}$$

EXAMPLE 2: Add the following fractions: $\frac{9}{16} + \frac{15}{16}$

Since each fraction is based upon the common unit fraction of $\frac{1}{16}$, we can add all the sixteenth's together. Visually, using our number line.

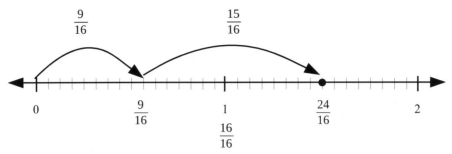

With the tape diagram we have

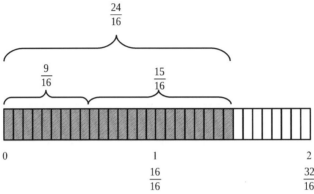

EXAMPLE 3: Add the following fractions: $\frac{3}{4}$ + $\frac{5}{8}$

Since one fraction is based upon the unit fraction of $\frac{1}{4}$, and the other is based on the unit fraction of $\frac{1}{8}$ we have to find a common equivalent fraction before we can add them together.

Visually, using our number line.

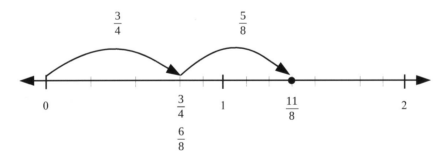

From this we see that $\frac{3}{4}$ is equivalent to $\frac{6}{8}$, so we can now add the fractions together as

$$\frac{6}{8} + \frac{5}{8} = \frac{11}{8}$$

With the tape diagram we have

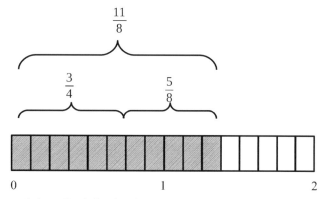

We can also use an area model to find the basic common unit fraction.

We first represent the two fractions.

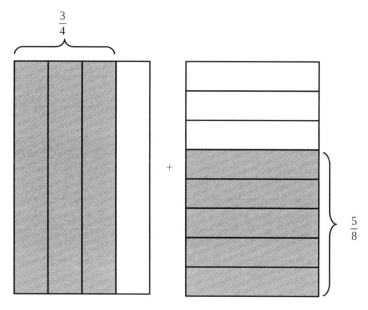

Next, we create a common unit fraction by dividing the ¾ in half, and add all the common 1/8 unit fractions together.

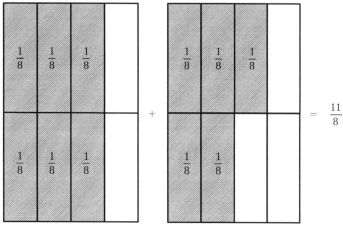

374

Finally we use our equivalent fraction approach without the visual aid.

In this problem we see that 4 and 8 are related, since $8 = 2 \times 4$. This means we only have to change one of the fractions into its equivalent form, and that is the $\frac{3}{4}$, which becomes,

$$\frac{3}{4} = \frac{3 \times 2}{4 \times 2} = \frac{6}{8}$$

Then we can write

$$\frac{3}{4} + \frac{5}{8} = \frac{6}{8} + \frac{5}{8} = \frac{11}{8}$$

We'll leave this answer as an improper fraction of $\frac{11}{8}$

EXAMPLE 4: Add the following fractions: $\frac{3}{7} + \frac{5}{6}$

In this problem we see that 7 and 6 are not related, meaning that they have no factors in common. Using our number line with two different scales (unit fractions) we can see the problem.

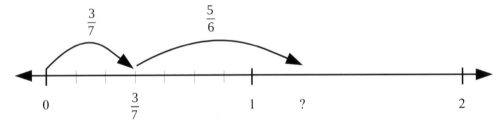

We are using two different unit fractions to add, so we don't know where to place our final point. Now we could work with the number line to find the common unit fraction, but we'll use the area model instead. (Note: the tape diagram is similar to the number line in its visualization of fractions.) If we use our area model we can easily see the common unit fraction as shown below. First we visual represent our two fractions we are adding.

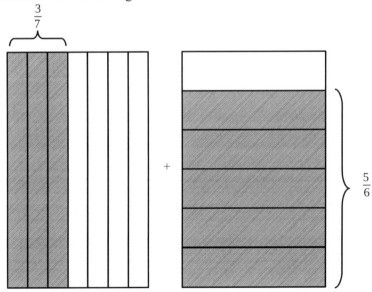

Next, if we divide the 3/7 into 6 equal parts as well as the 5/6 into 7 equal parts, we have the following.

Now we can visually see why this approach works, but it would be extremely cumbersome if we had to do this every time. Instead we rely on the arithmetic approach, using the fraction notation and writing equivalent fractions numerically, to do our calculations. We only use the visual approach as a means for understanding the concept of adding fractions with unlike denominators.
We finish this problem using the arithmetic approach.

We still need to find the common equivalent unit fraction we need so that we can find out how many equivalent unit fractions they have between the both of them.

> If the denominators do not have any factors in common, we simply multiply the denominators together to get the smallest common equivalent denominator.

In this case it is

$$7 \times 6 = 42$$

This means we have to change both of the fractions into their equivalent forms with 42 in the denominator.

$$\frac{3}{7} \text{ becomes } \frac{3}{7} = \frac{3 \times 6}{7 \times 6} = \frac{18}{42}$$

Notice, for this fraction we simply multiply by the opposite factor of 6.

$$\frac{5}{6} \text{ becomes } \frac{5}{6} = \frac{5 \times 7}{6 \times 7} = \frac{35}{42}$$

In this fraction we multiplied by the opposite factor of 7 to get our equivalent fraction.

Then we can write

$$\frac{3}{7} + \frac{5}{6} = \frac{18}{42} + \frac{35}{42} = \frac{53}{42}$$

We'll leave this answer as an improper fraction of $\frac{53}{42}$

EXAMPLE 5: Add the following fractions: $\frac{7}{8} + \frac{11}{12}$

In this problem we will only demonstrate the arithmetic approach.

We first observe that 8 and 12 have a common factor of 4, since

$$8 = 2 \times 4 \qquad \text{and} \qquad 12 = 3 \times 4$$

We see that the uncommon factors are 2 and 3. The least common equivalent denominator is obtained by multiplying by the uncommon factors (2 and 3) and the common factor (4), or

$$2 \times 3 \times 4 = 24$$

> If the denominators do have factors in common, we simply multiply the uncommon factors by the common factors to get our common denominator.

This means we now have to change both of the fractions into their equivalent forms with 24 in the denominator.

$$\frac{7}{8} \text{ becomes } \frac{7}{8} = \frac{7 \times 3}{8 \times 3} = \frac{21}{24}$$

Notice, for this fraction we simply multiply by the opposite uncommon factor of 3.

$$\frac{11}{12} \text{ becomes } \frac{11}{12} = \frac{11 \times 2}{12 \times 2} = \frac{22}{24}$$

In this fraction we multiplied by the opposite uncommon factor of 2 to get our equivalent fraction.

Then we can write

$$\frac{7}{8} + \frac{11}{12} = \frac{21}{24} + \frac{22}{24} = \frac{43}{24}$$

We'll leave this answer as an improper fraction of $\frac{43}{24}$

Subtracting Fractions

We have spent a great deal of time showing how to do addition of fractions. This should now make the concept of subtraction easier. Subtraction of fractions is very similar to addition. Only now we remove common unit fractions instead of adding them.

Let's see this through a few examples.

EXAMPLE 1: Subtract the following fractions: $\frac{6}{7} - \frac{4}{7}$

In this case the denominators are the same, so we already know our unit fraction is 1/7. Then we can visually see what this is through our number line.

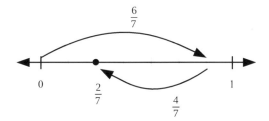

We first move 6/7 units to the right, then since we are subtracting we move 4/7 units to the left, leaving us at the location of 2/7.

This is similar to our tape diagram representation. We start with 6/7 and remove 4/7 of the tape.

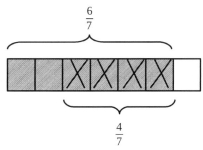

This leaves us with only 2/7 of the tape.

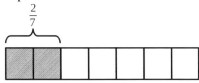

So our final answer is $\dfrac{2}{7}$

Arithmetically, to solve $\dfrac{6}{7} - \dfrac{4}{7}$ we notice that the denominators are the same, so we can simply subtract the numerators.

$$\frac{6}{7} - \frac{4}{7} = \frac{6-4}{7} = \frac{2}{7}$$

The above calculation is the standard algorithm, but the number line and the tape diagram explain why we can do this simple calculation.

EXAMPLE 2: Subtract the following fractions: $\frac{3}{4} - \frac{5}{8}$

We could show the same work visually using either the number line, tape diagram, or even the area model, but since we spent a lot of time doing this for addition, we will use the idea that it is the same, and with subtraction we are removing, rather than adding, the unit fractions.

In this case the denominators are not the same, so we must obtain a common denominator (equivalent unit fraction) for both. We notice that we can simply change the 3/4's and rewrite it as the equivalent fraction of $\frac{6}{8}$.

Thus, we can obtain:

$$\frac{3}{4} - \frac{5}{8} = \frac{6}{8} - \frac{5}{8} = \frac{6-5}{8} = \frac{1}{8}$$

EXERCISES A.3

Adding and Subtracting Fractions

Add the following fractions with the standard algorithm and visually show the process using a number line or tape diagram.

1. $\dfrac{3}{8} + \dfrac{5}{8}$ 3. $\dfrac{4}{9} + \dfrac{2}{9}$ 5. $\dfrac{3}{10} + \dfrac{2}{5}$ 7. $\dfrac{2}{9} + \dfrac{2}{3}$

2. $\dfrac{5}{12} + \dfrac{1}{12}$ 4. $\dfrac{4}{11} + \dfrac{3}{11}$ 6. $\dfrac{2}{7} + \dfrac{5}{14}$ 8. $\dfrac{1}{4} + \dfrac{5}{12}$

Add the following fractions with the standard algorithm and visually show the process using an area model.

9. $\dfrac{2}{5} + \dfrac{3}{7}$ 10. $\dfrac{2}{9} + \dfrac{4}{5}$ 11. $\dfrac{2}{3} + \dfrac{2}{7}$ 12. $\dfrac{1}{6} + \dfrac{3}{11}$

Add the following fractions with the standard algorithm

13. $\dfrac{3}{13} + \dfrac{5}{26}$ 15. $\dfrac{7}{18} + \dfrac{8}{9}$ 17. $\dfrac{9}{16} + \dfrac{7}{24}$ 19. $\dfrac{3}{16} + \dfrac{5}{12}$

14. $\dfrac{5}{21} + \dfrac{1}{7}$ 16. $\dfrac{3}{5} + \dfrac{4}{15}$ 18. $\dfrac{5}{8} + \dfrac{7}{18}$ 20. $\dfrac{9}{22} + \dfrac{8}{33}$

Subtract the following fractions with the standard algorithm and visually show the process using a number line or tape diagram.

21. $\dfrac{5}{8} - \dfrac{1}{8}$ 23. $\dfrac{4}{9} - \dfrac{2}{9}$ 25. $\dfrac{2}{5} - \dfrac{3}{10}$ 27. $\dfrac{7}{9} - \dfrac{2}{3}$

22. $\dfrac{5}{12} - \dfrac{1}{12}$ 24. $\dfrac{4}{11} - \dfrac{3}{11}$ 26. $\dfrac{4}{7} - \dfrac{5}{14}$ 28. $\dfrac{3}{4} - \dfrac{5}{12}$

Subtract the following fractions with the standard algorithm and visually show the process using an area model.

29. $\dfrac{3}{5} - \dfrac{2}{7}$ 30. $\dfrac{5}{9} - \dfrac{2}{5}$ 31. $\dfrac{2}{3} - \dfrac{2}{7}$ 32. $\dfrac{5}{6} - \dfrac{3}{11}$

Subtract the following fractions with the standard algorithm

33. $\dfrac{3}{13} - \dfrac{5}{26}$ 34. $\dfrac{10}{21} - \dfrac{1}{7}$ 35. $\dfrac{7}{18} - \dfrac{2}{9}$ 36. $\dfrac{3}{5} - \dfrac{4}{15}$

37. $\dfrac{9}{16} - \dfrac{5}{24}$ 38. $\dfrac{5}{8} - \dfrac{5}{18}$ 39. $\dfrac{7}{16} - \dfrac{5}{12}$ 40. $\dfrac{9}{22} - \dfrac{5}{33}$

Add or Subtract fractions applications

41. Kaitlyn read $\dfrac{2}{5}$ of a book on Monday, then read $\dfrac{1}{4}$ of the same book on Tuesday, and the remainder of the book on Wednesday, What part of the book did she read on Wednesday?

42. Ethan ate $\dfrac{1}{4}$ of a huge cookie his mother made for him. Two hours later, he ate $\dfrac{1}{3}$ of the same cookie. What part of the cookie is left?

43. A student did $\dfrac{3}{5}$ of her homework on Saturday morning and $\dfrac{1}{4}$ of her homework in the afternoon. She said she would do the remaining homework in the evening. What part of her homework does she still have to do?

44. A farmer stated that he and his crew harvested $\dfrac{5}{8}$ of his strawberry field on day 1. They harvested $\dfrac{1}{5}$ of the field on day 2. How much of the whole field does the farmer and his crew still need to harvest?

A.4 MULTIPLYING AND DIVIDING FRACTIONS

When we learned how to multiply whole numbers we thought of multiplication as repeated addition, and division as finding the number of times you were repeating the additions. Applying the operations of multiplication and division to fractions, we will use the same concepts. A number is a number, regardless of its form, so the operations should be the same conceptually. We start with multiplication.

Multiplying Fractions

Again, multiplication is the process of repeated additions. It is stated as $(\text{Number 1}) \times (\text{Number 2})$ and asks the question;

What number is obtained when (Number 2) is added to itself (Number 1) times?

Multiplying Whole Numbers Times Fractions

Let's start with the simplest form of multiplication, that of a whole number times a fraction. Now, we have already seen this in our multiplication sentence for adding unit fractions. For example we saw that

$$9 \times \dfrac{1}{16} = \dfrac{9}{16}$$

We essentially multiplied the 9 times the 1 in the numerator of the unit fraction. This was because 9/16 was just 9 copies of 1/16. The same holds true if the term we are multiplying is not a unit fraction. Let's illustrate this with some examples.

EXAMPLE 1: Multiply the following: $4 \times \dfrac{3}{5}$

Visual Models

What this is telling us is that we have 4 copies of 3/5. We can show this, with either our number line,

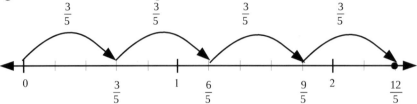

or our tape diagram

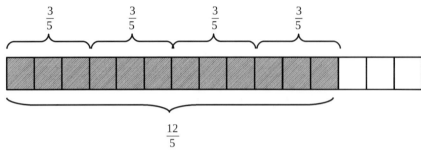

Standard Algorithm

Arithmetically we have the standard algorithm which says we just multiply the numerator by the whole number we are multiplying by.

$$4 \times \frac{3}{5} = \frac{4 \times 3}{5} = \frac{12}{5}$$

We have left our final answer as an improper fraction. We could change it to a mixed number and get $2 \frac{2}{5}$.

We should point out that the operation of multiplication is what we call commutative. That means $3 \times 9 = 9 \times 3$. We can reverse the order in which we multiply and get the same answer. The reason why we mention this now is because if we see the following multiplication problem $\frac{5}{7} \times 6$ we can always rewrite it as $6 \times \frac{5}{7}$ and use the same interpretation and approach for multiplying we just introduced. Both expressions mean that we have 6 copies of $\frac{5}{7}$.

Furthermore, we should also point out that if we don't use the above analogy, there is another analogy we will use later on in interpreting $\frac{5}{7} \times 6$. This expression also means that we are trying to found out what number is $\frac{5}{7}$ of 6. Depending upon the type of problem we are working on, one analogy or the other will be easier, so we need to understand both. Thus, we explain it in detail in what follows.

Multiplying Fractions Times Whole Numbers

EXAMPLE 2: As we just said multiplying a whole number by a fraction, such as $\frac{5}{7} \times 14$, is the same as asking the question; $\frac{5}{7}$ of 14 is equal to what number?

<u>Visual Model</u>

Let's consider the concept visually using our tape diagram. Here, we will take the number 14 and divide it into 7 equal parts and then just count 5 of them for $\frac{5}{7}$ of 14.

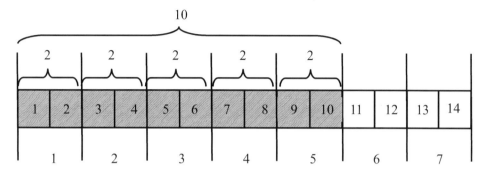

This shows us that $\frac{5}{7}$ of 14 is equal to 10.

<u>Standard Algorithm</u>

To compute using the standard algorithm we simply multiply the 5 times the 14 and divide by 7,

$$\frac{5}{7} \times 14 = \frac{5 \times 14}{7} = \frac{70}{7} = \frac{10 \times \cancel{7}}{\cancel{7}} = 10$$

Notice how we got the same result using our tape diagram.

EXAMPLE 3: $\frac{3}{5}$ of 35 is what number?

<u>Visual Model</u>

Using our tape diagram we interpret this as:

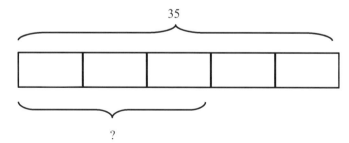

Thus, we must break 35 into 5 equal parts and count 3 of them.

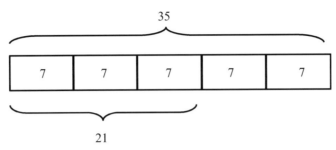

So we see our answer is 21.

<u>Standard Algorithm</u>
Again, if we use the standard algorithm we get:

$$\frac{3}{5} \times 35 = \frac{3 \times 35}{5} = \frac{115}{5} = \frac{21 \times 5}{5} = 21$$

In these examples the answer worked out to be a whole number. Most of the time the result is a fraction, though.

EXAMPLE 4: $\frac{2}{7}$ of 9 is what number?

<u>Standard Algorithm</u>
As we can see with this problem, our answer will not be a whole number. Let's start with the standard algorithm, and then try and justify it with another approach.

$$\frac{2}{7} \times 9 = \frac{2 \times 9}{7} = \frac{18}{7}$$

The final answer is the fraction $\frac{18}{7}$.

<u>Visual Model</u>
To justify this, consider the area model. We start with 9, and then break it into 7 equal parts, and then choose only two of the 7 parts (the shaded region below). Each smaller square is actually $\frac{1}{7}$. There are 18, $\frac{1}{7}$ squares shaded in for a total of $\frac{18}{7}$. Thus, the multiplication is just 18 repeated additions of the unit fraction $\frac{1}{7}$.

As we can see we get the same answer. This is our visual justification of the standard algorithm. This is why it works. For more complicated problems like this, however, we will use the standard algorithm.

$$\frac{2}{7} \text{ of } 9 = 18 \times \frac{1}{7} = \frac{18}{7}$$

Multiplying Fractions Times Fractions

We now consider multiplying a fraction by another fraction, such as $\frac{4}{5} \times \frac{2}{3}$. Again, this is still the multiplication operation, so the same ideas and concepts we've been discussing in this section still apply. Multiplication is still a repeated addition. What we are adding, and how many times, however, is not as obvious. In what follows we will attempt to explain it to you. Let's look at an example.

EXAMPLE 5: $\frac{4}{5}$ of $\frac{2}{3}$ is what number? Or $\frac{4}{5} \times \frac{2}{3}$ equals what number?

In this case we will start with an area interpretation before using the standard algorithm. We first want to understand the concept before we learn the short cut (standard algorithm) in finding our results. A conceptual understanding should always come first.

Visual Model

In considering this type of problem, $\frac{4}{5} \times \frac{2}{3}$, we will use the second interpretation of multiplying by a fraction in the beginning of this section. We have already shown that multiplying by a fraction is equivalent to saying we want to find the number that is $\frac{4}{5}$ of $\frac{2}{3}$. Recall that we have already shown that this is just another way to think of multiplication as repeated addition, but addition of specific unit fractions that we have to find.

384

For this problem we shall start with $\frac{2}{3}$

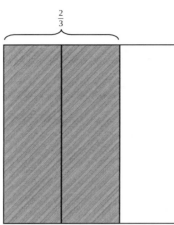

We then break the $\frac{2}{3}$ into fifths, and count only $\frac{4}{5}$ of the $\frac{2}{3}$ as shown below.

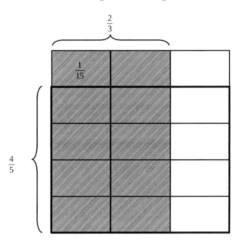

When we break the $\frac{2}{3}$ into fifths we see that the unit fraction we are adding is actually $\frac{1}{15}$.

If we count the overlap of the shaded region with the $\frac{4}{5}$ box we count 8 boxes. Thus, we are actually adding $\frac{1}{15}$ 8 times, for a final answer of $\frac{8}{15}$. Again, we can see that multiplication is still a repeated addition, even when the multipliers are fractions.

Standard Algorithm
Let's now see how this works for the standard algorithm.

$$\frac{4}{5} \times \frac{2}{3} = \frac{4 \times 2}{5 \times 3} = \frac{8}{15}$$

Again, the same final answer. You can see why standard algorithms were developed, as they make getting the final answer quite simple. The problem is when people use the standard algorithm and think this tells them what they are multiplying and what are the repeated unit fractions they are working with. It really doesn't. This is why so many students mess up the multiplication of fractions. They have learned so many standard algorithms that they all begin to merge together in their minds, so sometimes they mix them up because the concepts are not that obvious in the different algorithms.

Dividing Fractions

Division is the process of finding the number used in the repeated additions of our related multiplication problem. Division is sometimes called the reciprocal relation of multiplication, and we will show why.

The division problem:

$$(\text{Number 1}) \div (\text{Number 2})$$

is essentially asking the question; What number added to itself (Number 2) times equals (Number 1)? Sometimes we rephrase this question and ask; How many (Number 2's) are there in (Number 1)?

If you think about this carefully you will see that both questions are actually asking you to find the same thing, but only in a different way. Why have two ways to ask the question? Because sometimes one question gives you a better grasp on understanding what is being asked for a particular problem than the other question. We'll show this through examples below.

We'd also like to point out that these questions are addressing what the concept of division really is. Eventually we will end up with a standard algorithm to use, but we don't want to lose sight of the concept of what we are actually doing. The algorithm is not division, it is just a set of steps used to quickly find our answer, with no obvious connection to the concept of division. We stress this point so that you don't get confused later on. Don't mistake the standard algorithm for division. Division is a concept, the standard algorithm is not.

We'll start by showing examples of division, for different types of division problems involving fractions.

Dividing a Fraction by a Whole Number

Let's start with the simplest form of division, that of dividing a fraction by a whole number. For this type of problem, the question:

What number added to itself (Number 2) times equals (Number 1)?

makes more sense intuitively. Anytime the fraction is the first number and a whole number is the second number you should use this conceptual interpretation of the division problem.

EXAMPLE 1: Perform the division $\dfrac{3}{4} \div 6$

We are actually asking; What number added to itself 6 times is equal to $\dfrac{3}{4}$? This question makes more sense than asking; How many 6's are in $\dfrac{3}{4}$? They both, however, are asking the same question.

Visual Model

To find the answer visually, let's look at the area model. We start with $\frac{3}{4}$ as shown in the figure to the right.

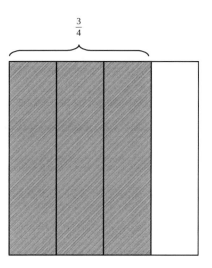

We then break (divide) it into 6 equal pieces, and figure out the size of an individual one sixth piece. We are looking for the amount in one part out of 6 of our 3/4, so we count how many pieces we have as shown below.

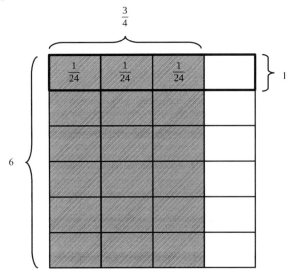

From the figure we see that when we break 3/4 into 6 pieces, each piece contains boxes that are 1/24 in size. Then we simply count the number of 1/24 boxes we get, which is 3, so our final answer is.

$$\frac{3}{24} = \frac{1 \times 3}{8 \times 3} = \frac{1}{8}$$

The answer is $\frac{3}{24}$ or $\frac{1}{8}$.

This means that if we add $\frac{3}{24}$ to itself 6 times we would have

$$\frac{18}{24} = \frac{3}{4}$$

We also see that the answer to the other question also works, but is harder to understand, i.e. there are $\frac{3}{24}$ 6's in 3/4.

Standard Algorithm

The standard algorithm is as follows. We start by changing all division problems to a multiplication by the reciprocal of the term we are dividing by. This makes sense for the problem above says that $\frac{3}{4} \div 6$ is equivalent to finding 1/6 of 3/4, or

$$\frac{3}{4} \div 6 = \frac{3}{4} \times \frac{1}{6}$$

Then we use the multiplication algorithm we developed earlier to obtain

$$\frac{3}{4} \times \frac{1}{6} = \frac{3 \times 1}{4 \times 6} = \frac{3}{24}.$$

Which we have shown is equivalent to $\frac{1}{8}$.

If we think about this, we can see that we were also looking for 1/6 of 3/4. This really is our multiplication problem in reverse. We want to know what number is 1/6 of 3/4 and that number is 1/8. This number is what we must add 6 times to get 3/4.

EXAMPLE 2:

Let's look at a slightly different example. Imagine that you have $\frac{2}{5}$ of a cup of frosting to share equally among three desserts. How would we write this as a division question? It is asking us to find $\frac{2}{5}$ of 3, or mathematically

$$\frac{2}{5} \div 3$$

Again, we are asking; What number added to itself 3 times is equal to $\frac{2}{5}$?

<u>Visual Model</u>
We can start by drawing a model of two-fifths shown to the right.

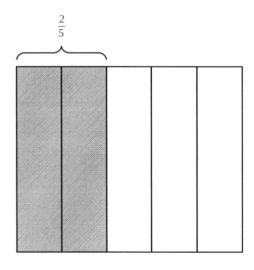

Next we show dividing two-fifths into three equal parts, and taking one of those 3 parts as our answer, shown below:

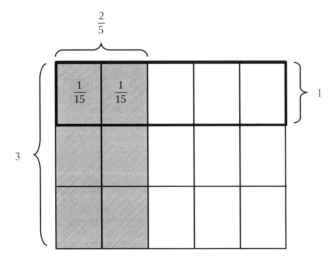

From the visual model, we can determine that $\frac{2}{5} \div 3 = \frac{2}{15}$.

$$\frac{2}{5} \div 3 = \frac{2}{5} \times \frac{1}{3} = \frac{2 \times 1}{5 \times 3} = \frac{2}{15}$$

Dividing a Whole Number by a Fraction

For these type of problems we are asking the question; How many (Number 2's) are in (Number 1)? Anytime a whole number comes first it may make sense to phrase the division problem in this way. Again, it makes more sense intuitively for certain problems. This is due to our intuitive bias to think of a number as being a whole number and not a fraction. Sometimes the second question; What number added to itself (Number 2) times equals (Number 1), will make more sense, though. You should look at both so that you better understand what you are being asked to find.

EXAMPLE 3: Perform the division $5 \div \frac{1}{2}$

This problem asks the question; How many $\frac{1}{2}$'s are there in 5?

Visual Model

We start with 5 units and divide each unit in half as shown below.

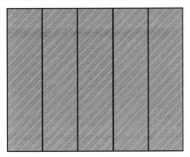

And then count the number of half boxes.

Our answer is $5 \div \frac{1}{2} = 10$

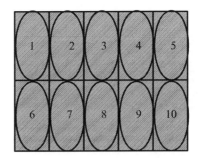

Standard Algorithm

Again, we just multiply 5 by the reciprocal of 1/2, which is 2.

$$5 \div \frac{1}{2} = 5 \times \frac{2}{1} = 10$$

EXAMPLE 4: Perform the division $3 \div \dfrac{2}{5}$

This problem asks the question; How many $\dfrac{2}{5}$'s are there in 3? or What number added to itself $\dfrac{2}{5}$ times equals 3. As you can see the second question is less intuitive.

<u>Visual Model</u>
We start with 3 units. divide it into fifths and count how many two–fifths we have

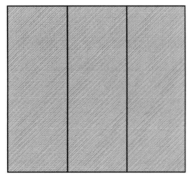
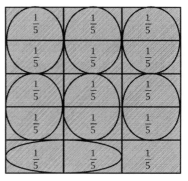

We can see we have 7 and one left over. The answer is $7 \dfrac{1}{2} = \dfrac{15}{2}$

<u>Standard Algorithm</u>

$$3 \div \frac{2}{5} = 3 \times \frac{5}{2} = \frac{3 \times 5}{2} = \frac{15}{2}$$

Dividing a Fraction by a Fraction
This is the final type of division problem we can expect. Since the second number is a fraction it makes more sense to ask the question;

How many (Number 2)'s are there in (Number 1)?

We start with an example where we divide a fraction by a fraction with the same denominator. We do this to develop an intuitive understanding before it becomes more abstract.

EXAMPLE 5: Perform the division $\dfrac{8}{9} \div \dfrac{2}{9}$

How many $\dfrac{2}{9}$'s are there in $\dfrac{8}{9}$?

Visual Model

We start with $\dfrac{8}{9}$ units

We then count how many $\dfrac{2}{9}$'s there are

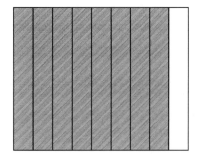

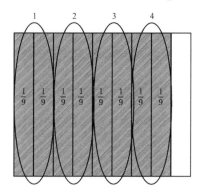

From our model we can see the answer is 4. Thus, $\dfrac{8}{9} \div \dfrac{2}{9} = 4$

Standard Algorithm

$$\dfrac{8}{9} \div \dfrac{2}{9} = \dfrac{8}{9} \times \dfrac{9}{2} = \dfrac{8 \times 9}{9 \times 2} = \dfrac{8}{2} = 4$$

EXAMPLE 6: Perform the division $\dfrac{9}{12} \div \dfrac{3}{12}$

How many $\dfrac{3}{12}$'s are there in $\dfrac{9}{12}$?

Visual Model

We start with $\dfrac{9}{12}$ units

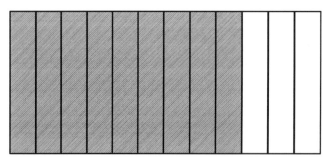

We then count how many $\dfrac{3}{12}$'s there are in this number.

From our model we can see the answer is 3. Thus, $\dfrac{9}{12} \div \dfrac{3}{12} = 3$

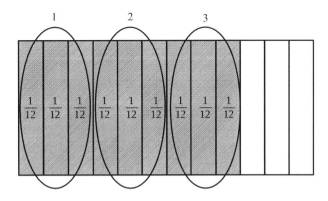

<u>Standard Algorithm</u>

$$\frac{9}{12} \div \frac{3}{12} \;=\; \frac{9}{12} \times \frac{12}{3} \;=\; \frac{9 \times \cancel{12}}{\cancel{12} \times 3} \;=\; \frac{9}{3} \;=\; 3$$

We now do two examples where the denominators are not the same.

EXAMPLE 7: Perform the division $\dfrac{5}{6} \div \dfrac{3}{8}$

How many $\dfrac{3}{8}$'s are there in $\dfrac{5}{6}$?

<u>Visual Model</u>

We start with $\dfrac{5}{6}$

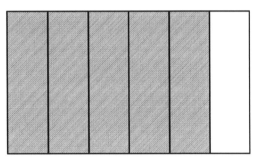

We need to count how many $\dfrac{3}{8}$'s there are in this number. To do this, we first need to construct $\dfrac{1}{8}$'s in our model. This can be a bit confusing, so bear with us.

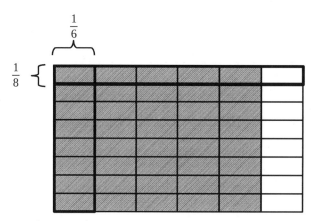

What we are showing here is $\frac{1}{8}$ and $\frac{1}{6}$ and where they overlap is actually $\frac{1}{48}$, but we don't really need this value. What we need to show is $\frac{3}{8}$ and count how many boxes of this size would be in a fully shaded $\frac{3}{8}$ amount. From what we see below, $\frac{3}{8}$ contains 18 of these boxes.

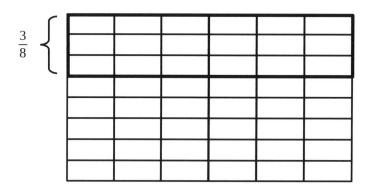

Thus we need to count 18 shaded boxes in our previous model, and then count how many groups of 18 can be formed.

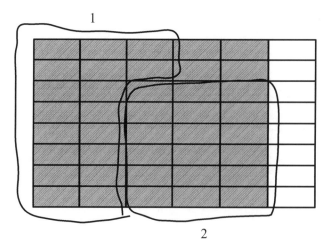

We see that we have two full groups and four left over. That means the last groups is 4 out of 18 or $\frac{4}{18}$. So, we have: $\frac{5}{6} \div \frac{3}{8} = \frac{5}{6} \times \frac{8}{3} = \frac{5 \times 8}{6 \times 3} = \frac{40}{18} = \frac{20}{9} = 2\frac{2}{9}$ or we count the number of eighteenths to get $\frac{40}{18} = \frac{20}{9}$.

<u>Standard Algorithm</u>

$$\frac{5}{6} \div \frac{3}{8} = \frac{5}{6} \times \frac{8}{3} = \frac{5 \times 8}{6 \times 3} = \frac{40}{18} = \frac{20}{9}, \text{ or } 2\frac{4}{18} = 2\frac{2}{9}$$

Obviously, the standard algorithm is far less work and we don't want to do the models for every problem. We are simply using them here to demonstrate that there is a concept behind the division algorithm, but you really can't see it. That is the down side of an algorithm, it hides the concept, but it makes the process to get a result so much easier.

EXERCISES A.4

Multiplying and Dividing Fractions

<u>Multiplication:</u> Solve using a visual model and the standard algorithm.

1. $\frac{1}{3}$ of 18

2. $\frac{1}{3}$ of 36

3. $\frac{3}{4} \times 24$

4. $\frac{3}{8} \times 24$

5. $\frac{4}{5} \times 25$

6. $\frac{1}{7} \times 140$

7. $\frac{1}{2} \times 8$

8. $8 \times \frac{1}{2}$

9. $\frac{3}{5} \times 10$

10. $10 \times \frac{3}{5}$

11. $14 \times \frac{3}{7}$

12. $\frac{3}{4} \times 36$

13. $\frac{1}{2}$ of $\frac{2}{2}$

14. $\frac{2}{3}$ of $\frac{1}{2}$

15. $\frac{3}{4}$ of $\frac{4}{5}$

16. $\frac{2}{5}$ of $\frac{2}{3}$

17. $\frac{1}{2} \times \frac{4}{5}$

18. $\frac{2}{3} \times \frac{1}{4}$

Set up and solve with a visual model

19. $\frac{2}{3}$ of a number is 10. What's the number?

20. $\frac{3}{4}$ of a number is 24. What's the number?

Solve using the standard algorithm

21. $\frac{3}{4} \times \frac{2}{3}$

22. $\frac{4}{5} \times \frac{5}{8}$

23. $\frac{3}{4} \times \frac{5}{6}$

24. $\frac{2}{3} \times \frac{6}{7}$

25. $\frac{4}{9} \times \frac{3}{10}$

26. $\frac{3}{11} \times \frac{7}{9}$

27. $\frac{11}{12} \times \frac{13}{5}$

Applications of Multiplying Fractions. Solve using whatever technique you wish.

28. There are 48 students going on a field trip. One-fourth are girls. How many boys are going on the trip?

29. Abbie spent $\frac{5}{8}$ of her money and saved the rest. If she spent $45, how much money did she have at first?

30. A marching band is rehearsing in rectangular formation. $\frac{1}{5}$ of the marching band members play percussion instruments. $\frac{1}{2}$ of the percussionists play the snare drum. What fraction of all the band members play the snare drum?

31. Phillip's family traveled $\frac{3}{10}$ of the distance to his grandmother's house on Saturday. They traveled $\frac{4}{7}$ of the remaining distance on Sunday. What fraction of the total distance to his grandmother's house was traveled on Sunday?

32. Santino bought a $\frac{3}{4}$ pound bag of chocolate chips. He used $\frac{2}{3}$ of the bag while baking. How many pounds of chocolate chips did he use while baking?

33. Farmer Dave harvested his corn. He stored $\frac{5}{9}$ of his corn in one large silo and $\frac{3}{4}$ of the remaining corn in a small silo. The rest was taken to market to be sold.
    a. What fraction of the corn was stored in the small silo?
    b. If he harvested 18 tons of corn, how many tons did he take to market?

34. $\frac{5}{8}$ of the songs on Harrison's music player are hip-hop. $\frac{1}{3}$ of the remaining songs are rhythm and blues. What fraction of all the songs are rhythm and blues?

35. Three-fifths of the students in a room are girls. One-third of the girls have blond hair. One-half of the boys have brown hair.
    a. What fraction of all the students are girls with blond hair?
    b. What fraction of all the students are boys without brown hair?

Division: Solve using a visual model to support your answer.

36. $\frac{3}{5} \div 6$	40. $4 \div \frac{2}{5}$	44. $\frac{4}{5} \div \frac{2}{5}$	48. $\frac{11}{9} \div \frac{3}{9}$
37. $\frac{2}{3} \div 4$	41. $3 \div \frac{2}{3}$	45. $\frac{9}{4} \div \frac{3}{4}$	49. $\frac{3}{4} \div \frac{2}{3}$
38. $\frac{5}{6} \div 3$	42. $7 \div \frac{3}{8}$	46. $\frac{7}{8} \div \frac{2}{8}$	50. $\frac{3}{5} \div \frac{6}{8}$
39. $\frac{4}{5} \div 2$	43. $9 \div \frac{1}{6}$	47. $\frac{13}{10} \div \frac{2}{10}$	

Solve using the standard algorithm

51. $\dfrac{7}{9} \div \dfrac{8}{11}$

52. $\dfrac{5}{11} \div \dfrac{3}{7}$

53. $\dfrac{3}{7} \div \dfrac{5}{16}$

54. $\dfrac{11}{2} \div \dfrac{4}{3}$

55. $\dfrac{9}{13} \div \dfrac{3}{10}$

56. $\dfrac{6}{5} \div \dfrac{1}{7}$

57. $\dfrac{1}{9} \div \dfrac{2}{5}$

58. $\dfrac{12}{5} \div \dfrac{7}{5}$

Applications for division of fractions

59. Mother has 20 pounds of salad. She wants each person to have $\dfrac{1}{4}$ pounds of salad. How many people can she feed with the 20 pounds of salad that she has?

60. I bought 10-lbs of sugar to make cakes. Each cake needs $\dfrac{3}{4}$ cups of sugar. Assuming that the 10-lbs of sugar is approximately 20 cups, how many cakes can I make with the 10 -lbs of sugar?

61. A piece of ribbon $\dfrac{18}{5}$ m long is cut into 12 shorter pieces of equal length. What is the length of each short piece?

A.5 DECIMAL FRACTIONS (DECIMALS)

In the previous sections we saw that fractions can be considered as numbers, since they enable us to measure in–between values. However, the notation that we used for fractions, did not make them appear to be actual numbers, but rather pairs of numbers used to represent parts of a whole. Then add to that there were infinitely many ways to write the same numbers using equivalent fractions, and you are bound to be confused. Finally, the rules for arithmetic are really quite complicated, so that the standard algorithms can confuse even the brightest students. This is the major reason why fractions are so difficult for many students to understand well enough to be able to master.

In this section we will show that fractions can be represented in an alternative way, so that they behave more like the natural and whole numbers we are familiar with. Instead of being written using the fraction notation of having a specific number for the numerator and one for the denominator, where we can have infinitely many different ways of writing the same value, we can standardize what we use for the denominators and just vary the numerator. This is really just an extension of the place–value numeral system we developed in Chapter 1, and makes many of the operations on fractions easier to work with and understand. This new notation is called decimal fractions or decimals for short.

Decimal Notation

In decimals the denominators are always powers of 10. for example:

$$\dfrac{1}{10} \ \text{or one-tenth}, \quad \dfrac{1}{100} \ \text{or one-hundredth}, \quad \dfrac{1}{1000} \ \text{or one-thousandth, etc.}$$

We remove the denominator from our representation, since we will always use the same denominators with our decimal fractions, and just the numerator term will change.

For example, to write the combined sum of fractions we simply write this in the following way:

$$\frac{3}{10} + \frac{5}{100} + \frac{7}{1000}$$

First, we need to identify that the number we are writing is a fraction. For this we introduce the decimal point notation. Anytime we see a period in a number, we are to interpret whatever comes after the period to the right as a fraction less than one. The period is called a decimal point.

Finally, we only write the numerator terms of the fractions, since the denominators are already known. Thus, we can write the sum of the three fractions above as

$$0.357$$

Where it is to be understood that the period, or decimal point, means that the numbers following this point are all numerators for fractions with their denominators coming in increasing powers of 10 as shown above.

Now, for numbers that have a whole part greater than one, we simply write the whole number part of the number to the left of the decimal point, and the fraction part to the right of the decimal.
As an example, the number 524.75 is understood to be

$$500 + 20 + 4 + \frac{7}{10} + \frac{5}{100}$$

Using this new notation, we can also see the connection to the place–values we used earlier.

Alternatively we could write this number as:

$$5 \times 100 + 2 \times 10 + 4 \times 1 + 7 \times \frac{1}{10} + 5 \times \frac{1}{100}$$

with the values after the times symbols being the place–values.

This means we can extend our place values to numbers to the right of the decimal point representing fractions with powers of 10 in their denominators. The place–values are now:

...	Hundreds	Tens	Ones	.	Tenths	Hundredths	Thousandths	...
...	100	10	1	.	1/10	1/100	1/1000	...

We explore this new notation further in the next section by also introducing the metric system of units. The metric system is a measurement system based on the same concept of multiplying and dividing by powers of 10 to get new unit types.

We now provide a few examples of writing decimals in expanded notation.

EXAMPLE 1: Write the decimal in expanded form, 0.59

$$0.59 = 5 \times \frac{1}{10} + 9 \times \frac{1}{100}$$

EXAMPLE 2: Write the decimal in expanded form, 1.703

$$1.703 = 1 + 7 \times \frac{1}{10} + 0 \times \frac{1}{100} + 3 \times \frac{1}{1000}$$

EXAMPLE 3: Write the decimal in expanded form, 36.34

$$36.34 = 3 \times 10 + 6 + 3 \times \frac{1}{10} + 4 \times \frac{1}{100}$$

EXAMPLE 4: Write the decimal in expanded form, 237.4981

$$237.4981 = 2 \times 100 + 3 \times 10 + 7 + 4 \times \frac{1}{10} + 9 \times \frac{1}{100} + 8 \times \frac{1}{1,000} + 1 \times \frac{1}{10,000}$$

Comparing Decimals

Another necessary skill when working with decimals is to determine their order and to be able to place them correctly on a number line. In this section we show how to determine if a decimal number is either less than ($<$) or greater than ($>$) another decimal number.

When comparing two decimals you need to compare them place–value by place–value starting with the largest place–value from the left. If the digits in the same place–values are the same then the numbers are the same. If, however, you hit a place–value where one digit is less than the other similarly placed digit, then that decimal is less than the decimal you are comparing it to. For example:

$$21.3174 < 21.32$$

The numbers are the same in the 10's, one's and tenths, but differ in the hundredth's place, Since 2 is greater than 1, 21.32 is larger than 21.3174, even though the second number has more digits.

Consider the following examples.

EXAMPLE 1: Compare 103.576 to 3.578
In this case it is obvious that $103.576 > 3.578$ since $100 > 3$.

EXAMPLE 2: Compare 0.049 to 0.05
The numbers are the same up until the hundredth's place. Since 4 is less than 5 we see that $0.049 < 0.05$

EXAMPLE 3: Compare 5.07081 and 5.1

The numbers are the same up until the tenth's place. Since 0 is less than 1 we see that 5.07081 < 5.1

EXAMPLE 4: Order the set of numbers and then place them correctly on a number line:
$$\{4.9, 5.07081, 5.1, 5.01, 5.7003, 5.68, 5.09\}$$

Following the previous examples we can order these numbers as follows:

$$4.9 < 5.01, 5.07081 < 5.09 < 5.1 < 5.68 < 5.7003$$

Putting them correctly on a number line we have:

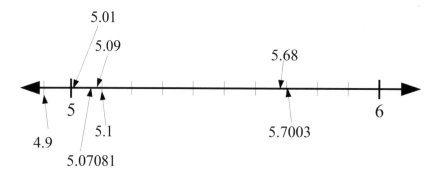

Adding Decimals

You can add decimals in two different ways. The first approach to adding decimals is to add the two decimals according to their place–values and whenever you obtain a 10 or larger in a place value you are adding, you carry the amount over to the next place–value. Exactly how you learned to add whole numbers.

EXAMPLE 1: Here is an example of adding 5.6 and 4.53

Convert both to their place–value form:

$$5.6 = 5 + 6 \times \frac{1}{10} \text{ and } 4.53 = 4 + 5 \times \frac{1}{10} + 3 \times \frac{1}{100}$$

Next we add the two and combine like place–value terms

$$5.6 + 4.53 = (5 + 4) + (5 + 6)\times\frac{1}{10} + (3 + 0)\times\frac{1}{100}$$

$$= 9 + 11\times\frac{1}{10} + 3\times\frac{1}{100}$$

$$= (9 + 1) + 1\times\frac{1}{10} + 3\times\frac{1}{100}$$

$$= 10 + 1\times\frac{1}{10} + 3\times\frac{1}{100}$$

$$= 10.13$$

EXAMPLE 2: Consider another example of adding 6.8 to 5.7

$$6.8 + 5.7 = (6 + 5) + (8 + 7)\times\frac{1}{10}$$

$$= 11 + 15\times\frac{1}{10}$$

$$= (11 + 1) + 5\times\frac{1}{10}$$

$$= 12 + 5\times\frac{1}{10}$$

$$= 12.5$$

The second approach is by converting the decimals to fractions and adding the like whole numbers and fractions.

EXAMPLE 3: Consider the example of adding 5.6 and 4.53, that we did earlier

$$5.6 + 4.53 = 5 + \frac{6}{10} + 4 + \frac{5}{10} + \frac{3}{100}$$

$$= (5 + 4) + (\frac{6}{10} + \frac{5}{10}) + \frac{3}{100}$$

$$= 9 + \frac{11}{10} + \frac{3}{100}$$

$$= 9 + 1 + \frac{1}{10} + \frac{3}{100}$$

$$= 10 + \frac{1}{10} + \frac{3}{100}$$

$$= 10.13$$

400

Multiplying Decimals

EXAMPLE 1: We shall demonstrate multiplying using the expanded notation.

Multiply 3.2 × 2.4.

In expanded notation, $3.2 = 3\times1 + 2\times \dfrac{1}{10}$ and $2.4 = 2\times1 + 4\times \dfrac{1}{10}$.

Recall how we multiply two numbers using partial products using a box, we are going to show how this multiplication is done using that method.

	2	$\dfrac{4}{10}$
$\dfrac{2}{10}$	$\dfrac{4}{10}$	$\dfrac{8}{100}$
3	6	$\dfrac{12}{10}$

$$\begin{array}{r} 2 \bullet 4 \\ \times \quad 3 \bullet 2 \\ \hline 6 \bullet \\ 1 \bullet 2 \end{array}$$

3 ones × 2 ones $3 \times 2 = 6$

3 ones × 4tenths $3\times \dfrac{4}{10} = \dfrac{12}{10}$

$\bullet 4$ 2 ones × 2 tenths $2 \times \dfrac{2}{10} = \dfrac{4}{10}$

$\bullet 08$ 2 tenths × 4 tenths $\dfrac{2}{10} \times \dfrac{4}{10} = \dfrac{8}{100}$

$$7 \bullet 68$$

Rounding Decimals

Before we multiply and divide decimal numbers, we will learn how to round numbers to the nearest place value indicated. For example, if we want to round 238 to the nearest tens, we want to determine if 238 is closer to 230 or 240. 230 and 240 are called the benchmark numbers. Below is the segment that shows the two benchmark numbers and the location of 238.

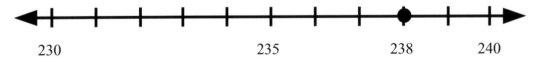

230   235   238   240

Notice that 235 is the midpoint between 230 and 240. We can see on the number line that 238 is closer to 240 or 24 tens, so we say that to round 238 to the nearest tens is 24 tens or 240.

Another example, round 2,318 to the nearest thousand, we first establish the benchmark numbers. They are 2,000 and 3,000.

2000   2318   2500   3000

Since 2,318 is closer to 2,000, we say that 2,318 rounded to the nearest thousand is 2,000.

Suppose we want to round 2,318 to the nearest hundred, what would be the two benchmark numbers that we will use? They are 23 hundreds or 2,300 and 24 hundreds or 2,400.

2300   2318   2350   2400

 Since 2,318 is closer to 2,300, we say that 2,318 rounded to the nearest hundred is 2,300.

To do one more, we round 2,318 to the nearest tens, the two benchmark numbers are 2,310 and 2,320.

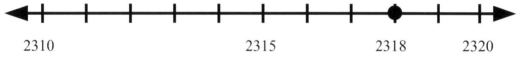

2310   2315   2318   2320

Since 2,318 is closer to 2,320 or 232 tens, we round 2,318 to the nearest tens which is 2,320.

If we want to round 2,315 to the nearest tens, we will adopt the rule to round it to the higher tens which in this case is 2,320. There are some other rules that determine how to round this number but we will not use these rules.

Rounding decimal numbers to the nearest whole number, tenths, hundredths, and so on is the same as the way we round the whole numbers. We figure out which two benchmark numbers the number we want to round are and use the number line to place the benchmark numbers. Then place the number we want to round on that number line and determine which benchmark number it is closer to.

EXAMPLE 1: Round 3.84 to the nearest tenths.

In this example the two benchmark numbers are 3.8 and 3.9.

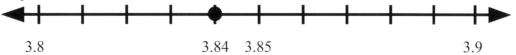

3.8   3.84   3.85   3.9

402

Since 3.84 is closer to 3.8, we round it to 3.8 or 3 and 8 tenths.

EXAMPLE 2: Round 3.84 to the nearest ones.

The two benchmark numbers are 3 and 4.

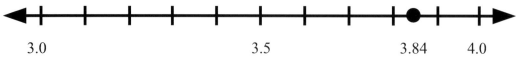

3.0   3.5   3.84   4.0

Since 3.84 is closer to 4, we round it to the nearest ones which is 4.

We could also do the rounding using a vertical line, but we'll stick to the horizontal number line.

EXAMPLE 3: Round 247.728 to the nearest tenths.

To round this number to the nearest tenths, the two benchmark numbers we will use are 247.7 and 247.8. The midpoint between these two numbers is 247.75.

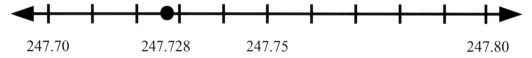

247.70   247.728   247.75   247.80

Since 247.728 is closer to 247.7, we round the number to 247.7.

EXAMPLE 4: Round 369.46 to the nearest tenths.

The two benchmark numbers that we use in this case are 369.4 and 369.5.

369.46

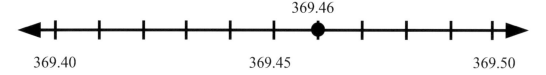

369.40   369.45   369.50

Since the number 369.46 is closer to 369.5, we round the number to its nearest tenths which is 369.5.

EXAMPLE 5: Round 369.46 to the nearest ones or whole number.

The two benchmark numbers in this case are 369 and 370.

369.46

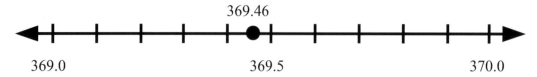

369.0   369.5   370.0

Since the number 369.46 is closer to 369, we round the number to the nearest whole number or ones which is 369.

Estimations with Decimals

Rounding numbers is used in estimating the operations of numbers.

EXAMPLE 1: Add 232 + 589.

We could easily add these two numbers using the standard algorithm. However, we could get a very quick estimate by rounding each of the addends to the nearest hundreds. 232 can be rounded to 200 while 589 can be rounded to 600 as it is closer to 600. Then we can add quickly 200 + 600 = 800 to see that our sum once we add these two numbers, it should be close to 800.

In fact when we add 232 + 589, we get 821.

EXAMPLE 2: Multiply 12.9 × 8.3.

We round 12.9 to ones which will be 13. We round 8.3 to the nearest ones which is 8.

Using 13 × 8, we get 104.

When we actually multiply the two numbers, 12.9 × 8.3, we get $\frac{129}{10} \times \frac{83}{10} = \frac{10{,}707}{100} = 107.07$

Note that the denominator is 100, that means the last digit of the product will be the hundredths.

We could round 12.9 to the nearest tens which is 10 and round 8.3 to the nearest tens too which is 10, the product 10×10= 100 which is not too far from the actual product too.

Estimation is very useful particularly in multiplication and division of decimal numbers.

EXAMPLE 3: Multiply 32.82 ×79.23.

We could first round each number to the nearest tens which will allow us to do a quick calculation.

32.82 can be rounded to 30 as it is closer to 30 and
79.23 can be rounded to 80 as it is closer to 80.

Then we can multiply 30×80 which is 2400. Note that this is an estimation.

When we multiply these two numbers, Rewriting 32.82 as a fraction, we have $32\frac{82}{100}$ which is $\frac{3282}{100}$ while 79.23 is $\frac{7923}{100}$.

Now, multiplying $\frac{3282}{100} \times \frac{7923}{100} = \frac{3282 \times 7923}{10{,}000} = \frac{26{,}003{,}286}{10{,}000} = 2{,}600.3286$

Note that the denominator is 10,000, so the last digit of the answer will be the ten thousandths.

Compare this answer from the estimate that we had which was 2,400.

EXERCISES A.5

Write the decimals in expanded form.

1. 0.035
2. 1.19
3. 16.0843
4. 25.0938
5. 11.002
6. 45.02023

Compare the decimals

7. 23.4, 23.39
8. 16.8, 16.09
9. 0.035,   0.03
10. 13.29, 3.3
11. 8.79,   7.8
12. 15.4309,   15.43

Order the set of numbers and then place them correctly on a number line:

13. {3.77, 3.02, 3.58, 3.95, 3.09, 3.8}

14. {0.23, 0.901, 0.3, 0.73, 0.703, 0.98}

15. {2.81, 3.0, 2.18, 2.06, 2.2, 2.005, 2.7, 2.71}

16. {11.9, 11.001, 11.71, 11.930, 11.554, 11.3, 11.06}

17. {1.7, 1.9, 1.07, 1.38, 1.83, 1.76, 1.67, 1.99, 2.0}

18. {5.05, 4.95, 5.15, 5.51, 5.85, 5.58, 5.976}

Add the decimals

19. 32.06 + 9.57
20. 16.2 + 3.89
21. 11.543 + 54.897
22. 0.395 + 2.785
23. 13.2 + 26.095
24. 8.07 + 12.387

Multiply the decimals

25. 2.5×3.7
26. 7.4×2.1
27. 1.21×3.07
28. 8.561×2.3
29. 7.93×5.6
30. 3.6×7.38

Round to the nearest whole number, tenth's, and hundredth's

31. 87.943
32. 389.854
33. 0.138
34. 2.793
35. 17.832
36. 20.097

37. Round the following numbers to the nearest tens.
    a) 376
    b) 9,636
    c) 126

38. Round the following numbers to the nearest hundreds.
    a) 3,786
    b) 12,349
    c) 9,023

39. Round the following numbers to the nearest tenths.
    a) 14.74
    b) 6.472
    c) 89.943

40. Round the following numbers to the nearest hundredths.
    a) 10.963
    b) 59.095
    c) 7.007

41. Round the following numbers to the nearest thousandths.
    a) 0.6718
    b) 15.7093
    c) 1.0086

Find the decimal values of the following fractions and round to the nearest hundredths, if necessary:

42. $\dfrac{1}{9}$ 43. $\dfrac{2}{5}$ 44. $\dfrac{1}{7}$ 45. $\dfrac{1}{11}$ 46. $\dfrac{3}{8}$ 47. $\dfrac{1}{6}$

Estimate each of the following correct to 1 significant digit, then perform the operation using your calculator correct to 3 significant digits and compare your answer to the estimate.

48. 28.8×2.1 49. $52.3 \div 4.8$ 50. 201.57×19.75 51. $201.57 \div 19.8$

Answers to Exercises

Chapter 1
Section 1.2

	Number	Word	Expanded form	Place-Value Chart
1.	29	twenty-nine	$20+9=$ $2\times10+9\times1$	Tens: 2, Ones: 9
2.	36	Thirty-six	$30+6=$ $3\times10+6\times1$	Tens: 3, Ones: 6
3.	125	One hundred twenty-five	$100+20+5=$ $1\times100+2\times10+5\times1$	Hundreds: 1, Tens: 2, Ones: 5
4.	237	Two hundred thirty-seven	$200+30+7=$ $2\times100+3\times10+7\times1$	Hundreds: 2, Tens: 3, Ones: 7
5.	3,458	Three thousand, four hundred fifty-eight	$3000+400+50+8=$ $3\times1000+4\times100+5\times10+8\times1$	Thousands: 3, Hundreds: 4, Tens: 5, Ones: 8

	Number	Word	Expanded form	Place-Value Chart
6.	7,239	Seven thousand, two hundred thirty-nine	7000+200+30+9 = 7×1000+2×100+3×10+ 9×1	Thousands 7 / Hundreds 2 / Tens 3 / Ones 9
7.	13,452	Thirteen thousand, four hundred fifty-two	10,000+3000+400+50+2 = 1×10000+3×1000+ 4×100+5×10+ 2×1	Ten thousands 1 / Thousands 3 / Hundreds 4 / Tens 5 / Ones 2
8.	79,872	Seventy-nine thousand, eight hundred seventy-two	70,000+9000+800+70+2 = 7×10000+9×1000+ 8×100+7×10+ 2×1	Ten thousands 7 / Thousands 9 / Hundreds 8 / Tens 7 / Ones 2
9.	326,913	Three hundred twenty-six thousand, nine hundred thirteen	300,000+20,000+6000+ 900+10+3 = 3×100000+2×10000+ 6×1000+9×100+ 1×10+3×1	Hundred thousands 3 / Ten thousands 2 / Thousands 6 / Hundreds 9 / Tens 1 / Ones 3

10. thousands   12. hundreds   14. millions
11. ten-thousands   13. ten-thousands   15. ten-thousands

Section 1.3

Adding Integers

| 1. 1 | 3. 20 | 5. 0 | 7. −45 | 9. 4 | 11. −14 |
| 2. −11 | 4. −12 | 6. −42 | 8. −100 | 10. −10 | 12. −4 |

Subtracting Integers

| 13. −6 | 15. 4 | 17. −46 | 19. 110 | 21. 4 | 23. −17 |
| 14. −3 | 16. 9 | 18. 35 | 20. −11 | 22. 2 | 24. −6 |

Adding and Subtracting Three or More Integers

| 25. 0 | 27. 0 | 29. 11 | 31. 6 | 33. −6 | 35. 9 |
| 26. −5 | 28. −14 | 30. −19 | 32. −3 | 34. 7 | 36. −11 |

Multiplying Integers

37. −56   38. 24   39. −104   40. −175   41. 48   42. −99   43. 144

44. -80 46. -45 48. 301   50. 65   52. 72   54. -42 56. 54
45. 126   47. 138   49. -196 51. 112   53. 240   55. -108

Dividing Integers

57. 2, $6 \times 2 = 12$
58. 2, $13 \times 2 = 26$
59. 8, $-9 \times 8 = -72$
60. -25, $-5 \times -25 = 125$
61. -11, $4 \times -11 = -44$

62. 18, $-4 \times 18 = -72$
63. 10, $-4 \times 10 = -40$
64. -5, $-7 \times -5 = 35$
65. -5, $-15 \times -5 = 75$
66. 4, $-14 \times 4 = -56$

67. -13, $5 \times -13 = -65$
68. -9, $-8 \times -9 = 72$
69. -6, $-23 \times -6 = 138$
70. 5, $-17 \times 5 = -85$

Section 1.4

Adding Fractions

1. $\dfrac{7}{8}$ 3. $\dfrac{7}{19}$ 5. $\dfrac{11}{26}$ 7. $\dfrac{23}{18}$ 9. $\dfrac{-13}{48}$ 10. $\dfrac{73}{72}$ 11. $\dfrac{-11}{48}$ 12. $\dfrac{43}{66}$

2. $\dfrac{17}{25}$ 4. $\dfrac{5}{9}$ 6. $\dfrac{8}{21}$ 8. $\dfrac{5}{15} = \dfrac{1}{5}$

Subtracting Fractions

13. $\dfrac{2}{10} = \dfrac{1}{5}$ 15. $\dfrac{8}{25}$ 17. $\dfrac{11}{26}$ 18. $\dfrac{-13}{21}$ 19. $\dfrac{3}{18} = \dfrac{1}{6}$ 21. $\dfrac{37}{48}$ 23. $\dfrac{1}{48}$ 24. $\dfrac{-37}{66}$

14. $\dfrac{4}{8} = \dfrac{1}{2}$ 16. $\dfrac{3}{15} = \dfrac{1}{5}$ 20. $\dfrac{5}{15} = \dfrac{1}{3}$ 22. $\dfrac{25}{72}$

Add or Subtract Fractions Applications

25. $\dfrac{7}{20}$ 26. $\dfrac{5}{12}$ 27. $\dfrac{3}{20}$ 28. $\dfrac{33}{40}$

Multiplying Fractions

29. 6   31. -18 33. 20   35. -4 37. 6   39. 6
30. 12   32. 9   34. 20   36. 4   38. 6   40. 27

41. $\dfrac{1}{2}$ 43. $\dfrac{-3}{5}$ 45. $\dfrac{3}{10}$ 47. $\dfrac{1}{2}$ 49. $\dfrac{-5}{8}$ 51. $\dfrac{2}{15}$

42. $\dfrac{1}{3}$ 44. $\dfrac{-4}{15}$ 46. $\dfrac{-1}{6}$ 48. $\dfrac{1}{2}$ 50. $\dfrac{4}{7}$ 52. $\dfrac{7}{33}$

Applications of Multiplying Fractions

53. 36 boys   56. $\dfrac{4}{10} = \dfrac{2}{5}$ 58. a. $\dfrac{1}{3}$, b. 2 tons of corn   60. a. $\dfrac{1}{5}$, b. $\dfrac{1}{5}$
54. \$72

55. $\dfrac{1}{10}$ 57. $\dfrac{1}{2}$ pound   59. $\dfrac{1}{8}$

Division of Fractions

61. $\dfrac{77}{72}$ 62. $\dfrac{-35}{33}$ 63. $\dfrac{48}{35}$ 64. $\dfrac{33}{8}$ 65. $\dfrac{-30}{13}$ 66. $\dfrac{42}{5}$ 67. $\dfrac{5}{18}$ 68. $\dfrac{12}{7}$

Applications of Division of Fractions

69. 80 people   70. about 26 cakes   71. $\dfrac{3}{10} m$

408

Section 1.5

Decimal Fractions

1. $0 \times \dfrac{1}{10} + 3 \times \dfrac{1}{100} + 5 \times \dfrac{1}{1000}$

2. $1 \times 1 + 1 \times \dfrac{1}{10} + 9 \times \dfrac{1}{100}$

3. $1 \times 10 + 6 \times 1 + 0 \times \dfrac{1}{10} + 8 \times \dfrac{1}{100} + 4 \times \dfrac{1}{1000} + 3 \times \dfrac{1}{10,000}$

4. $2 \times 10 + 5 \times 1 + 0 \times \dfrac{1}{10} + 9 \times \dfrac{1}{100} + 3 \times \dfrac{1}{1000} + 8 \times \dfrac{1}{10,000}$

5. $1 \times 10 + 1 \times 1 + 0 \times \dfrac{1}{10} + 0 \times \dfrac{1}{100} + 2 \times \dfrac{1}{1000}$

6. $4 \times 10 + 5 \times 1 + 0 \times \dfrac{1}{10} + 2 \times \dfrac{1}{100} + 0 \times \dfrac{1}{1000} + 2 \times \dfrac{1}{10,000} + 3 \times \dfrac{1}{100,000}$

7. hundredths 8. tenths   9. thousandths   10. hundredths   11. tenths   12. ten-thousandths

Percents

Express percents as fractions in simplest form

13. $\dfrac{19}{100}$ 14. $\dfrac{1}{2}$ 15. $\dfrac{38}{100} = \dfrac{19}{50}$ 16. $\dfrac{9}{10}$ 17. $\dfrac{5}{8}$ 18. $\dfrac{5}{4}$ 19. $\dfrac{61}{200}$ 20. $\dfrac{2}{3}$

Express Percents as Decimals

21. 0.5   22. 0.3   23. 0.275   24. 0.2125   25. 0.45   26. 0.625

Express Fraction as Decimal and then Percent

27. 0.4, 40%   28. 0.875, 87.5%   29. 0.075, 7.5%   30. 2.4, 240%

Express Decimal as Percent and then Fraction

31. 40%, $\dfrac{2}{5}$ 32. 38%, $\dfrac{19}{50}$ 33. 87.5%, $\dfrac{7}{8}$ 34. 192%, $\dfrac{48}{25}$

Applications of Percents

35. 36%

36. 30% is rent, 5% is for gasoline.

37. 25% of the class received an A, 75% did not.

38. 20.16% aluminum, 79.84% chlorine

39. 34.34% iron, 65.66% chlorine

40. 45.95% barium, 54.05% silicon

41. 28.17% silver, 20.24% nitrogen, 51.59% oxygen

42. $\dfrac{21}{70} = 0.3 = 30\%$

43. $\dfrac{37}{148} = 0.25 = 25\%$

More Applications

44. total is 40, 37.5%female, 62.5% male   46. 28%

45. 57%   47. 55%

Sec. 1.6
Properties of Real Numbers
1. Associative law of addition
2. Associative law of mult.
3. Multiplicative inverse
4. Multiplicative identity
5. Commutative law of addition
6. Additive inverse
7. Commutative law of mult.
8. Multiplicative identity
9. Commutative law of addition
10. Additive inverse
11. Multiplicative inverse
12. Additive identity
13. Distributive law
14. Distributive law
15. Associative law of addition

Sec. 1.7
Distributive Property

1. $-6(3+8) = -6\cdot3 + -6\cdot8 = -18 + -48 = -66; \quad -6(3+8) = -6\cdot11 = -66$
2. $4\cdot(6-15) = 4\cdot6 + 4\cdot-15 = 24 + -60 = -36; \quad 4\cdot(6-15) = 4\cdot-9 = -36$
3. $-4\cdot(3-8) = -4\cdot3 - -4\cdot8 = -12 + 4\cdot8 = -12 + 32 = 20; \quad -4\cdot(3-8) = -4\cdot(-5) = 20$
4. $\dfrac{3}{5}\cdot(9-4) = \dfrac{3}{5}\cdot9 + \dfrac{3}{5}\cdot(-4) = \dfrac{27}{5} + \dfrac{-12}{5} = \dfrac{15}{5} = 3 \quad ; \quad \dfrac{3}{5}\cdot(9-4) = \dfrac{3}{5}\cdot5 = 3$

Numbers 1-4 were done for you. Now try numbers 5 - 21 on your own. Answers are below.

5. 20
6. $\dfrac{-25}{9}$
7. $\dfrac{32}{3}$
8. -60
9. 18
10. 24
11. 28
12. 20
13. 0
14. 28
15. 30
16. 8
17. $^-9$
18. $^-9$
19. $\dfrac{24}{5}$
20. $\dfrac{20}{3}$
21. $\dfrac{10}{9}$

Sec. 1.8
Order of Operations

1. 55
2. 11
3. 27
4. 6
5. 16
6. 17
7. -9
8. 3
9. 32
10. -5
11. -18
12. -8
13. -6
14. 8
15. 12
16. 1
17. 13
18. 15
19. -10
20. 18
21. 5
22. 7
23. 8
24. 11
25. 12
26. 18
27. 4
28. 0

Order of Operations – Intermediate and Advanced

29. 7
30. 11
31. 13
32. -1
33. -5
34. 15
35. 24
36. -15
37. -12
38. 3
39. 2
40. -1
41. 8
42. 2
43. 30
44. 7
45. -1
46. 17
47. -4

Chapter 1 Practice Test

1. -5
2. 7
3. $\dfrac{5}{4}$
4. $\dfrac{9}{16}$
5. $\dfrac{-4}{15}$
6. -42
7. 40
8. -90
9. -225
10. 2
11. $\dfrac{5}{4}$
12. $\dfrac{-4}{3}$
13. $\dfrac{-2}{3}$
14. $\dfrac{1}{14}$
15. $\dfrac{-15}{2}$
16. $\dfrac{3}{2}$

17. $\dfrac{2}{3} = \dfrac{4}{6} = \dfrac{20}{30}$ (answers may vary)

18. $\dfrac{3}{8} = \dfrac{6}{16} = \dfrac{30}{80}$ (answers may vary)

19. $9 \times 10,000 + 3 \times 1,000 + 0 \times 100 + 5 \times 10 + 9 \times 1$

20. $2 \times 10 + 1 \times 1 + 3 \times \dfrac{1}{10} + 2 \times \dfrac{1}{100} + 5 \times \dfrac{1}{1000}$

21. $1\times100+9\times10+7\times1+0\times\dfrac{1}{10}+1\times\dfrac{1}{100}+3\times\dfrac{1}{1000}+4\times\dfrac{1}{10,000}$

22. millions 　　　　23. thousandths 　　　　24. ten-thousandths

25. $\dfrac{1}{20}$

26. $\dfrac{19}{50}$

27. $\dfrac{11}{10}$

28. 0.17

29. 1.05

30. 0.02

31. 0.80, 80%

32. 0.75, 75%

33. 15%, $\dfrac{3}{20}$

34. 4%, $\dfrac{1}{25}$

35. 250%, $\dfrac{5}{2}$

36. 55

37. 11

38. 27

39. 6

40. -2

41. 16

42. 20

43. 3

44. 7

45. $^-5$

46. 7

47. 4

48. $\dfrac{1}{12}$

49. 5%

Chapter 2
Section 2.1
Write Each Using Exponents

1. 10^7 　　　2. 10^3 　　　3. 10^6 　　　4. 10^7 　　　5. 10^{11} 　　　6. 10^4

Write Without Exponents and Multiply Out

7. $10\times10\times10=1000$ 　8. $10\times10\times10\times10\times10=100,000$ 　9. $^-10\times^-10\times^-10\times^-10=10,000$

10. $10\cdot10\cdot10\cdot10\cdot10\cdot10\cdot10\cdot10\cdot10=1,000,000,000$

11. $10\cdot10\cdot10\cdot10\cdot10=100,000$

12. $^-10\times^-10\times^-10\times^-10\times^-10\times^-10=1,000,000$

Write in Expanded Exponent Notation Form

13. $7\times10^1+3\times10^0$

14. $5\times10^1+6\times10^0$

15. $1\times10^2+4\times10^1+7\times10^0$

16. $2\times10^2+8\times10^1+9\times10^0$

17. $5\times10^3+0\times10^2+2\times10^1+4\times10^0$

18. $8\times10^3+2\times10^2+9\times10^1+0\times10^0$

19. $1\times10^4+3\times10^3+3\times10^2+9\times10^1+0$

20. $1\times10^5+3\times10^4+0\times10^3+4\times10^2+5\times10^1+0$

21. $5\times10^6+6\times10^5+8\times10^4+9\times10^3+0\times10^2+0\times10^1+3$

22. $2\times10^4+1\times10^3+0\times10^2+0\times10^1+0$

23. $2\times10^5+0\times10^4+0\times10^3+3\times10^2+9\times10^1+5$

24. $7\times10^4+0\times10^3+4\times10^2+0\times10^1+2$

25. $9\times10^5+7\times10^4+5\times10^3+3\times10^2+8\times10^1+9$

26. $8\times10^3+3\times10^2+8\times10^1+1$

27. $2\times10^9+4\times10^8+5\times10^7+0\times10^6+0\times10^5+0\times10^4+0\times10^3+0\times10^2+0\times10^1+0\times10^0$

28. $1\times10^{11}+5\times10^{10}+0\times10^9+0\times10^8+0\times10^7+3\times10^6+0\times10^5+0\times10^4+1\times10^3+2\times10^2+0\times10^1+2\times10^0$

Section 2.2
Rewrite Using Exponents

1. 3^7

2. 7^3

3. -2^7 or $(^-2)^7$

4. -3^6 or $(^-3)^6$

5. $\left(\dfrac{3}{5}\right)^3$

6. $\left(\dfrac{1}{2}\right)^5$

7. $\left(\dfrac{2}{3}\right)^7$

Rewrite Without Exponents and Multiply Out

8. $4\times4\times4=64$

9. $2\times2\times2\times2\times2=32$

10. $-3\times-3\times-3=-27$

11. $-2\times-2\times-2\times-2=16$

12. $-3\times3\times3\times3=-81$

13. $-5\times5=-25$

14. $^-6\times^-6=36$

15. $^-4\times^-4=16$

16. $\left(\dfrac{2}{3}\right) \times \left(\dfrac{2}{3}\right) \times \left(\dfrac{2}{3}\right) \times \left(\dfrac{2}{3}\right) = \dfrac{16}{81}$ 18. $\left(\dfrac{3}{4}\right) \times \left(\dfrac{3}{4}\right) \times \left(\dfrac{3}{4}\right) = \dfrac{27}{64}$ 21. $\left(\dfrac{1}{4}\right) \times \left(\dfrac{1}{4}\right) \times \left(\dfrac{1}{4}\right) = \dfrac{1}{64}$

17. $\left(\dfrac{1}{3}\right) \times \left(\dfrac{1}{3}\right) = \dfrac{1}{9}$ 19. $-\,4 \times 4 \times 4 = -\,64$ 22. $1 \times 1 \times 1 \times 1 \times 1 \times 1 \times 1 \times 1 = 1$

20. $^-3 \times\, ^-3 \times\, ^-3 \times\, ^-3 = 81$

23. $-1 \times 1 \times 1 \times 1 \times 1 \times 1 \times 1 \times 1 \times 1 \times 1 \times 1 \times 1 = -1$

24. $\left(\dfrac{5}{2}\right) \times \left(\dfrac{5}{2}\right) \times \left(\dfrac{5}{2}\right) \times \left(\dfrac{5}{2}\right) = \dfrac{625}{16}$ 25. $\left(-\dfrac{2}{7}\right)\left(-\dfrac{2}{7}\right) = \dfrac{4}{49}$

Section 2.3

Basic Problems

1. $2^5 = 32$ 3. $3^6 = 729$ 5. $2^8 = 256$ 7. $5^3 = 125$ 9. 1 11. $4^2 = 16$

2. 1 4. $15^1 = 15$ 6. 16 8. $^-8$ 10. $3^6 = 729$ 12. $6^2 = 36$

Intermediate to Advanced Problems

13. $^-81$ 18. $(^-4)^5 = ^-1024$ 21. $\dfrac{16}{25}$ 25. $\dfrac{16}{49}$ 28. $(^-5)^3 = ^-125$ 31. 1

14. 49 19. $^-5^4 = 625$ 22. $\dfrac{27}{8}$ 26. $\dfrac{15}{19}$ 29. $\dfrac{1}{7^2} = \dfrac{1}{49}$ 32. $\dfrac{1}{6^2} = \dfrac{1}{36}$

15. $^-8$ 20. $\left(\dfrac{2}{3}\right)^6 = \dfrac{64}{729}$ 23. $5^0 = 1$ 27. $(^-3)^6 = 729$ 30. $\dfrac{-27}{8}$ 33. $4^0 = 1$

16. $^-125$ 24. $3^1 = 3$ 34. $5^2 = 25$

17. $3^0 = 1$

Section 2.4

Simplify by Removing Negative Exponents

1. $\dfrac{1}{10}$

2. $\dfrac{1}{10^3}$

3. $\dfrac{1}{10^2}$

4. 10^2

5. 10^5

6. 10^1

7. 10^8

8. $10^{5-8} = 10^{-3} = \dfrac{1}{10^3}$

9. $10^{2-6} = 10^{-4} = \dfrac{1}{10^4}$

10. $10^{2-(-2)} = 10^{2+2} = 10^4$

11. $10^{1-(-3)} = 10^{1+3} = 10^4$

12. $10^{-4-(-4)} = 10^{-4+4} = 10^0$

13. $10^{-3-3} = 10^{-6} = \dfrac{1}{10^6}$

14. $10^{1-(-1)} = 10^{1+1} = 10^2$

15. $10^{5+(-4)} = 10^1$

16. $10^{-6+(-7)} = 10^{-13} = \dfrac{1}{10^{13}}$

17. $10^{-1+4} = 10^3$

18. $10^{-2+(-6)} = 10^{-8} = \dfrac{1}{10^8}$

19. $10^{5+(-8)} = 10^{-3} = \dfrac{1}{10^3}$

20. $10^{-3+(-1)} = 10^{-4} = \dfrac{1}{10^4}$

21. $10^{-6+8} = 10^2$

22. $10^{2+(-7)} = 10^{-5} = \dfrac{1}{10^5}$

23. $10^{-6 \cdot (-2)} = 10^{12}$

24. $10^{-4 \cdot (-3)} = 10^{12}$

25. $10^{5 \cdot (-3)} = 10^{-15} = \dfrac{1}{10^{15}}$

26. $10^{-3 \cdot (6)} = 10^{-18} = \dfrac{1}{10^{18}}$

27. $10^{8 \cdot (-2)} = 10^{-16} = \dfrac{1}{10^{16}}$

Write Each Decimal Number in Expanded Exponent Notation Form

28. $4 \times 10^0 + 1 \times 10^{-1} + 2 \times 10^{-2} + 3 \times 10^{-3}$

29. $0 \times 10^{-1} + 1 \times 10^{-2} + 7 \times 10^{-3}$

30. $2 \times 10^1 + 1 \times 10^0 + 0 \times 10^{-1} + 0 \times 10^{-2} + 5 \times 10^{-3}$

31. $9 \times 10^0 + 9 \times 10^{-1} + 8 \times 10^{-2} + 7 \times 10^{-3}$

32. $7 \times 10^{-1} + 8 \times 10^{-2} + 4 \times 10^{-3} + 3 \times 10^{-4}$

33. $5 \times 10^0 + 3 \times 10^{-1} + 8 \times 10^{-2} + 2 \times 10^{-3} + 7 \times 10^{-4}$

412
34. $3 \times 10^{-3} + 0 \times 10^{-4} + 5 \times 10^{-5}$

35. $3 \times 10^{1} + 0 \times 10^{0} + 0 \times 10^{-1} + 0 \times 10^{-2} + 1 \times 10^{-3} + 0 \times 10^{-4} + 3 \times 10^{-5}$

Sec. 2.5
Square Roots of Real Numbers

1. 6
2. 11
3. 12
4. 10
5. $^{-}7$
6. 13
7. $^{-}14$
8. undefined
9. undefined
10. undefined
11. $^{-}4$
12. $\frac{2}{3}$
13. $\frac{21}{5}$
14. $\frac{15}{17}$

Find the Square Root Using Calculator

15. 4.47
16. 8.54
17. $^{-}6.71$
18. undefined
19. 0.89
20. $^{-}1.41$
21. 0.2
22. 0.08
23. 1.55
24. 3.65

Section 2.6

1. $^{-}17$
2. 6
3. 23
4. 3
5. 29
6. 3
7. 28
8. 5
9. 0
10. $^{-}1$
11. 8
12. $^{-}20$
13. 26
14. $^{-}12$
15. $^{-}5$
16. 3
17. 29
18. 1
19. 19
20. $^{-}18$
21. 4
22. 20
23. 5
24. 9.84

25. $\frac{119}{3} \approx 39.7$ rounded to the nearest tenths

26. 10
27. 1.4
28. $^{-}3.5$

Chapter 2: Practice Test

1. 10^5
2. 10^7
3. 100,000,000
4. $^{-}100$
5. 10,000

6. $5 \times 10^4 + 6 \times 10^3 + 0 \times 10^2 + 9 \times 10^1 + 8$
7. $8 \times 10^5 + 7 \times 10^4 + 3 \times 10^3 + 1 \times 10^2 + 5 \times 10^1 + 2$

8. 5^4
9. $^{-}7^5$ or $(^{-}7)^5$
10. $\left(\frac{3}{5}\right)^4$
11. 125
12. 81
13. $^{-}16$
14. 256
15. $\frac{27}{125}$
16. $3^6 = 729$
17. 4^{30}
18. $7^2 = 49$
19. $\frac{9}{25}$
20. $^{-}1$
21. $\frac{1}{10}$
22. $10^3 = 1000$
23. 10

24. $1 \times 10^1 + 2 \times 10^0 + 1 \times 10^{-1} + 7 \times 10^{-2}$
25. $4 \times 10^1 + 5 \times 10^0 + 3 \times 10^{-1} + 7 \times 10^{-2} + 5 \times 10^{-3}$

26. 8
27. $^{-}6$
28. 12
29. undefined
30. 5.74
31. 6.32
32. 36
33. $^{-}3$
34. $^{-}12$
35. $\frac{^{-}116}{3} \approx ^{-}38.67$
36. 5
37. $\frac{62}{21} \approx 2.95$
38. $\frac{^{-}7}{6} \approx -1.17$

Chapter 3
Section 3.1
Combining Like Terms

1. 2x
2. 25n
3. $^{-}2$x
4. 8a
5. 5x
6. 12x
7. $3v + 3$
8. $8r - 4$
9. $^{-}15$x
10. $^{-}25$n
11. $2x + 13$
12. $9t - 12$
13. 0
14. $22c - 13$
15. $- 6x + 4$

16. $18x^2 + 16x - 3$
17. $^{-}5x^3 + 4x^2 + 7x$
18. $- 4x^3 + 4x^2 - 10x + 11$

19. $3x - 16$
20. $4x^2 + x - 13$
21. $7x^3 - 2x^2 - 7$
22. $4x^2 - 15x - 5$
23. $^-3x^2 - 7x - 5$
24. $16x^3 - 5x^2 - 3x - 1$

25. $\frac{7}{6}x^2 + x + 6$
26. $2x^3 - \frac{1}{3}x^2 - 2x + 1$
27. $\frac{2}{3}x^3 - \frac{2}{9}x^2 + \frac{5}{18}$

28. $\frac{1}{2}x^3 - \frac{5}{12}x^2 + 7x + 1$
29. $\frac{3}{2}x^3 - \frac{19}{15}x^2 + \frac{2}{9}x + 11$
30. $\frac{-5}{2}x^3 + \frac{1}{6}x^2 - 5x + \frac{9}{2}$

Section 3.2
Distributive Property – Basic Problems
1. $-15x - 10$
2. $20 + 16x$
3. $-8x - 36$
4. $30a + 12$
5. $-7b + 35$
6. $-10a + 15$
7. $27x + 18$
8. $12n + 8$
9. $6 - 3x$
10. $9r + 15$
11. $-21x - 49$
12. $-8 + 32v$
13. $-5 + 5c$
14. $8 + 8x$
15. $52 - 13r$
16. $4x + 8$
17. $\frac{1}{5}x - 2$
18. $-\frac{3}{2}x - \frac{1}{3}$
19. $-12x + \frac{3}{2}$

Distributive Property – Intermediate and Advanced Problems
20. $10x - 18$
21. $4x + 15$
22. $27x + 8$
23. $7x + 3$
24. $-6x + 8$
25. $-9x + 6$
26. $11 + 7x$
27. -33
28. $2 + x$
29. $-10x - 5$
30. $x - 16$
31. $12 - 7x$
32. -5
33. $-15x + 18$
34. $6x - 7$
35. $-6x + 12$
36. $-8x - 31$
37. 4
38. $\frac{51}{2}x + 2$
39. $4x^2 - 12x + 20$
40. $12x^2 + 24x - 12$
41. $-6x^2 + 8x + 10$
42. $-5x^2 + 5x + 30$
43. $4x^2 - 8x + 6$

Section 3.3
Terminology

	Degree	Number of terms	Leading coefficient	In descending order
1.	second	Binomial	2	Already in descending order
2.	zeroth	Monomial	-6	Only one term
3.	first	Binomial	-3	Already in descending order
4.	first	Binomial	-2	$-2w + 12$
5.	ninth	Polynomial	4	$4x^9 - 3x^6 - 4x + 1$
6.	ninth	Polynomial	-8	$-8m^9 + 16m^3 - 3m + 12$
7.	zeroth	Monomial	-2	Only one term – constant
8.	second	Trinomial	5	Already in descending order
9.	fourth	Polynomial	3	$3p^4 - 5p^3 + 2p - 7$
10.	seventh	Binomial	-13	$-13x^7 + 5x^3$
11.	second	Binomial	-1	$-z^2 + 1$
12.	second	Trinomial	-1	$-x^2 + 7x + 3$

	Degree	Number of terms	Leading coefficient	In descending order
13.	eighth	Monomial	-14	Only one term
14.	first	Monomial	3	Only one term
15.	second	Binomial	3	Already in descending order
16.	fourth	Trinomial	-56	$-56d^4-16d^2+42d$
17.	fifth	Polynomial	-2	$-2v^5-3v^4+v-4$
18.	fourth	Trinomial	13	Already in descending order
19.	fourth	Binomial	-6	Already in descending order
20.	first	Binomial	2	Already in descending order
21.	first	Binomial	-3	$-3x+1$
22.	eighth	Monomial	64	Only one term
23.	first	Binomial	$\dfrac{1}{2}$	$\dfrac{1}{2}x-5$
24.	first	Binomial	2	$2x+\dfrac{3}{4}$ Already in descending order
25.	fifth	Polynomial	15	$15w^5-3w^3+2w-5$
26.	sixth	Trinomial	$\dfrac{-2}{3}$	$\dfrac{-2}{3}x^6+4x^3+15$
27.	third	Polynomial	-1	$-x^3+24x^2+22x-13$
28.	second	Binomial	$\dfrac{-1}{2}$	$\dfrac{-1}{2}w^2+1$
29.	third	Polynomial	$\dfrac{2}{3}$	$\dfrac{2}{3}b^3+44b^2+b+13$

Section 3.4
Adding and Subtracting Polynomials

1. $5x+1$
2. $-2x-1$
3. $8x-6$
4. $8x^2+3x$
5. $3x^3+3x^2-x$
6. $-x^2-3x+12$
7. $-2a^2-3a$
8. $3w^2-7w+12$
9. $-6n^5+5n^4-3$
10. $14a$
11. -8
12. $6b^2$
13. 0
14. $6w^5-4w$
15. r^2+r-9
16. $-14n^2+10n$
17. $-3p^3+3p^2-5p-1$
18. $3x^2-16$
19. $4x^7-19x^4+24x$
20. $2x^2-3x+15$
21. $-5x^2+9x+10$
22. $-3t^4-2t^3-t-2$
23. $4x^4-2x^2-6x+6$
24. $-6x^4+12x^2$

Section 3.5
Multiplying Polynomials

1. $6x^2-3x$
2. $-20x^3-8x^2$
3. $7x^3-14x^4$
4. $-2a^3-8a$
5. $-7a-49$
6. $n^2+3n-18$
7. $6x^2+14x+8$
8. $p^2+9p+18$

9. $x^2+10x+21$
10. $2c^2-5c-3$
11. $3x^2+16x-35$
12. $5w^2-8w-4$
13. $56t^2-119t+63$
14. $36b^2-45b-4$
15. $7x^2-31x+12$

16. $10a^2-68a+90$
17. $3x^2-14x+16$
18. x^2-1
19. $4a^2-9$
20. $36w^2-16$
21. $25r^2-81$
22. $4x^2+4x+1$

23. a^2+6a+9
24. $4c^2+24c+36$
25. $9w^2+30w+25$
26. x^2-2x+1
27. $9x^2-36x+36$
28. $4w^2-12w+9$
29. $25a^2-70a+49$

30. $81x^2-144x+64$
31. $4x^3-12x^2+80x$
32. $2x^4-8x^3+14x^2$
33. $9x^5-15x^4-27x^3$
34. $-18x^4+48x^3-30x^2$
35. $-5x^5+7x^4+6x^3$

Section 3.6

Greatest Common Factor- Basic Problems

1. $3(3x+4)$
2. $7(3x+1)$
3. $15(c+4)$
4. $8(x+3)$
5. $3(w-3)$
6. $16(2x-1)$
7. $6(3y-1)$
8. $14(n-2)$

Greatest Common Factor- Intermediate and Advanced Problems

9. $3(x^2+2x+3)$
10. $7(2b^2+b+3)$
11. $5(x^2+5x+3)$
12. $6(6a^2+3a+1)$
13. $3x^2(2x-1)$
14. $5x^2(3x^2-2)$
15. $8x^2(3x^3+1)$
16. $9a^2(a-3)$
17. $13p(p^2-2)$
18. $7k^4(1+7k^4)$
19. $5x(x^2+4x-2)$
20. $16t^2(t^2-2t+1)$
21. $5y^3(1+5y^3+3y^6)$

Trinomials

22. $(x+2)(x+1)$
23. $(a+5)(a+1)$
24. $(y+3)(y+1)$
25. $(z-3)(z+2)$
26. $(m+5)(m+3)$
27. $(x-5)(x+3)$
28. $(n-9)(n+2)$
29. $(c-5)(c+2)$
30. $(x-3)(x-3)$
31. $(v-6)(v-1)$
32. $(w-5)(w-4)$
33. $(z-9)(z-4)$
34. $(k+6)(k-4)$
35. $(x+6)(x-2)$
36. $(b-2)(b-3)$
37. $(z-8)(z-2)$

Difference of Squares

38. $(b+1)(b-1)$
39. $(x+3)(x-3)$
40. $(w+5)(w-5)$
41. $(a+6)(a-6)$
42. $(y+7)(y-7)$
43. $(2z+3)(2z-3)$
44. $(5x+6)(5x-6)$
45. $(4b+3)(4b-3)$
46. $(3x+8)(3x-8)$
47. $(2r+1)(2r-1)$
48. $(3c+2)(3c-2)$
49. $(5x+4)(5x-4)$
50. $(x^2+7)(x^2-7)$
51. $(2x^2+3)(2x^2-3)$
52. prime
53. prime
54. $(5x^3+2)(5x^3-2)$
55. $(x^3+6)(x^3-6)$
56. prime
57. $(x^3+8)(x^3-8)$
58. $(3x^2+5)(3x^2-5)$
59. $(x^3+1)(x^3-1)$
60. $(4x^2+7)(4x^2-7)$
61. prime

Section 3.7

1. –5
2. –27
3. 33
4. 21
5. 27
6. 34
7. –7
8. 4
9. 41
10. 27
11. 0
12. 0
13. 19
14. –11
15. –5
16. –10
17. 2
18. 10
19. 2
20. 12
21. –5
22. 1
23. –1
24. 4
25. – 11
26. –1
27. 2
28. 15
29. –10
30. –37
31. 0
32. 110
33. 1
34. –4
35. 21
36. –14
37. –1
38. –1
39. –17
40. –22
41. 7
42. 30
43. 6
44. 4

Chapter 3 Practice Test

1. $2x+1$
2. $25n+3$
3. $-3x-9$
4. $-2a-6$
5. $-15x-10$
6. $8-4x$
7. $-17-16x$
8. $-5x-34$
9. $32a+9$
10. $2c-5$

11. Second degree binomial; leading coefficient is 2; in descending powers: $2x^2-x$
12. zeroth degree monomial; leading coefficient is -6 ; There is only one term.
13. First degree binomial; leading coefficient is -3 ; already in descending powers
14. Ninth degree polynomial; leading coefficient is 4 ; in descending powers: $4x^9-3x^6-4x+1$

15. Second degree trinomial; leading coefficient is 5; in descending powers: $5x^2+2x-3$

16. Fourth degree polynomial; leading coefficient is 3; already in descending powers.

17. $5x+1$

18. $-2x-1$

19. $8x-6$

20. $8x^2+3x$

21. $6b^2$

22. 0

23. r^2+r-9

24. $-14n^2+10n$

25. $6x^2-3x$

26. $-20x^3-8x^2$

27. $7x^3-14x^4$

28. $-2a^3-8a$

29. $-7a-49$

30. $n^2+3n-18$

31. $81x^2-144x+64$

32. $7x^2-31x+12$

33. $10a^2-68a+90$

34. $3x^2-14x+16$

35. $25w^2-9$

36. $3(3x+4)$

37. $7b(3b+1)$

38. $(4w+5)(4w-5)$

39. $(z-3)(z-2)$

40. $(b+3)(b+3)$

41. $15(c+4)$

42. $(y+3)(y+1)$

43. $3(w-3)$

44. $16z^3(2z-1)$

45. $(x-5)(x-3)$

46. $(a+6)(a-6)$

47. $4(b^2+1)$

48. $(6t+3)(6t-3)$

49. $-8(x+3)$

50. -5

51. -27

52. -29

53. -4

54. -8

55. 17

56. 23

57. -12

58. 0

Chapter 4

Section 4.1

1. Answers will vary. Some examples: length, area, volume, mass, density, weight, energy, force, time, temperature

2. and 3.

Different unit names for the same attribute		
Attribute	**Fundamental units**	**Derived units**
length	meters, inches	board-feet, centimeters, feet, kilometers, light-years, microns, mils, miles, nanometers, yards
mass	grams	centigrams, kilograms, metric tons, micrograms, milligrams
time	seconds	centuries, days, decades, hours, millennia, milliseconds, minutes, nanoseconds
temperature	°Celsius, °Fahrenheit	centigrade (which is just another name for celsius)
volume	liters, fluid ounces	cubic centimeters, cubic yards, cubic meters, cups, gallons, milliliters, pints, quarts
density		grams per cubic centimeter, grams per liter, grams per milliliter, kilograms per cubic meter, pounds per cubic foot
velocity		feet per second, kilometers per hour, meters per second, miles per hour

Section 4.3

Conversions Within the Metric System

1. 22,000 mm
2. 0.275 kg
3. 0.15 m
4. 15 cm
5. 34 mm
6. 49,000 μm
7. 3,200,000 m
8. 0.068 km
9. 22,000 mL
10. 0.026 μm
11. 0.9 cm
12. 2 mm

Conversions Within the English System

13. 10,560 ft
14. 13.5 ft
15. 3.75 gallons
16. 6600 ft
17. 216 inches
18. 2640 yards
19. 2.3125 quarts
20. 90 inches

Conversions Between the English and Metric Systems

21. 55.88 cm
22. 3.2186 km
23. 23.6128 miles
24. 3.077 miles

25. 72.26 m
26. 1.5892 ounces
27. 18.939 L
28. 27.126 inches
29. 1958.16 ft

30. 68.58 m
31. 2040.606 mL
32. 86.18 kg
33. 0.2198 ft
34. 52.25 kg

35. 0.09252 inches
36. 7.756 inches
37. 52480 ft or
    52,494.874 ft
38. 16.9 ounces

39. 99.2 lb
40. 3681.72 minutes

Temperature Conversions
41. –13 °F
42. 100.4 °F

43. 25.6 °C
44. –23.3 °C

45. 34.35 °C, 93.83 °F
46. 369.261 K

47. 5 °C, 278.15 K
48. 25 °C, 298.15 K

Applications
49. 1.6975 lb
50. 36,008.0875 km
51. 3.937 in

52. -63.15 °C, -81.67 °F
53. 180.34 cm, 1.8034 m
54. 42.22 °C

55. 2.63×10^{13} RBCs
56. 516.67 μm

57. 0.3 mm

Section 4.4

1. 21 cm
2. 8.4 kg
3. 8.7 mg
4. 210.6 km
5. 1.2 nm
6. 4.02 m
7. 3.3 kg
8. 4.378 L

9. 2.046 L
10. 2.04 mm
11. 2.07 cm
12. 13.257 m
13. 36 m^2
14. 15.6 kg^2
15. 65 mm^2
16. 66 cm^3

17. 7.5 mm^3
18. 2.4 cm^3
19. 6.6 m^3
20. 29.4 kg m/s^2
21. 39.2 kg m/s^2
22. 5
23. 0.7
24. 1.2

25. 38
26. 2 kg/m^3
27. 2 lb/in^3
28. 5.5
29. 7.1 mm
30. 5 cm^2
31. 4 L
32. 0.03 cm

33. 320 kg^2
34. 72
35. 42 kg
36. 85 g
37. 5 g/cm^3
38. 3 kg/m^3
39. 0.3 mm
40. 500 μm

Chapter 4 Practice Test

1. 0.2198 ft
2. 52.25 kg
3. 1.6 m
4. 0.0925in
5. 96.1 °C
6. -13 °F

7. 55.88 cm
8. 3.2186 km
9. 23.6128 miles
10. 3.08 miles
11. 21 cm
12. 8.4 kg

13. 8.7 mg
14. 210.6 km
15. 1.2 nm
16. 85 g
17. 5 g/cm^3
18. 3 kg/m^3

19. 0.3 mm
20. 500 μm
21. .0003937 in
22. 531.43 μm
23. 24,700,000,000,000 RBCs
24. 0.2725 mm

Chapter 5
Section 5.2

1. length, dim 1, cm
2. volume, dim 3, ft^3
3. area, dim 2, in^2
4. length, dim 1, mm
5. volume, dim 3, L
6. area, dim 2,, ft^2
7. length, dim 1, km
8. volume, dim 3, m^3

9. volume, dim 3, ft^3
10. length (diagonal), dim 1, in
11. length, dim 1, in
12. volume, dim 3, ft^3
13. area, dim 2, cm^2
14. area, dim 2, in^2
15. length dim 1, m
16. volume dim 3. cubic inches

17. area dim 2, square feet
18. volume dim 3, cubic feet
19. length dim 1, miles
20. area dim 2, square meters
21. volume dim 3, liters
22. volume dim 3, cubic feet

Section 5.3

1. a) 10.8 cm, 7.29 cm^2
   b) 64 in, 220 in^2
   c) 56 cm, 144 cm^2
   d) 11.2 π in ≈ 35.17 in
      31.36 π in^2 ≈ 98.47 in^2

2. 144 in^2
3. 7 cm
4. (720 + 72 π) in^2 ≈ 946.08 in^2
5. 8,000 cm^3
6. 4,752 cm^3

7. 250 π in^3 ≈ 785 in^3
8. 26,244 π in^3 ≈ 82,406.16 in^3
9. 36 π ft^2 ≈ 113.1 ft^2
10. 2.25 ft^3 = 3,888 in^3
11. 3 cm^2 = 300 mm^2

12. 22,400 cm² = 2.24 m²
13. 30 cm³
14. 225 π mm² ≈ 706.86 mm²
15. 36 π cm² ≈ 113.1cm²
16. 19.125 cm²
17. 25.67 cm²

18. 6.8 π cm ≈ 21.352 cm
19. 10.4 π in ≈ 32.656 in
20. 7.84 π cm² ≈ 24.6176 cm²
21. ≈ 28.26 cm
22. 25.625 π in³ ≈ 80.46 in³
23. 1218.816 π cm³

≈ 3,827.08 cm³
24. ≈ 1.6 cm
25. ≈ 588.7 in³
26. ≈ 540.3 cm³
27. 4 in
28. 21,952 cm³, 4704 cm²
29. 128 π cm³ ≈ 402.176 cm³

Section 5.4

1. B=65°
2. A=40°, B= 140°
3. A=40°
4. A=60°
5. A=118°, B=118°, C=62°
6. A=15°, B=165°, C=165°
7. A=155°
8. A=85°
9. A=73.6°
10. A=49.1°
11. A=26.5°
12. A=137.8°
13. A=55.7°
14. A=115.9°
15. A=62.5°
16. A=28.1°
17. A=75.3°
18. A=13.8°, B= 166.2°, C=166.2°
19. A= 81.4°
20. A=27.1°

Section 5.5

1. 6,912 in³
2. 210 mm²
3. 7,350 mm³
4. 1,224 in²
5. 462 in³
6. 50 cm³
7. 11,356.236 cm³
8. 610.237 in³
9. ≈ 3.47 ft²
10. 6 ft³
11. 0.75 m³
12. 37,000,000 cm³
13. ≈ 4.33 gal
14. 580,000 mm³
15. 0.000005 L
16. ≈ 0.000058 ft³
17. 0.332 L
18. ≈ 14.9 gal
19. 5,184 in²
20. 16,000 mm³
21. 250 cm³

Section 5.6

1. use π or 2 decimal places
2. use π or 2 decimal places
3. use π or 2 decimal places
4. 5 or 6 decimal places
5. 2 or 3 decimal places
6. 4 or 5 decimal places
7. 2 decimal places (given in problem)
8. 5 or 6 decimal places
9. use π or 2 decimal places
10. 4 π cm² (just use π symbol)
11. 4 or 5 decimal places
12. 5 or 6 decimal places
13. 6 or 7 decimal places
14. 2 or 3 decimal places
15. use π or 2 decimal places
16. 2 or 3 decimal places
17. use π or 2 decimal places
18. use π symbol
19. use π symbol
20. use π symbol
21. 3 or 4 decimal places

Chapter 5 Practice Test

1. length, dim 1, miles
2. volume, dim 3, cubic feet
3. area, dim 2, square inches
4. length, dim 1, meters
5. 36.424 cm
6. 8 ft² or 1,152 in²
7. 7.5 m²
8. 1,820 cm³
9. 7,335.04 cm³
10. 103°
11. 70°
12. A = 42.7°
13. A = 123.5°
14. 1,617 in³
15. 324 in²
16. 0.038 cm³
17. 0.054 L
18. ≈ 589.9 cm³
19. 3441.6 in²
20. three or four
21. π or two
22. two or three
23. four or five
24. π

Chapter 6
Section 6.2a

1. x=9
2. a=4
3. w=10
4. y= -19
5. p=21
6. n= -12
7. r=8
8. t=26
9. a= -3
10. v=20
11. x=3
12. w= -8
13. y=5
14. b= -2
15. a= -2
16. s=0
17. p= -3
18. t= -12
19. x=0
20. x=50
21. c= -17
22. m= -28
23. y=81
24. a=0
25. b=0
26. $y=\dfrac{2}{3}$
27. v= -2
28. x= -72
29. k=33
30. b=13
31. $x=\dfrac{7}{5}$
32. p= -2
33. no solution
34. all real numbers
35. no solution
36. no solution

Section 6.2b

1. $a=3$
2. $x=5$
3. $x=\dfrac{7}{5}$
4. $p=3$
5. $x=\dfrac{13}{2}$
6. $p=-\dfrac{5}{4}$
7. $u=0$
8. $c=0$
9. $z=10$
10. $a=-\dfrac{7}{3}$
11. $w=\dfrac{20}{3}$
12. $x=-4$
13. $b=13$
14. $a=-4$
15. $p=-\dfrac{23}{5}$
16. $x=\dfrac{4}{5}$
17. $c=\dfrac{25}{4}$
18. $x=\dfrac{13}{2}$
19. $p=\dfrac{13}{2}$
20. $b=-\dfrac{26}{7}$
21. $n=-60$
22. $s=-21$
23. $u=4$
24. $y=100$
25. $p=55$
26. $r=-40$
27. $u=80$
28. $x=20$
29. $z=-5$
30. $v=3$
31. $z=14$
32. $r=0$

Section 6.2c

1. $x=-2$
2. $a=1$
3. $y=2$
4. $c=-4$
5. no solution
6. no solution
7. $x=7$
8. $x=-\dfrac{17}{2}$
9. $y=\dfrac{8}{7}$
10. $s=-3$
11. $n=-\dfrac{15}{2}$
12. $r=-\dfrac{8}{5}$
13. $m=1$
14. $p=-\dfrac{5}{3}$
15. $n=-\dfrac{9}{5}$
16. $w=-\dfrac{26}{15}$
17. $u=-\dfrac{16}{5}$
18. $a=-\dfrac{6}{19}$
19. $a=-\dfrac{1}{18}$
20. $k=\dfrac{10}{13}$
21. $t=\dfrac{1}{15}$
22. $x=-\dfrac{61}{13}$
23. $y=-29$
24. $p=\dfrac{9}{7}$
25. no solution
26. $x=7$

Section 6.3

1. n is the number, 6 + 3n= 33, The number is 9.
2. n is the number, 6n – 5 = 7, The number is 2.
3. n is the number, 25 + 8n = 1, The number is -3.
4. n is the number, 2n – 7 = –15, The number is -4.
5. L is the length and W is the width, L = 3W, 6W+2W = 64, The width of the rectangle is 8 inches. The length of the rectangle is 24 inches.
6. L is the length and W is the width, L + 2W, 4W+2W = 150, The width of the rectangle is 25 feet. The length of the rectangle is 50 feet.
7. L is the length of each side, 3L = 36, The length of each side is 12 inches.
8. L is the length of each side, 3L = 12, The length of each side is 4 feet.
9. A is the smallest angle, B is the next largest, and C is the largest angle, 3A+2A+A =180, The angles of the triangle are: A=30º, B=60º, and C=90º.
10. A is the smallest angle, B is the next largest, and C is the largest angle, 5A+3A+A =180, The angles of the triangle are: A=20º, B=60º, and C=100º.
11. n is the first number and n+1 is the next number, n+ (n+1)= 23, The numbers are 11 and 12.
12. n is the first number and n+1 is the next number, n+ (n+1)= 39, The numbers are 19 and 20.
13. n is the first number and m=2n is the second number, n+ 2n= 30, The numbers are 10 and 20.
14. n is the first number and m=3n is the second number, n+ 3n= 8, The numbers are 2 and 6.
15. n is the smallest number and m=2n is the next smallest number and p is the largest number p=3n, n+ 2n+3n= 36, The numbers are n=6, 1m=2, and p=18.
16. C is the cost of a chair and T = C+60 is the cost of a table, 2(C+60)+ 3C=745, C=$125, T= $185
17. L is the length and W is the width, L = W+5, 2W + 2(W+5)=38. The width of the rectangle is 7 feet and the length of the rectangle is 12 feet.
18. L is the length and W is the width, W = L– 12, 2L + 2(L -12)=156. The width of the rectangle is

33 meters and the length of the rectangle is 45 meters.

19. C is the cost of a chair and T = C+80 is the cost of a table, (C+80)+ 6C=745, C=$95, T= $175

20. L is the length and W is the width, L=2.5W, 2W+2(2.5W) = 189, The width of the rectangle is 27 cm. The length of the rectangle is 67.5 cm.

21. L is the length of a rung, 9L=15, The length of each rung is 5/3 feet, or approximately 1.67 feet.

22. L is the length and W is the width, W= 1/3 L, 2(1/3 L)+2L = 15, The width of the rectangle is 15/8 feet. The length of the rectangle is 45/8 feet.

Section 6.4

1. $x = 2, x = -2$
2. $x = 3, x = -3$
3. $x = 2/5, x = -2/5$
4. $x = 0$
5. no solution
6. $x = 4, x = -4$
7. $x = 0, x = 3$
8. $x = -3, x = 7/3$
9. no solution
10. $x = -2/3, x = 5/3$

Section 6.5

1. $x = 1, x = -2$
2. $a = -4, a = 5$
3. $b = -4, b = 1/2$
4. $y = 7, y = -1/2$
5. $p = -12, p = 5/3$
6. $x = 5, x = -2$
7. $y = 3$
8. $c = -8, c = 2$
9. $w = 6, w = 2$
10. $x = 3, x = -1/2$
11. $m = -2, m = -5/3$
12. $p = -3/2, p = 4/3$
13. $y = 4, y = -7/5$
14. $z = 2, z = -7/4$
15. $x = -3/2$
16. $r = 3, r = -2$
17. $w = 2, w = -5$
18. $u = 1, u = 4$
19. $c = 2, c = -12$
20. $y = 8, y = -2$
21. $x = 4, x = -3$
22. $w = -7, w = -3$
23. $a = 4, a = 7$
24. $p = 1, p = -5$
25. $u = 4, u = 2$
26. $z = 0, z = 5$
27. $a = 0, a = -3/2$
28. $x = 3, x = -6$
29. $y = 7, y = -2$
30. $r = 3, r = 1/2$
31. $a = -4, a = -2/3$
32. $b = -1, b = -2$
33. $v = 3$
34. $x = 7, x = 1$
35. $m = -4$

Section 6.6

Graph the inequalities

1.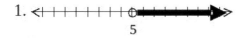
   5

2.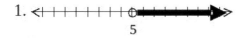
   - 2

3.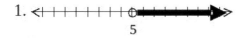
   7

4.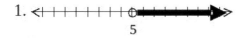
   0

5.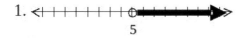
   -5

6.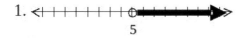
   6

7.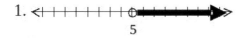
   -7

8.
   5

9.
   -3

10.
   9

11.
   0

12.
   -6

13.
   10

14.
   - 8

One-step Inequalities

15.

16.

17.

18.

19.

20.

21.

22.

23.

24.

25.

26.

27.

28.

Two-step Inequalities

29.

30.

31.

32.

33.

34.

35.

36.

37.

38.

39.

40.

41.

42.

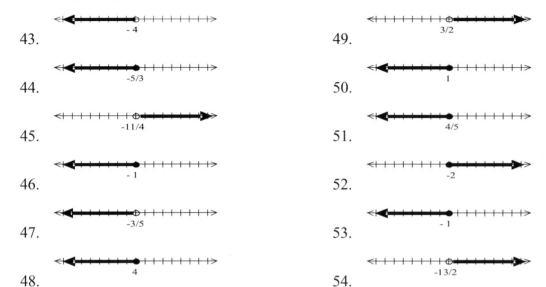

43.

44.

45.

46.

47.

48.

49.

50.

51.

52.

53.

54.

Absolute value Inequalities

55. $-2 \le x \le 2$ 57. $-3 < x < 11$ 59. $x \le -2$ or $x \ge 2/5$ 61. $x < 1$ or $x > 5$

56. $x < 0$ or $x > 5$ 58. $x < -3$ or $x > 3$ 60. $-5/3 \le x \le 1/3$ 62. $-3/4 \le x \le 13/4$

Section 6.7

Applications of Inequalities

1. $x + \$1,025 + \$689 \le \$3,700$, $x \le \$1,986$, Javier has $19826 or less to spend for everything else on his vacation.

2. $x\,2,250\,\text{lbs} \le 19,500\,\text{lbs}$, $x \le 8.67\,\text{skids}$ The driver can no more than 8 full skids on the truck safely.

3. $x + 325 \ge 875$, $x \ge 550$ The quality control inspects has to inspect at least 550 products in the remainder of her shift.

4. $x + \$3,800 \ge \$6,500$, $x \ge \$2,700$ The bakery must sell at least another $2,7000 of goods to make a profit for the week.

5. $|x - \$28.00| \le \$7.50$
    $-\$7.50 \le x - \$28.00 \le \$7.50$, $\$20.50 \le x \le \$35.50$ The range of hourly wages at a construction company, ranges from $20.50 up to $35.50.

6. $|x - 14.75\,\text{oz}| \le 0.65\,\text{oz}$
    $-0.65\,\text{oz} \le x - 14.75\,\text{oz} \le 0.65\,\text{oz}$, $14.1\,\text{oz} \le x \le 15.4\,\text{oz}$ The range of the amount of cereal in a 14.75 oz box of cereal is from 14.1 to 15.4 ounces.

7. $|x - 74\,\text{in}| \le 2 \times 2.9\,\text{in}$
    $-5.8\,\text{in} \le x - 74\,\text{in} \le 5.8\,\text{in}$, $68.2\,\text{in} \le x \le 79.8\,\text{in}$ The range of heights of 95% of male members of a basketball club is from 68.2 to 79.8 inches.

8. $|x - \$122| \le 2 \times \$10.50$
    $-\$21 \le x - \$122 \le \$21$, $\$101 \le x \le \$143$. The amount of money spent per child for 95% of the children going back to school ranges from $101-$143.

Chapter 6 Practice Test

1. z = 0	3. x = –6	5. x = 28	7. x = 8/3
2. x = 11	4. y = –3	6. y = – 40	8. no solution

9. n is the number, 3n – 5= –17, The number is –4

10. L is the length and W is the width, L = 3W, 6W+2W = 96, The width of the rectangle is 12 inches. The length of the rectangle is 36 inches.

11. L is the length and W is the width, L = 5W, 10W+2W = 120, The width of the rectangle is 10 feet. The length of the rectangle is 50 feet

12. L is the length and W is the width, L = W+5, 2(W+5)+2W = 154, The width of the rectangle is 36 cm. The length of the rectangle is 41 cm

13. C is the cost of a chair and T = C+150 is the cost of a table, (C+150)+ 4C=525, C=$75, T= $225

14. x = –3, x= 12	17. x = –2, x = 1/3	20. x = 4	23. $-7 < x > 1$
15. x = –14/3, x = 14/3	18. x = 2, x = 3	21. $x > -7/3$	24. $x \geq 4$ or $x \leq -3$
16. no solution	19. x = –2, x = 5	22. $x \geq 5/2$	

25. $x + \$500 + \$2,100 + \$1,400 \leq \$6,500$, $x \leq \$2,500$ Kaitlyn needs to see at least %2,500 more merchandise to meet her quota.

26. $\begin{array}{c} |x-36\,oz| \leq 0.75\,oz \\ -0.75\,oz \leq x - 36\,oz \leq 0.75\,oz \end{array}$, $35.25\,oz \leq x \leq 36.75\,oz$ The range of the amount of flour in a 36oz package is from 35.25 to 36.75 ounces.

Chapter 7
Section 7.1

1. $\dfrac{32}{25} = 1.28 = \dfrac{128}{100} = 128\%$

2. $\dfrac{36}{12} = 3 = \dfrac{300}{100} = 300\%$

3. $\dfrac{16}{19} \approx 0.84 = \dfrac{84}{100} = 84\%$

4. $\dfrac{45}{10} = 4.5 = \dfrac{450}{100} = 450\%$

5. $\dfrac{175}{25} = 7 = \dfrac{700}{100} = 700\%$

6. $\dfrac{28}{56} = 0.5 = \dfrac{50}{100} = 50\%$

7. $\dfrac{120}{10} = 12 = \dfrac{1200}{100} = 1200\%$

8. $\dfrac{4}{3} \approx 1.33 = \dfrac{133}{100} = 133\%$

Section 7.2

1. 57.5 mph
2. approx 62 mph
3. approx –0.017 °F/ft
4. –0.005°F/ft
5. approx 71.4 beats/min
6. 80 beats/min
7. 1.625 mL/hour
8. 1.9 mL/hour
9. approx 0.031 mL/min
10. 346.15ft/min, 1.92 yds/sec
11. 0.00025km/sec, 900 m/hr
12. 0.324 km/min; approx 5.4 m/sec
13. 80 L/hr
14. 1100 ft/min

15. 31.875m/min
16. 90 heartbeats/min
17. 96 beats/min
18. 300 miles/hr
19. approx 21.82 mi/hr
20. 45000m/hr
21. 440 m/min
22. 54 cm/hr
23. 24cm/hr
24. 60 m
25. 192.5 miles
26. 3750 cm
27. 768 feet
28. 1280 m

29. 2240 miles
30. 140 km
31. 412.5 miles
32. 311.4 feet
33. 0.1 km
34. 400 km
35. 4500 m
36. 40500 cm
37. 420 feet
38. 62.5 m
39. 270 miles
40. 4875 cm
41. 2160 feet
42. 4741 m

43. 281.25 miles   44. 2325 miles   45. 7020 feet

Section 7.4

1. yes	4. no	7. yes	10. no	13. no
2. yes	5. no	8. yes	11. yes	14. yes
3. no	6. yes	9. no	12. yes	15. no

Section 7.5

1. $x=\dfrac{12}{5}$

2. $b=\dfrac{12}{7}$

3. $n=\dfrac{9}{2}$

4. $a=\dfrac{40}{3}$

5. $d=\dfrac{60}{7}$

6. $y=\dfrac{56}{9}$

7. $n=\dfrac{3}{2}$

8. $x=4$

9. $r=\dfrac{35}{2}$

10. $s=\dfrac{3}{2}$

11. $w=\dfrac{9}{8}$

12. $p=\dfrac{40}{9}$

13. $x=-\dfrac{2}{7}$

14. $m=\dfrac{63}{5}$

15. $a=-\dfrac{22}{5}$

16. $p=-\dfrac{27}{2}$

17. $w=10$

18. $r=5$

19. $v=-\dfrac{38}{7}$

20. $x=-10$

21. $n=-\dfrac{15}{11}$

22. $p=\dfrac{47}{5}$

23. $c=-43$

24. $b=\dfrac{12}{11}$

25. $y=\dfrac{21}{4}$

26. 3.8 hours

27. approx 251.1 miles

28. 622.5 km

29. 2250 km

30. 2625 m

31. approx 66.8 sec

Section 7.6

Note: If appropriate, all answers are rounded to the nearest hundredth, unless otherwise noted.

1. I can buy 28 pencils (must round down assuming only whole pencils can be purchased).

2. I can buy 21 batteries.
6. It took Juan 1.35 seconds.

3. I can buy 8.73 pounds of nails.
7. It took Jessica 5.6 seconds.

4. I can buy 39.27 pounds of chlorine.
8. It took Justin 2.59 hours.

5. It took Shirley 4.62 hours.
9. It will take 50 minutes to fill the entire tank.

10. It will take 8/3 hours (or approximately 2.67 hours) to fill the entire tank.

11. It will take 5.4 hours to fill the entire tank.
17. There are 280 g of nitrogen.

12. It will take 12 minutes to fill the entire tank.
18. There are 1676.29 g of potassium.

13. It will take 0.84 seconds for a heartbeat to occur.
19. There are 0.10 g of chlorine.

14. It will take 1 second for a heartbeat to occur.
20. There are 3.86 g of carbon.

15. There are 2222.22 g of oxygen.
21. There are 0.19 g of hydrogen.

16. There are 486.28 g of chlorine.

Section 7.7

Note: all percent answers rounded to nearest percent

1. 7%	6. 3%	11. $23	16. 15%	21. $24.93, $313.93
2. 8%	7. $1.6 million	12. 1600 kilometers	17. 31%	22. $7.68, $96.68
3. $80	8. 16%	13. 12%	18. 120%	
4. 20%	9. $22.86	14. $111.60	19. 30%	
5. 4%	10. 6%	15. $125	20. 64%	

Chapter 7 Practice Test

1. 48 miles per hour

2. –0.02°F/ft

3. 7.5 m/sec

4. $\approx 92.31\dfrac{\text{beats}}{\text{minute}}$

5. $\approx 0.07\dfrac{\text{mL}}{\text{minutes}}$

6. $\approx 266.7\dfrac{\text{ft.}}{\text{minutes}}$

7. $\approx 54.4\dfrac{\text{miles}}{\text{hour}}$

8. 45.6 miles

9. 1080 feet

10. 26.5 miles

11. 178.5 feet

425

12. yes
13. no
14. yes
15. yes

16. $x = \dfrac{10}{7}$

17. $r = \dfrac{8}{5}$

18. $y = \dfrac{3}{2}$

19. $x = \dfrac{26}{3}$

20. 52.5 sec.
21. approx 1.4 hours
22. 112.5 minutes

23. 9% (rounded to nearest percent)
24. 35% (rounded to nearest percent)
25. I pay $11.64 in sales tax. The final cost of the item is $146.64.
26. 10%

Chapter 8
Section 8.1

1.

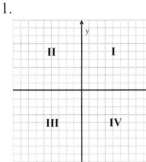

2.

$A = (-4, 4)$, $B = (0, 3)$,

$C = (3, 2)$, $D = (4, 0)$,

$E = (5, -5)$, $F = (0, 0)$,

$G = (0, -2)$, $H = (-1, -4)$,

$I = (-3, 0)$

3.

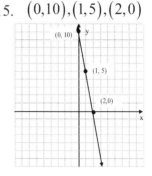

4. $(-1, 1)$
5. $(0, 3), (2, 11), (-1, -1)$

6. $(3, -3), (-3, -7)$
7. $(-2, -1)$

8. $(2, 5)$
9. $(0, 6), (-1, -8)$

Problems 10-15, answers may vary

10. $(0, -6), (1, -5), (2, -4)$

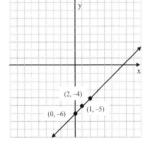

11. $(0, 7), (1, 4), (2, 1)$

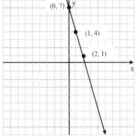

12. $(0, 3), (1, 2), (2, 1)$

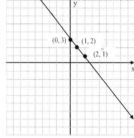

13. $(0, -1), (1, 1), (2, 3)$

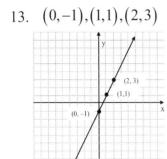

14. $(0, 6), (1, 4), (2, 2)$

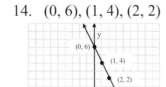

15. $(0, 10), (1, 5), (2, 0)$

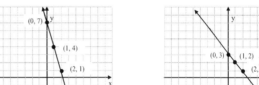

Section 8.2
Problems 1-8, answers may vary

426

1. (0,5), (1,8), (−1, 2)

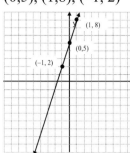

2. (0, −2), (1, −3), (−1, −1)

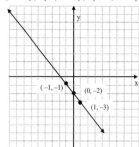

3. (0, −7), (1, −6), (−1, −8)

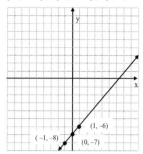

4. (0, 11), (1, 9), (−1, 13)

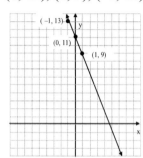

5. (0, −8), (1, −3), (−1, −13)

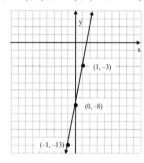

6. (0, 3), (1, 3/2), (−1, 9/2)

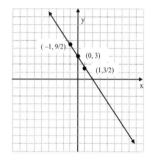

7. (0,4), (1, 5), (−1, 3)

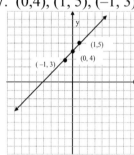

8. (0, 8), (1, 12), (− 1, 4)

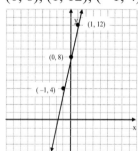

9. x-int (−5/3,0), y-int (0, 5)

10. x-int(7,0), y-int (0, −7)

11. x-int (2,0), y-int (0,3)

12. x-int (−4,0), y-int (0, 4)

13. x-int (−2,0), y-int (0, 8)

14. x-int (11/2, 0), y-int (0, 11)

15. x-int (3,0), y-int (0, −5)

16. x-int (2,0), y-int (0, −7)

17. 2
18. − 5
19. − 1/3
20. −1

21. 8/3
22. 3/2
23. −11/7
24. 13

25. −3
26. 1/2
27. 2
28. −5/2

29. 0
30. 0
31. Undefined
32. Undefined

33. m = 4
34. m = −9/7
35. m = −2/3
36. m = −4/7

37. m = −16/9
38. m = 9/14
39. m = 5/16

Section 8.3

1. 8
2. −3
3. −3/5
4. 4/3

5. 2
6. −1
7. 1
8. Undefined

9. −12
10. 0
11. 3/2
12. Undefined

13. 1
14. 2
15. −2
16. −5/2

17. 3/2
18. −1
19. 1/2
20. −2/3

427

Section 8.4

Please note that all the graphs below should contain arrows in both directions on the lines of the equations

1.
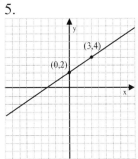

5.
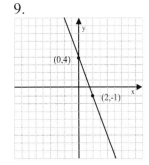

9.
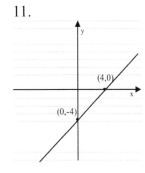

13.
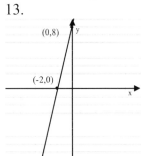

2.
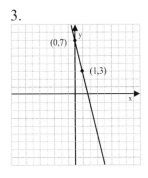

6.
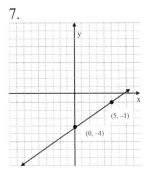

10.
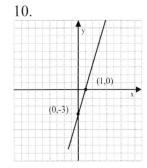

14.
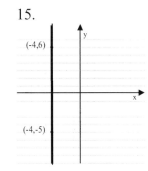

3.
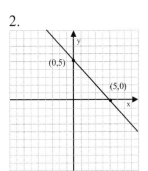

7.
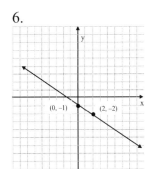

11.
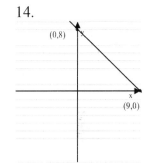

15.
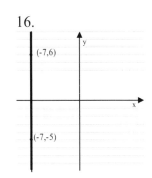

4.
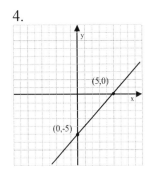

8.
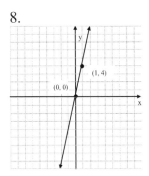

12.
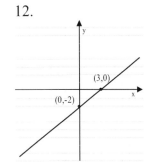

16.

428

17.

21.

25.

29.

18.

22.

26.

30.

19.

23.

27.

31.

20.

24.

28.

32..

33.

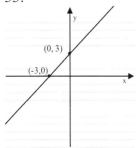

Section 8.5

1. Independent: chlorophyll by-products; Dependent: pigments
2. Independent: light brightness; Dependent: moth behavior
3. Independent: temperature; Dependent: vapor pressure
4. Independent: temperature; Dependent: rate of photosynthesis
5. Independent: ambient temperature; Dependent: animal metabolism
6. Independent: exposure time; Dependent: effectiveness of disinfectant
7. Independent: time; Dependent: height of ball
8. Independent: humidity; Dependent: transpiration
9. Independent: amount of gas; Dependent: how far you go
10. Independent: gravity; Dependent: root growth
11. Independent: temperature; Dependent: rate of carbon dioxide produced
12. Independent: salinity of water; Dependent: hatching of brine shrimp
13. Independent: ratio of S/V; Dependent: diffusion efficiency
14. Independent: number of carrots; Dependent: vision
15. independent: humidity; Dependent: height of redwoods

Answers will vary!

16. $m \approx 1/2$
17. $m \approx -1$
18. $m \approx 0$
19. $m \approx 2$

20. $m \approx -5/4$
21. $m \approx -3/7$
22. $m \approx 2/5$
23. $m \approx 5/2$

24. $m \approx 6/11$
25. $m \approx -8/3$
26. $m \approx 6/5$
27. $m \approx -1$

28. $m \approx 0$
29. $m \approx -9/10$
30. $m \approx -3/5$
31. $m \approx 13/10$

32. $m \approx -1$
33. $m = 1$
34. $m \approx -2$
35. $m \approx 3/4$

36. $m \approx 68 \text{ km}/(\text{mpc s}) \quad \pm 5 \text{ km}/(\text{mpc s})$

37. $m \approx -0.5 \text{ mortality}/\text{year} \quad \pm 0.1 \text{ mortality}/\text{year}$, 2011 ± 1 year

38. $m \approx -11 \text{ sunspots/year} \quad \pm 1$, decreasing at a rate of 11 sunspots per year

39. $m \approx 0.24 \text{ \%GDP/year} \quad \pm 0.02$, increasing at a rate of 0.24%GDP per year

40. $m \approx 0.007 \text{ }^{\circ}\text{F/year} \quad \pm 0.002$, increasing at a rate of 0.007 ºF per year

Section 8.6

1. a) ≈ 0.14, b) ≈ 0.175, c) ≈ 0.415, d) ≈ 0.62
2. a) $\approx 120^{\circ}$, b) $\approx 160^{\circ}$, c) $\approx 10^{\circ}$, d) $\approx 65^{\circ}$
3. a) $\approx 25^{\circ}F$, b) $\approx 54^{\circ}F$, c) $\approx 73^{\circ}F$, d) $\approx 0^{\circ}F$, e) $\approx 90^{\circ}F$,
   f),It decreases, g), It increases, h) It increases
4. a) $-5 \text{ } m$, b), $-3 \text{ } m$ c) $-4 \text{ } m$
5. a) $\approx 5 \text{ m}$, b) $\approx 2 \text{ m}$, c) $\approx 18 \text{ m}$, d) $\approx 6 \text{ m}$

430

6. a) ≈ 4.7 miles , b) ≈ 7.5 miles , c) ≈ 1.7 hours , d) ≈ 0.5 hours

7. a) ≈ 7 gallons , b) ≈ 5 gallons , c) ≈ 90 miles , d) ≈ 330 miles e) No

Chapter 8 Practice Test

1. A = (0,4), B = (–3, 3), C = (3,0), D = (4, 4), E = (0, –4), F = (0, 0), G = (4, –2), H = (–2, 0), I = (–3, –5)

2.

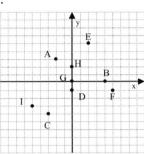

3. $m = 2$
4. $m = -1/3$
5. $m = 8/3$

6. a yes, b no, c yes, d no   9. $m = 1/2$
7. $m = 8$ 10. $m = -2/5$
8. $m = 0$

11.

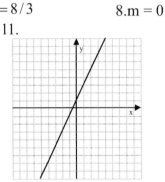

12.

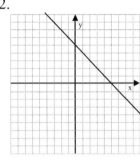

13.

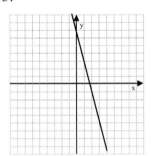

14.

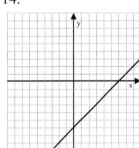

15.

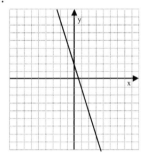

16.

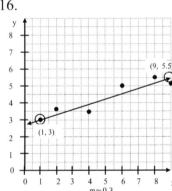

$m \approx 0.3$

17.

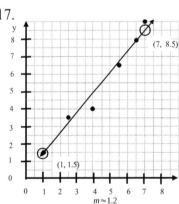

$m \approx 1.2$

18.

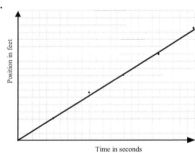

$m \approx 8/13 \approx 0.615$ ft / sec , Velocity

19. a) 0.075   b) 0.0425   c) 0.21   d) 0.70

Chapter 9
Section 9.1
1. $x = 2, y = 1$
2. $x = 0, y = 6$
3. $x = 3, y = -4$
4. $x = 0, y = 0$
5. $x = 5, y = -5$
6. $x = 4, y = 0$
7. $x = 1, y = -6$
8. $x = -4, y = 3$
9. $x = -3, y = -3$
10. $x = -4, y = 7$
11. $x = 0, y = 5$
12. $x = -3, y = 6$
13. $x = -2, y = 0$
14. $x = 6, y = 1$
15. $x = 7, y = -2$
16. $x = 0, y = -3$
17. $x = 6, y = 2$
18. Infinitely many solutions; same lines
19. Infinitely many solutions; same lines
20. $x = 2, y = 6$
21. No solution
22. Infinitely many solutions; same lines
23. No solution

Section 9.2
1. $x = 6, y = -6$
2. $x = 3, y = 2$
3. $x = 12/7, y = 9$
4. $x = -20, y = 6$
5. $x = 6, y = -9$
6. $x = -1, y = 19/7$
7. $x = 4/3, y = 2$
8. $x = 2, y = -7$
9. No solution
10. $x = 2, y = 9$
11. $x = -6, y = -3$
12. $x = 13/4, y = 5$
13. $x = -14, y = -6$
14. $x = -2, y = 9/7$
15. $x = -2, y = 1$
16. $x = -1, y = 0$
17. $x = -8, y = -10$
18. Infinitely many solutions; same lines
19. Infinitely many solutions; same lines
20. No solution
21. No solution
22. $x = 2, y = 3$
23. $x = -6/7, y = -1$
24. $x = -3, y = 3/4$

Section 9.3
1. $x = 1, y = 1$
2. $x = -3, y = -2$
3. $x = -5, y = -2$
4. $x = -4, y = -10$
5. $x = 14, y = 5$
6. $x = 4, y = 0$
7. $x = 2, y = 0$
8. $x = 2, y = -3$
9. $x = 1, y = 2$
10. $x = 5, y = 2$
11. $x = 3, y = -3$
12. $x = 6, y = 7$
13. $x = 2, y = -1$
14. $x = 7, y = 4$
15. $x = -6, y = -81$
16. $x = -1, y = 3$
17. $x = -5, y = 15$
18. $x = 2, y = 0$
19. No solution
20. No solution
21. Infinitely many solutions; same lines
22. Infinitely many solutions; same lines
23. $x = 3, y = 8$
24. $x = -8, y = 25$
25. $x = 2, y = -7$
26. $x = -7, y = -10$
27. $x = 3, y = -5$
28. $x = 16/5, y = 1/5$
29. $x = -3, y = 3$

Section 9.4
1. 7.5 ml of the 10% solution, and 12.5 ml of the 50% solution
2. 60/23 ≈ 2.6 servings of chicken, and 90/23 ≈ 3.9 servings of corn
3. $12 per cat, and $24 per dog
4. 175/13 ≈ 13.46 lbs of peanuts, and 150/13 ≈ 11.54 lbs of cashews
5. 7/4 = 1.75 cups of spaghetti & meatballs, and 14/4 = 4.75 cups of salad
6. 50 lbs of Columbian, and 50 lbs of French roast

Chapter 9 Practice Test
1. $x = 3, y = 4$
2. $x = -2, y = -3$
3. $x = 6, y = -6$
4. $x = 3, y = 2$
5. $x = 12/7, y = 9$
6. $x = 2, y = 4$
7. $x = -3, y = 3/4$
8. $x = 1, y = 1$
9. $x = 3, y = 0$
10. $x = -5, y = -2$
11. $x = -2, y = 2$
12. $x = 4, y = 0$
13. 10 ml of 20%, 20 ml of 50%
14. $4 per bird, $16 per cat
15. 40 lbs Columbian, 60 lbs French Roast

Chapter 10
Section 10.1

1. 5.28045×10^9 5. 5.34×10^{-8} 8. -2.3005×10^{12} 11. -9.9999×10^{10} 14. 3.3302×10^{17}

2. 9.67×10^{-5} 6. 1.0×10^{11} 9. 2.4×10^1 12. -3.33×10^{-4} 15. -4.59×10^2

3. 1.0923×10^{11} 7. -7.95×10^{-7} 10. -1.0×10^{-1} 13. 5.0×10^{-4} 16. -2.349×10^{10}

4. 5.5×10^4

17. 3,000

18. -0.00000019

19. 0.000000042

20. 7,200,000,000,000

21. -0.0002

22. 0.00000000000045

23. $-87,900,000,000$

24. 38,900,000

25. 334,500,000

26. -0.000000000534

27. 0.0000003256

28. 0.0000000025

29. $-3,100,000$

30. 7,700,000,000,000,000,000

31. 1,853.2

32. 0.371

33. -0.231

34. 5,500,000

35. -5.3

36. 2

37. 6.9×10^7 43. $\approx 3.96 \times 10^5$ 49. 2.0×10^{-12} 55. 8.1×10^9 61. $\approx 8.96 \times 10^2$

38. -9.5×10^{-10} 44. $\approx -1.56 \times 10^{17}$ 50. 7.0×10^{20} 56. $\approx -3.4 \times 10^{-24}$ 62. 5.125×10^1

39. $\approx 2.1 \times 10^2$ 45. 1.1×10^6 51. $\approx 1.9 \times 10^{-11}$ 57. 2.0×10^{-9} 63. 1.728×10^0

40. $\approx 3.2 \times 10^{23}$ 46. -2.6×10^3 52. 2.2×10^{13} 58. $\approx 5.4 \times 10^{-12}$ 64. $1.0 \times 10^{-\frac{3}{2}} = 1$

41. -6 47. 3.0×10^{19} 53. $\approx -1.8 \times 10^{-13}$ 59. $\approx 6.68 \times 10^{-39}$

42. $\approx 4.0 \times 10^{-27}$ 48. $\approx 3.3 \times 10^{-1}$ 54. 6.6×10^{-14} 60. 6.4×10^{13}

Section 10.2

1. $2.8 \times 10^3 \, \text{kg} / \text{m}^3$

2. $1.28 \times 10^{11} \, \text{bits} / \text{s}$

3. 2.47×10^{13} RBCs

4. $5.0 \times 10^{17} \, \text{cycles} / \text{sec}$
   $1.7 \times 10^{25} \, \text{m}$, $6.0 \times 10^{-9} \, \text{m}$

5. $\approx 9.9 \times 10^{17} \, \text{kg}$

6. $\approx 6.7 \times 10^2 \, \text{cal} / \text{m}^2 \text{s}$

7. a. $\approx 2.17 \times 10^1 \, \text{people} / \text{square mile}$ b. $\approx 8.6 \times 10^0 \, \text{people} / \text{square mile}$

   c. $\approx 8.1 \times 10^1 \, \text{people} / \text{square mile}$ d. $\approx 3.5 \times 10^2 \, \text{people} / \text{square mile}$

   e. $\approx 8.7 \times 10^2 \, \text{people} / \text{square mile}$ f. $\approx 4.3 \times 10^4 \, \text{people} / \text{square mile}$

8. $\approx 2.5 \times 10^{12} \pi (\text{light yr})^3$, $\approx 7.9 \times 10^{12} (\text{light yr})^3$, $\approx 2.1 \times 10^{66} \pi \, \text{cm}^3$, $= 6.7 \times 10^{66} \, \text{cm}^3$

9. $6.0 \times 10^{44} \, \text{g}$, $6.0 \times 10^{41} \, \text{kg}$, not including planets and other dark matter

10. a. 4.6×10^9 gal , $1.7 \times 10^{10} \text{L}$ b. 7.7×10^9 gal , $2.9 \times 10^{10} \text{L}$ c. 1.2×10^{10} gal , $4.5 \times 10^{10} \text{L}$
    d. 1.7×10^{10} gal , $6.4 \times 10^{10} \text{L}$

11. a. $\approx 9.0 \times 10^{-23}$ g/cm^3 , b. $\approx 9.0 \times 10^{-20}$ kg/cm^3

12. $\approx 2.3 \times 10^{10}$ kWh   14. 3.99×10^{23} km , 2.48×10^{23} miles   15. 1.26×10^9 kg , 2.8×10^9 lb

13. 3.0×10^8 m/s

16. a. 3.011×10^{23} molecules , b. $\approx 1.662 \times 10^{24}$ molecules , c. $\approx 1.280 \times 10^{23}$ molecules

17. $\approx 1.98 \times 10^{20} \, \dfrac{\text{kg} \cdot \text{m}}{\text{s}^2}$

Section 10.3

1. Quantitative – discrete
2. Quantitative – continuous
3. Qualitative
4. Quantitative – continuous
5. Quantitative – discrete
6. Qualitative
7. Quantitative – discrete
8. Qualitative
9. Qualitative
10. Quantitative – discrete
11. Qualitative
12. Quantitative – discrete

13. Quantitative – discrete   14. Qualitative

15.

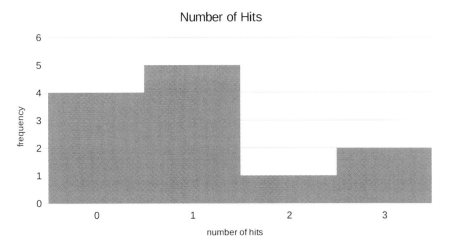

16.

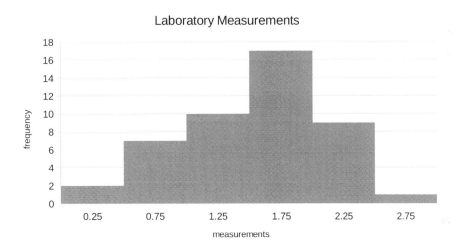

17. mean = 19.77; median = 20; mode = 20; range = 8; variance = 4.64; standard deviation = 2.15
18. mean = 2.74; median = 2.65; mode = 2.6; range = 0.8; variance = 0.06; standard deviation = 0.25
19. mean = 181.4; median = 186.5; mode = none; range = 116; variance = 831.44; standard deviation = 28.83
20. mean = 66.62; median = 65; mode = 65; range = 30; variance = 87.62; standard deviation = 9.36
21. mean = 1.91; median = 1.8; mode = 1.3; range = 4.9; variance = 1.49; standard deviation = 1.22
22. mean = 425.92; median = 432.5; mode = none; range = 582; variance = 24862.57; standard deviation = 157.68
23. mean = 1.08; median = 1; mode = 1; range = 3; variance = 1.08; standard deviation = 1.04

Section 10.4
Note: all answers rounded to nearest hundredth unless otherwise noted
1. °F = 23
2. A = $64,085.85
3. P = 2921.80 J/m^3
4. I = 11.2 Volts/Ohms
5. W = 6050 (Amps)^2 Ohms
6. D = 25 g/cm^3
7. z = 3
8. % Weight Change = 8
9. M = $1753.77
10. R_T = 6 Ohms
11. V_S = 113.10 in^3
12. A = 28.27 m^2
13. K = 294
14. V_{cyl} = 98.17 cm^3
15. p = 48 J/m^3

16. A = $34,125   19. C = 21.99 in   22. $\bar{x}$=21 g

17. Total Magnification = 1000 x 20. A_T = 20 m^2 23. σ = 1.5 hrs

18. Image size = 12,500 μm   21. A = 50.27 cm^2

24. P = 0.14   26. z = 1.4   28. τ = 3,628,873,693 yrs.   30. density = 0.2 g/cm^3

25. Q = 1544 J   27. V_2 = 3.02 m^3 29. R = 2136751.02

31. a) d=0.92 g/mL; b) d=12.03 g/mL; c) v=26.91 mL; d) v=54.17 mL; e) m=579 g; f) m=15.8 g

32. D = 50 parsecs   33. H = 4666.67 feet

34. a) P = 59 watts/m^2; b) P = 2793 watts/m^2; c) P = 10843.88 watts/m^2; d) D = 172.13 watts/m^2; e) P = 535.5 watts/m^2

35. Acceleration = 0.25 ft/s^2 37. W = 132000 pounds · feet

36. V = 40 mph; V = 58.67 feet/second

38. a) R_{Sch} = 30 km; b) R_{Sch} = 300 km; c) R_{Sch} = 0.3 km; d) R_{Sch} = 1.5 km

39. density = 1.03 g/cm^3 43. f = 0.2 seconds^{-1} 47. MAP = 91 mmHg

40. mass = 154.5 g   44. T = 8 seconds   48. pulse pressure = 36 mmHg

41. d = 0.75 m   45. A = 0.25 m

42. λ = 2.7 m   46. h = 7.3 m

49. partial pressure = 416 mmHg (rounded to nearest whole number)

50. C_2 = 0.91 mg/ml

51. density = 0.002 g/cm^3 (rounded to nearest thousandth)

52. $°C = \dfrac{°F - 32}{1.8}$ 56. $V = \dfrac{M}{D}$ 61. $h = \dfrac{V_{cyl}}{\pi r^2}$ 65. $m = \dfrac{Q}{c\,\Delta t}$

53. $T = \dfrac{PV}{nR}$ 57. $\mu = x - z\,\sigma$ 62. $w = \dfrac{p - 2l}{2}$ 66. $\bar{x} = z\left(\dfrac{\sigma}{\sqrt{n}}\right) + \mu$

54. $E = IR$ 58. $b = \dfrac{2A_T}{h}$ 63. $t = \dfrac{A - A_0}{A_0 r}$ 67. $P_2 = \dfrac{P_1 V_1}{V_2}$

55. $R = \dfrac{W}{I^2}$ 59. $r = \sqrt{\dfrac{A}{\pi}}$ 64. $r = \dfrac{C}{2\pi}$ 68. TV = 1000 ml

60. $°C = K - 273$

Chapter 10 Practice Test

1. 5.28045×10^9 5. 5.3×10^{-8} 10. 6.9×10^7

2. 9.67×10^{-5} 6. 3,000   11. -9.5×10^{-10} 14. $2.8 \times 10^3 \dfrac{kg}{m^3}$

3. 1.0923×10^{11} 7. −0.00000019   12. 2.142×10^2

4. 5.5×10^4 8. 0.000000042   13. 6.6×10^{-14} 15. Qualitative

9. 7,200,000,000,000

16. Quantitative and discrete   17. Quantitative and continuous   18. Qualitative

19. mean = 87.387, median = 63.85, mode = none, range = 216.988, variance = 5677.4, standard deviation = 75.3

20. mean = 9.14, median = 9, mode = 15, range = 12, variance = 21.54, standard deviation = 4.64

21. 757,382,400 seconds or $\approx 7.57 \times 10^8$ seconds

22. 4,320,000,000 cells per day or $\approx 4.32 \times 10^9$ cells per day

23. 50   27. 3   29. $n = \dfrac{PV}{RT}$ 30. $R = \dfrac{E}{I}$ 32. $l = \dfrac{p - 2w}{2}$

24. ≈ 13.34 28. $°C = \dfrac{°F - 32}{1.8}$

25. 3

26. 3125   31. $\sigma = \dfrac{x - \mu}{z}$ 33. $D = \dfrac{C}{\pi}$

Appendix A
Section A.1

1.

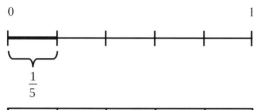

2.

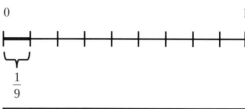

3.

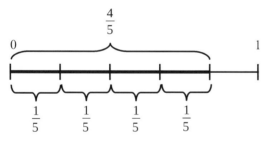

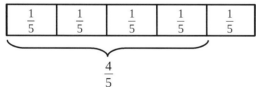

4.

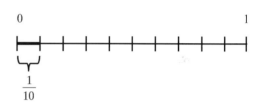

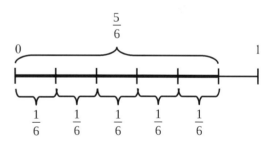

5.

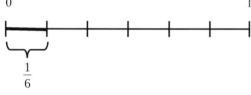

$$\frac{1}{5}+\frac{1}{5}+\frac{1}{5}+\frac{1}{5} = \frac{4}{5}$$

$$4\times\frac{1}{5} = \frac{4}{5}$$

6.

$$\frac{1}{6}+\frac{1}{6}+\frac{1}{6}+\frac{1}{6}+\frac{1}{6} = \frac{5}{6}$$

$$5\times\frac{1}{6} = \frac{5}{6}$$

7.

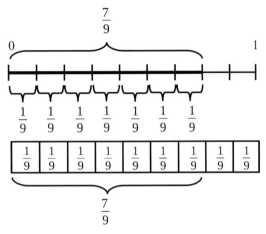

$$\frac{1}{9}+\frac{1}{9}+\frac{1}{9}+\frac{1}{9}+\frac{1}{9}+\frac{1}{9}+\frac{1}{9} = \frac{7}{9}$$

$$7\times\frac{1}{9} = \frac{7}{9}$$

8.

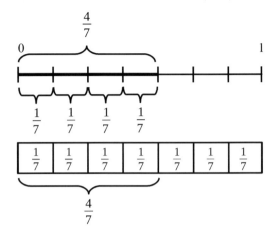

$$\frac{1}{7}+\frac{1}{7}+\frac{1}{7}+\frac{1}{7} = \frac{4}{7}$$

$$4\times\frac{1}{7} = \frac{4}{7}$$

9.

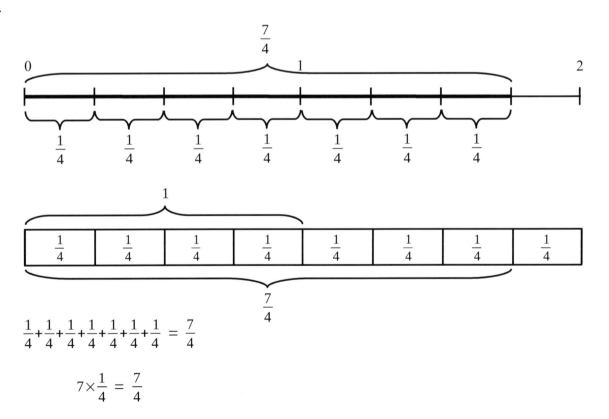

$$\frac{1}{4}+\frac{1}{4}+\frac{1}{4}+\frac{1}{4}+\frac{1}{4}+\frac{1}{4}+\frac{1}{4} = \frac{7}{4}$$

$$7\times\frac{1}{4} = \frac{7}{4}$$

10.

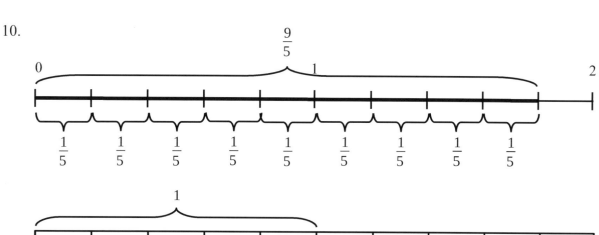

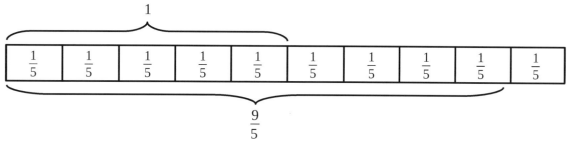

$$\frac{1}{5}+\frac{1}{5}+\frac{1}{5}+\frac{1}{5}+\frac{1}{5}+\frac{1}{5}+\frac{1}{5}+\frac{1}{5}+\frac{1}{5} = \frac{9}{5}$$

$$9\times\frac{1}{5} = \frac{9}{5}$$

11.

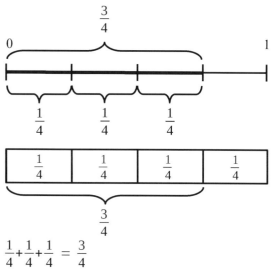

$$\frac{1}{4}+\frac{1}{4}+\frac{1}{4} = \frac{3}{4}$$

$$3\times\frac{1}{4} = \frac{3}{4}$$

12.

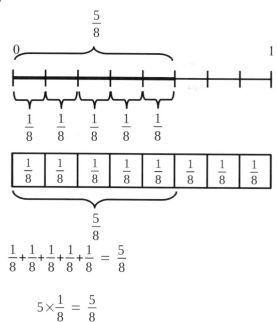

$$\frac{1}{8}+\frac{1}{8}+\frac{1}{8}+\frac{1}{8}+\frac{1}{8} = \frac{5}{8}$$

$$5\times\frac{1}{8} = \frac{5}{8}$$

438

13.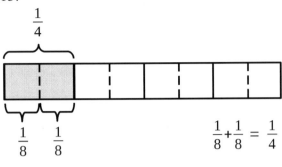

$$\frac{1}{8} + \frac{1}{8} = \frac{1}{4}$$

$$2 \times \frac{1}{8} = \frac{1}{4}$$

14.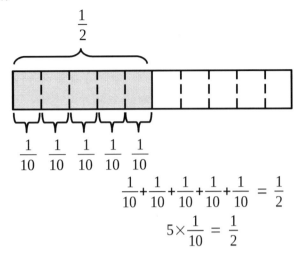

$$\frac{1}{10} + \frac{1}{10} + \frac{1}{10} + \frac{1}{10} + \frac{1}{10} = \frac{1}{2}$$

$$5 \times \frac{1}{10} = \frac{1}{2}$$

15.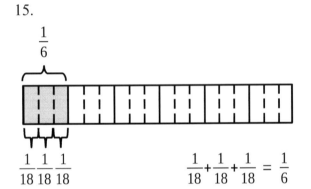

$$\frac{1}{18} + \frac{1}{18} + \frac{1}{18} = \frac{1}{6}$$

$$3 \times \frac{1}{18} = \frac{1}{6}$$

16.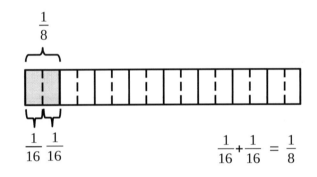

$$\frac{1}{16} + \frac{1}{16} = \frac{1}{8}$$

$$2 \times \frac{1}{16} = \frac{1}{8}$$

17.

$$\frac{1}{5} + \frac{1}{5} + \frac{1}{5} + \frac{1}{5} = \frac{4}{5}$$

$$4 \times \frac{1}{5} = \frac{4}{5}$$

18.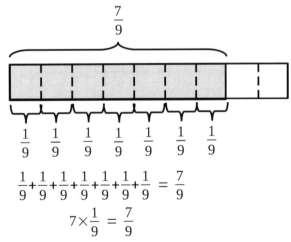

$$\frac{1}{9} + \frac{1}{9} + \frac{1}{9} + \frac{1}{9} + \frac{1}{9} + \frac{1}{9} + \frac{1}{9} = \frac{7}{9}$$

$$7 \times \frac{1}{9} = \frac{7}{9}$$

19.

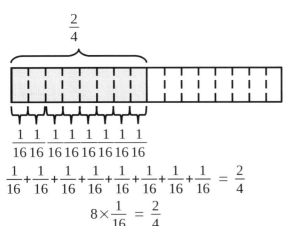

$$\frac{1}{16}+\frac{1}{16}+\frac{1}{16}+\frac{1}{16}+\frac{1}{16}+\frac{1}{16}+\frac{1}{16}+\frac{1}{16} = \frac{2}{4}$$

$$8\times\frac{1}{16} = \frac{2}{4}$$

20.

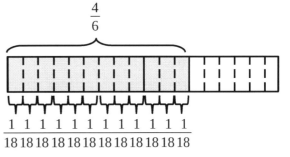

$$\frac{1}{18}+\frac{1}{18}+\frac{1}{18}+\frac{1}{18}+\frac{1}{18}+\frac{1}{18}+\frac{1}{18}+\frac{1}{18}+\frac{1}{18}+\frac{1}{18}+\frac{1}{18}+\frac{1}{18} = \frac{4}{6}$$

$$12\times\frac{1}{18} = \frac{4}{6}$$

21.

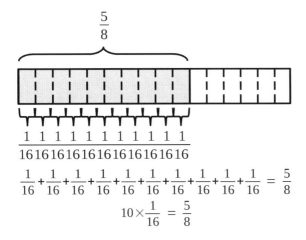

$$\frac{1}{16}+\frac{1}{16}+\frac{1}{16}+\frac{1}{16}+\frac{1}{16}+\frac{1}{16}+\frac{1}{16}+\frac{1}{16}+\frac{1}{16}+\frac{1}{16} = \frac{5}{8}$$

$$10\times\frac{1}{16} = \frac{5}{8}$$

22.

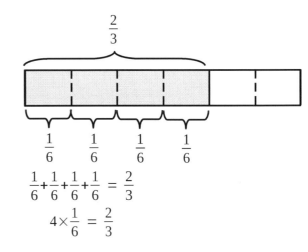

$$\frac{1}{6}+\frac{1}{6}+\frac{1}{6}+\frac{1}{6} = \frac{2}{3}$$

$$4\times\frac{1}{6} = \frac{2}{3}$$

23.

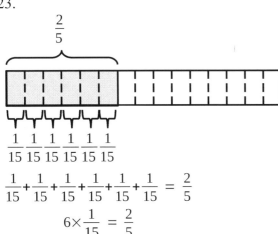

$$\frac{1}{15}+\frac{1}{15}+\frac{1}{15}+\frac{1}{15}+\frac{1}{15}+\frac{1}{15} = \frac{2}{5}$$

$$6\times\frac{1}{15} = \frac{2}{5}$$

24.

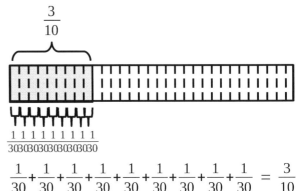

$$\frac{1}{30}+\frac{1}{30}+\frac{1}{30}+\frac{1}{30}+\frac{1}{30}+\frac{1}{30}+\frac{1}{30}+\frac{1}{30}+\frac{1}{30} = \frac{3}{10}$$

$$9\times\frac{1}{30} = \frac{3}{10}$$

Section A.2

1. $3 = \dfrac{9}{3}$, $4 = \dfrac{12}{3}$, $5 = \dfrac{15}{3}$, $6 = \dfrac{18}{3}$

2. $3 = \dfrac{12}{4}$, $4 = \dfrac{16}{4}$, $5 = \dfrac{20}{4}$, $6 = \dfrac{24}{4}$

3. $3 = \dfrac{15}{5}$, $4 = \dfrac{20}{5}$, $5 = \dfrac{25}{5}$, $6 = \dfrac{30}{5}$

4. $3 = \dfrac{18}{6}$, $4 = \dfrac{24}{6}$, $5 = \dfrac{30}{6}$, $6 = \dfrac{36}{6}$

5. $\dfrac{13}{2}$ and $\dfrac{23}{2}$

6. $\dfrac{17}{2}$ and $\dfrac{27}{2}$

7.

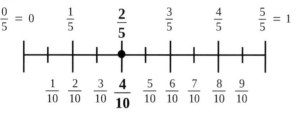

$$\frac{2}{5}=\frac{2\times 2}{5\times 2}=\frac{4}{10}$$

8.

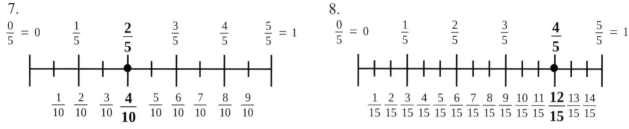

$$\frac{4}{5}=\frac{4\times 3}{5\times 3}=\frac{12}{15}$$

9.

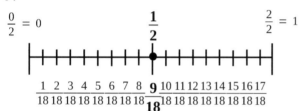

$$\frac{9}{18}=\frac{9\div 9}{18\div 9}=\frac{1}{2}$$

10.

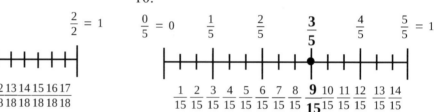

$$\frac{9}{15}=\frac{9\div 3}{15\div 3}=\frac{3}{5}$$

11.

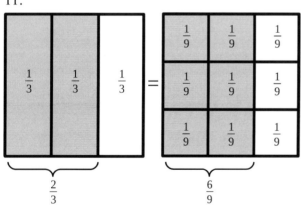

12.

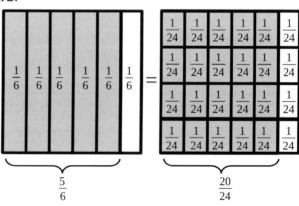

13.

$\frac{1}{7}$ $\frac{1}{7}$ $\frac{1}{7}$ $\frac{1}{7}$ $\frac{1}{7}$ $\frac{1}{7}$ $\frac{1}{7}$ $=$

| $\frac{1}{14}$ | $\frac{1}{14}$ | $\frac{1}{14}$ | $\frac{1}{14}$ | $\frac{1}{14}$ | $\frac{1}{14}$ | $\frac{1}{14}$ |
| $\frac{1}{14}$ | $\frac{1}{14}$ | $\frac{1}{14}$ | $\frac{1}{14}$ | $\frac{1}{14}$ | $\frac{1}{14}$ | $\frac{1}{14}$ |

$\frac{4}{7}$ $\frac{8}{14}$

14.

$\frac{1}{4}$ $\frac{1}{4}$ $\frac{1}{4}$ $\frac{1}{4}$ $=$

$\frac{1}{16}$	$\frac{1}{16}$	$\frac{1}{16}$	$\frac{1}{16}$
$\frac{1}{16}$	$\frac{1}{16}$	$\frac{1}{16}$	$\frac{1}{16}$
$\frac{1}{16}$	$\frac{1}{16}$	$\frac{1}{16}$	$\frac{1}{16}$
$\frac{1}{16}$	$\frac{1}{16}$	$\frac{1}{16}$	$\frac{1}{16}$

$\frac{3}{4}$ $\frac{12}{16}$

15.

$\frac{1}{3}$ $\frac{1}{3}$ $\frac{1}{3}$ $=$

1/18	1/18	1/18
1/18	1/18	1/18
1/18	1/18	1/18
1/18	1/18	1/18
1/18	1/18	1/18
1/18	1/18	1/18

$\frac{2}{3}$ $\frac{12}{18}$

16.

$\frac{1}{5}$ $\frac{1}{5}$ $\frac{1}{5}$ $\frac{1}{5}$ $\frac{1}{5}$ $=$

$\frac{1}{15}$	$\frac{1}{15}$	$\frac{1}{15}$	$\frac{1}{15}$	$\frac{1}{15}$
$\frac{1}{15}$	$\frac{1}{15}$	$\frac{1}{15}$	$\frac{1}{15}$	$\frac{1}{15}$
$\frac{1}{15}$	$\frac{1}{15}$	$\frac{1}{15}$	$\frac{1}{15}$	$\frac{1}{15}$

$\frac{4}{5}$ $\frac{12}{15}$

Section A.3

1.

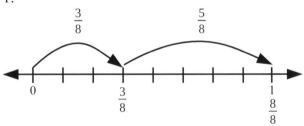

2.

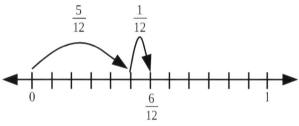

3.

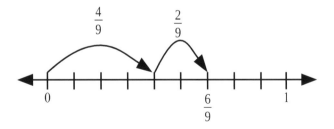

4.

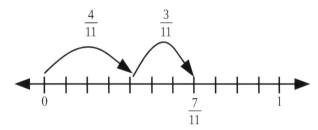

5.

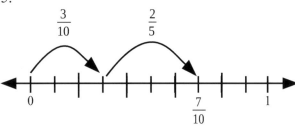

6.

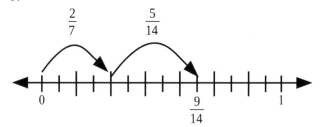

7.

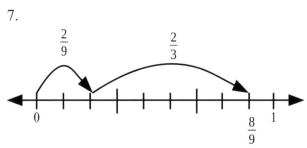

8.

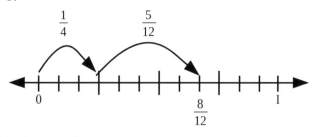

9.

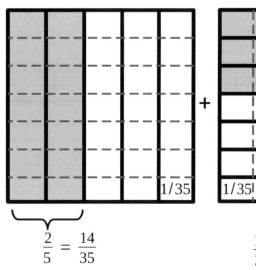

$$\frac{3}{7} = \frac{15}{35}$$

$$\frac{2}{5} = \frac{14}{35}$$

$$\frac{14}{35} + \frac{15}{35} = \frac{29}{35}$$

10.

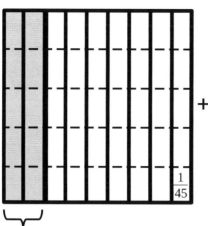

$$\frac{2}{9} = \frac{10}{45}$$

$$\frac{4}{5} = \frac{36}{45}$$

$$\frac{10}{45} + \frac{36}{45} = \frac{46}{45}$$

11.

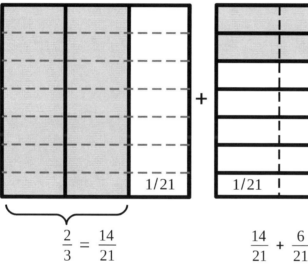

$$\frac{2}{7} = \frac{6}{21}$$

$$\frac{2}{3} = \frac{14}{21}$$

$$\frac{14}{21} + \frac{6}{21} = \frac{20}{21}$$

12.

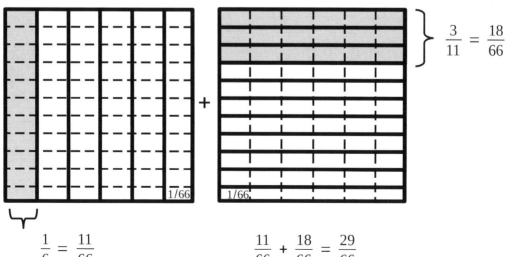

$$\frac{3}{11} = \frac{18}{66}$$

$$\frac{1}{6} = \frac{11}{66}$$

$$\frac{11}{66} + \frac{18}{66} = \frac{29}{66}$$

13. $\dfrac{11}{26}$ 14. $\dfrac{8}{21}$ 15. $\dfrac{23}{18}$ 16. $\dfrac{13}{15}$ 17. $\dfrac{41}{48}$ 18. $\dfrac{73}{72}$ 19. $\dfrac{29}{48}$ 20. $\dfrac{43}{66}$

21.

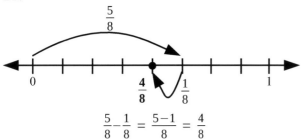

$$\frac{5}{8} - \frac{1}{8} = \frac{5-1}{8} = \frac{4}{8}$$

22.

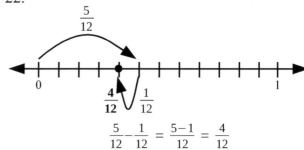

$$\frac{5}{12} - \frac{1}{12} = \frac{5-1}{12} = \frac{4}{12}$$

23.

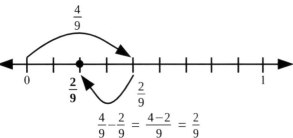

$$\frac{4}{9} - \frac{2}{9} = \frac{4-2}{9} = \frac{2}{9}$$

24.

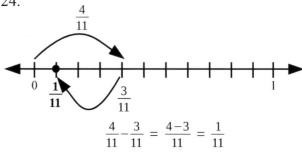

$$\frac{4}{11} - \frac{3}{11} = \frac{4-3}{11} = \frac{1}{11}$$

25.

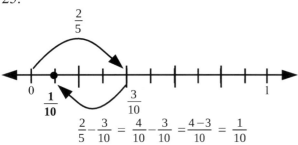

$$\frac{2}{5} - \frac{3}{10} = \frac{4}{10} - \frac{3}{10} = \frac{4-3}{10} = \frac{1}{10}$$

26.

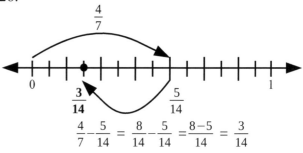

$$\frac{4}{7} - \frac{5}{14} = \frac{8}{14} - \frac{5}{14} = \frac{8-5}{14} = \frac{3}{14}$$

27.

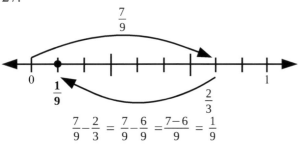

$$\frac{7}{9} - \frac{2}{3} = \frac{7}{9} - \frac{6}{9} = \frac{7-6}{9} = \frac{1}{9}$$

28.

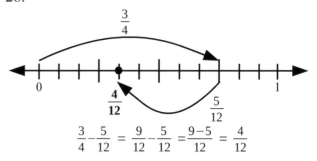

$$\frac{3}{4} - \frac{5}{12} = \frac{9}{12} - \frac{5}{12} = \frac{9-5}{12} = \frac{4}{12}$$

29.

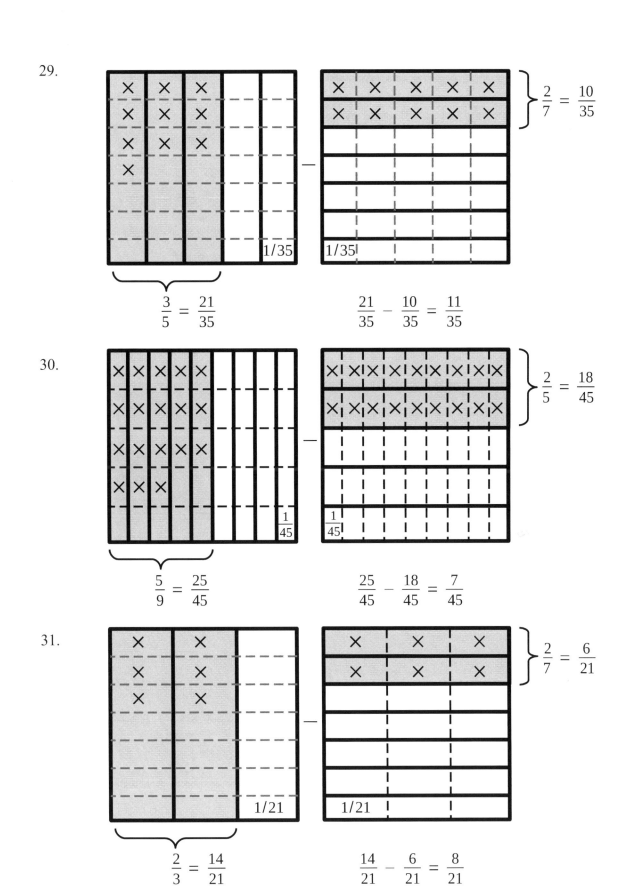

$$\frac{3}{5} = \frac{21}{35}$$

$$\frac{2}{7} = \frac{10}{35}$$

$$\frac{21}{35} - \frac{10}{35} = \frac{11}{35}$$

30.

$$\frac{5}{9} = \frac{25}{45}$$

$$\frac{2}{5} = \frac{18}{45}$$

$$\frac{25}{45} - \frac{18}{45} = \frac{7}{45}$$

31.

$$\frac{2}{3} = \frac{14}{21}$$

$$\frac{2}{7} = \frac{6}{21}$$

$$\frac{14}{21} - \frac{6}{21} = \frac{8}{21}$$

32.

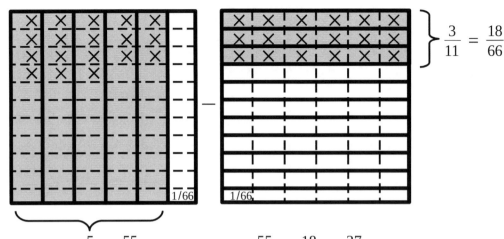

$$\frac{3}{11} = \frac{18}{66}$$

$$\frac{5}{6} = \frac{55}{66}$$

$$\frac{55}{66} - \frac{18}{66} = \frac{37}{66}$$

33. $\frac{1}{26}$ 35. $\frac{3}{18} = \frac{1}{6}$ 37. $\frac{17}{48}$ 39. $\frac{1}{48}$ 41. $\frac{7}{20}$ 43. $\frac{3}{20}$

34. $\frac{7}{21} = \frac{1}{3}$ 36. $\frac{5}{15} = \frac{1}{3}$ 38. $\frac{25}{72}$ 40. $\frac{17}{66}$ 42. $\frac{5}{12}$ 44. $\frac{7}{40}$

Section A.4

1.

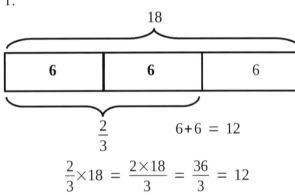

$$\frac{2}{3} \times 18 = \frac{2 \times 18}{3} = \frac{36}{3} = 12$$

2.

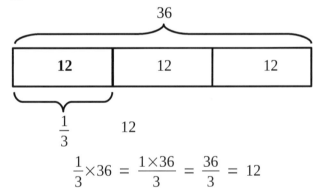

$$\frac{1}{3} \times 36 = \frac{1 \times 36}{3} = \frac{36}{3} = 12$$

3.

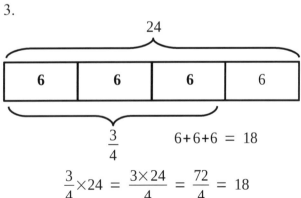

$$\frac{3}{4} \times 24 = \frac{3 \times 24}{4} = \frac{72}{4} = 18$$

4.

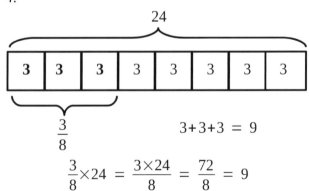

$$\frac{3}{8} \times 24 = \frac{3 \times 24}{8} = \frac{72}{8} = 9$$

447

5.

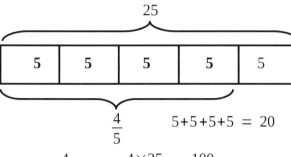

25

| **5** | **5** | **5** | **5** | 5 |

$\frac{4}{5}$

$5+5+5+5 = 20$

$$\frac{4}{5}\times25 = \frac{4\times25}{5} = \frac{100}{5} = 20$$

6.

140

| **20** | 20 | 20 | 20 | 20 | 20 | 20 |

$\frac{1}{7}$

20

$$\frac{1}{7}\times140 = \frac{1\times140}{7} = \frac{140}{7} = 20$$

7.

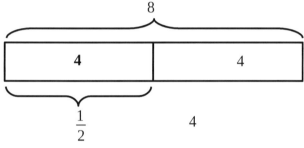

8

| **4** | 4 |

$\frac{1}{2}$

4

$$\frac{1}{2}\times8 = \frac{1\times8}{2} = \frac{8}{2} = 4$$

8.

1	2	3	4	5	6	7	8
$\frac{1}{2}$	$\frac{1}{2}$	$\frac{1}{2}$	$\frac{1}{2}$	$\frac{1}{2}$	$\frac{1}{2}$	$\frac{1}{2}$	$\frac{1}{2}$

$$\frac{1}{2}+\frac{1}{2}+\frac{1}{2}+\frac{1}{2}+\frac{1}{2}+\frac{1}{2}+\frac{1}{2}+\frac{1}{2} = 4$$

$$\frac{8\times1}{2} = \frac{8\times1}{2} = \frac{8}{2} = 4$$

9.

10

| **2** | **2** | **2** | 2 | 2 |

$\frac{3}{5}$

$2+2+2 = 6$

$$\frac{3}{5}\times10 = \frac{3\times10}{5} = \frac{30}{5} = 6$$

10.

1	2	3	4	5	6	7	8	9	10
$\frac{3}{5}$	$\frac{3}{5}$	$\frac{3}{5}$	$\frac{3}{5}$	$\frac{3}{5}$	$\frac{3}{5}$	$\frac{3}{5}$	$\frac{3}{5}$	$\frac{3}{5}$	$\frac{3}{5}$

$$\frac{3}{5}+\frac{3}{5}+\frac{3}{5}+\frac{3}{5}+\frac{3}{5}+\frac{3}{5}+\frac{3}{5}+\frac{3}{5}+\frac{3}{5}+\frac{3}{5} = 6$$

$$\frac{10\times3}{5} = \frac{10\times3}{5} = \frac{30}{5} = 6$$

11.

1	2	3	4	5	6	7	8	9	10	11	12	13	14
$\frac{3}{7}$	$\frac{3}{7}$	$\frac{3}{7}$	$\frac{3}{7}$	$\frac{3}{7}$	$\frac{3}{7}$	$\frac{3}{7}$	$\frac{3}{7}$	$\frac{3}{7}$	$\frac{3}{7}$	$\frac{3}{7}$	$\frac{3}{7}$	$\frac{3}{7}$	$\frac{3}{7}$

$$\frac{3}{7}+\frac{3}{7}+\frac{3}{7}+\frac{3}{7}+\frac{3}{7}+\frac{3}{7}+\frac{3}{7}+\frac{3}{7}+\frac{3}{7}+\frac{3}{7}+\frac{3}{7}+\frac{3}{7}+\frac{3}{7}+\frac{3}{7} = 6$$

$$\frac{14\times3}{7} = \frac{14\times3}{7} = \frac{42}{7} = 6$$

12.

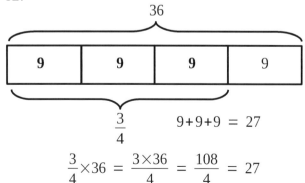

36

| **9** | **9** | **9** | 9 |

$\frac{3}{4}$

$9+9+9 = 27$

$$\frac{3}{4}\times36 = \frac{3\times36}{4} = \frac{108}{4} = 27$$

13.

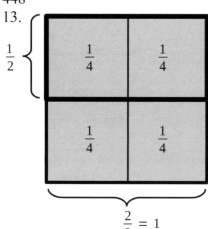

$$\frac{2}{2} = 1$$

$$\frac{1}{2} \times \frac{2}{2} = \frac{1 \times 2}{2 \times 2} = \frac{2}{4} = \frac{2 \div 2}{4 \div 2} = \frac{1}{2}$$

14.

$$\frac{2}{3} \times \frac{1}{2} = \frac{2 \times 1}{3 \times 2} = \frac{2}{6} = \frac{2 \div 2}{3 \div 2} = \frac{1}{3}$$

15.

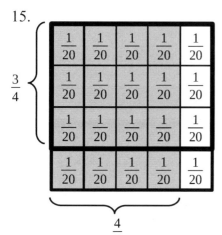

$$\frac{4}{5}$$

$$\frac{3}{4} \times \frac{4}{5} = \frac{3 \times 4}{4 \times 5} = \frac{12}{20} = \frac{12 \div 4}{20 \div 4} = \frac{3}{5}$$

16.

$$\frac{2}{3}$$

$$\frac{2}{5} \times \frac{2}{3} = \frac{2 \times 2}{5 \times 3} = \frac{4}{15}$$

17.

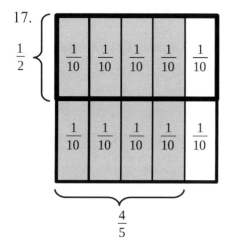

$$\frac{4}{5}$$

$$\frac{1}{2} \times \frac{4}{5} = \frac{1 \times 4}{2 \times 5} = \frac{4}{10} = \frac{4 \div 2}{10 \div 2} = \frac{2}{5}$$

18.

$$\frac{1}{4}$$

$$\frac{2}{3} \times \frac{1}{4} = \frac{2 \times 1}{3 \times 4} = \frac{2}{12} = \frac{2 \div 2}{12 \div 2} = \frac{1}{6}$$

19.

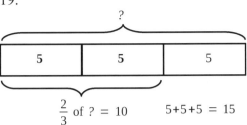

$\frac{2}{3}$ of ? = 10   5+5+5 = 15

20.

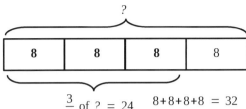

$\frac{3}{4}$ of ? = 24   8+8+8+8 = 32

21. $\frac{1}{2}$

22. $\frac{1}{2}$

23. $\frac{5}{8}$

24. $\frac{4}{7}$

25. $\frac{2}{15}$

26. $\frac{7}{33}$

27. $\frac{143}{60}$

28. 36
29. 72

30. $\frac{1}{10}$

31. $\frac{4}{10} = \frac{2}{5}$

32. $\frac{1}{2}$

33. a) 1/3, b) 2 tons

34. $\frac{1}{8}$

35. a) 1/5 b) 1/5

36.

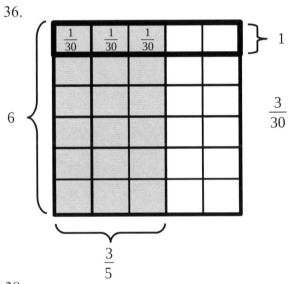

37.

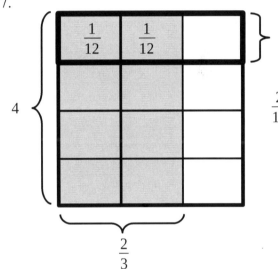

38.

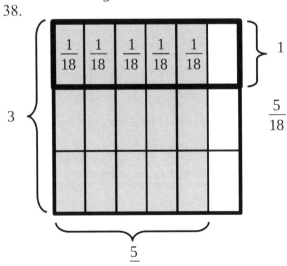

39.

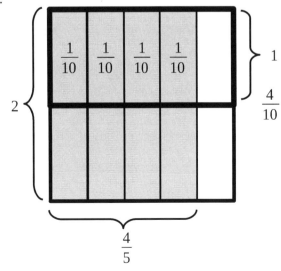

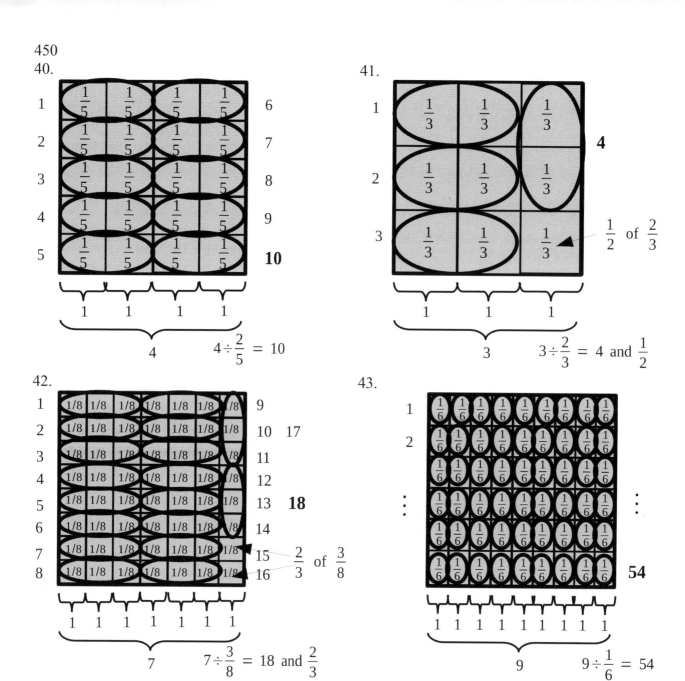

40.

$$4 \div \frac{2}{5} = 10$$

41.

$$3 \div \frac{2}{3} = 4 \text{ and } \frac{1}{2}$$

42.

$$7 \div \frac{3}{8} = 18 \text{ and } \frac{2}{3}$$

43.

$$9 \div \frac{1}{6} = 54$$

44.

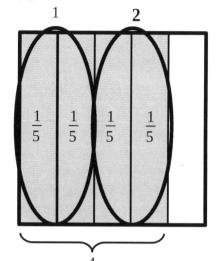

$$\frac{4}{5} \div \frac{2}{5} = \frac{4}{5} \times \frac{5}{2} = \frac{4 \times 5}{5 \times 2}$$

$$= \frac{20}{10} = \frac{20 \div 10}{10 \div 10} = 2$$

46.

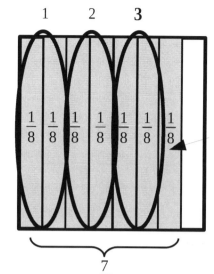

$\frac{1}{2}$ of $\frac{2}{8}$

$$\frac{7}{8} \div \frac{2}{8} = \frac{7}{8} \times \frac{8}{2} = \frac{7 \times 8}{8 \times 2} = \frac{56}{16}$$

$$= \frac{56 \div 8}{16 \div 8} = \frac{7}{2} = 3 \text{ and } \frac{1}{2}$$

45.

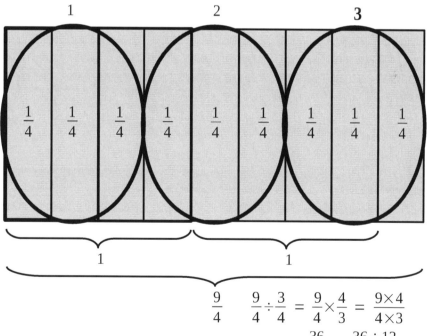

$\frac{9}{4}$

$$\frac{9}{4} \div \frac{3}{4} = \frac{9}{4} \times \frac{4}{3} = \frac{9 \times 4}{4 \times 3}$$

$$= \frac{36}{12} = \frac{36 \div 12}{12 \div 12} = 3$$

452

47.

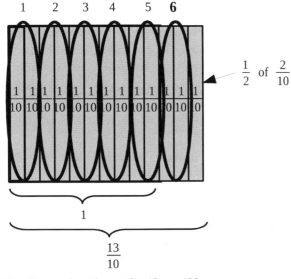

$$\frac{13}{10} \div \frac{2}{10} = \frac{13}{10} \times \frac{10}{2} = \frac{13 \times 10}{10 \times 2} = \frac{130}{20}$$
$$= \frac{130 \div 10}{20 \div 10} = \frac{13}{2} = 6 \text{ and } \frac{1}{2}$$

48.

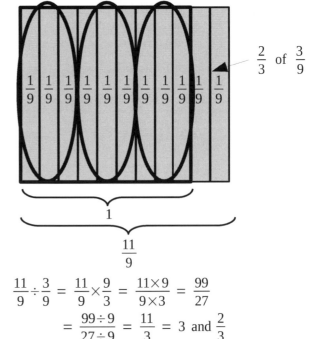

$$\frac{11}{9} \div \frac{3}{9} = \frac{11}{9} \times \frac{9}{3} = \frac{11 \times 9}{9 \times 3} = \frac{99}{27}$$
$$= \frac{99 \div 9}{27 \div 9} = \frac{11}{3} = 3 \text{ and } \frac{2}{3}$$

49.

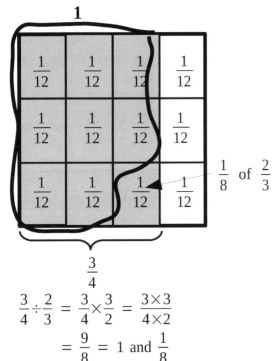

$$\frac{3}{4} \div \frac{2}{3} = \frac{3}{4} \times \frac{3}{2} = \frac{3 \times 3}{4 \times 2}$$
$$= \frac{9}{8} = 1 \text{ and } \frac{1}{8}$$

50.

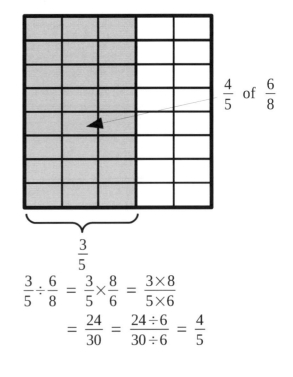

$$\frac{3}{5} \div \frac{6}{8} = \frac{3}{5} \times \frac{8}{6} = \frac{3 \times 8}{5 \times 6}$$
$$= \frac{24}{30} = \frac{24 \div 6}{30 \div 6} = \frac{4}{5}$$

51. $\dfrac{77}{72}$ 52. $\dfrac{35}{33}$ 53. $\dfrac{48}{35}$ 54. $\dfrac{33}{8}$ 55. $\dfrac{30}{13}$ 56. $\dfrac{42}{5}$ 57. $\dfrac{5}{18}$ 58. $\dfrac{12}{7}$

59. 80 people

60. Approximately 26 cakes. There is not enough for 27 cakes. $20\,\text{cups}\div\dfrac{3}{4}\,\text{cups}=20\times\dfrac{4}{3}=26\dfrac{2}{3}$

61. $\dfrac{3}{10}$ meter

Section A.5

1. $3\times10^{-2}+5\times10^{-3}$ 2. $1\times10^{0}+1\times10^{-1}+9\times10^{-2}$
3. $1\times10^{1}+6\times10^{0}+0\times10^{-1}+8\times10^{-2}+4\times10^{-3}+3\times10^{-4}$
4. $2\times10^{1}+5\times10^{0}+0\times10^{-1}+9\times10^{-2}+3\times10^{-3}+8\times10^{-4}$
5. $1\times10^{1}+1\times10^{0}+0\times10^{-1}+0\times10^{-2}+2\times10^{-3}$
6. $4\times10^{1}+5\times10^{0}+0\times10^{-1}+2\times10^{-2}+0\times10^{-3}+2\times10^{-4}+3\times10^{-5}$

7. 23.4 $>$ 23.39	9. 0.035 $>$ 0.03	11. 8.79 $>$ 7.8
8. 16.8 $>$ 16.09	10. 13.29 $>$ 3.3	12. 15.4309 $>$ 15.43

13. {3.02, 3.09, 3.58, 3.77, 3.8, 3.95}

14. {0.23, 0.3, 0.703, 0.73, 0.901, 0.98}

15. {2.005, 2.06, 2.18, 2.2, 2.7, 2.71, 2.81, 3.0,}

16. {11.001, 11.06, 11.3, 11.554, 11.71, 11.9, 11.930}

17. {1.07, 1.38, 1.67, 1.7, 1.76, 1.83, 1.9, 1.99, 2.0}

18. {4.95, 5.05, 5.15, 5.51, 5.58, 5.85, 5.976}

19. 41.63	25. 9.25	31. 88, 87.9, 87.94
20. 20.09	26. 15.54	32. 390, 389.9, 389.85
21. 66.440	27. 3.7147	33. 0, 0.1, 0.14
22. 3.180	28. 19.6903	34. 3, 2.8, 2.79
23. 39.295	29. 44.408	35. 18, 17.8, 17.83
24. 20.457	30. 26.568	36. 20, 20.1, 20.10

37. a) 370, b) 9,640, c) 130   40. a) 10.96, b) 59.10, c) 7.01
38. a) 3,800, b) 12,300, c) 9,000   41. a) 0.672, b) 15.709, c) 1.009
39. a) 14.7, b) 6.5, c) 89.9

42. 0.11   43. 0.40   44. 0.14   45. 0.09   46. 0.38   47. 0.17

48. 28.8 can be estimated to 30 and 2.1 can be estimated to 2. 30×2=60 , 28.8×2.1=60.48 which is
    60.5 rounded to the nearest tenths.

49. 52.3 can be estimated to 50 and 4.8 can be estimated to 5. $50\div5=10$.
    $52.3\div4.8=10.89583$ which is 10.9 rounded to the nearest tenths.

50. 201.57 can be estimated to 200 and 19.75 can be estimated to 20. 200×20=4000 .
    201.57×19.75=3981.0075 which is 3981.0 rounded to the nearest tenths.

51. 201.57 can be estimated to 200 and 19.8 can be estimated to 20. $200\div20=10$
    201.57÷19.8=10.2   rounded to the nearest tenths.

Measurement Conversion Formulas

From Unit	To Unit	Conversion Factor
pm pg pL	m g L	$\dfrac{1\text{m }(g,L)}{1{,}000{,}000{,}000{,}000 \text{ pm}(pg,pL)}$
nm ng nL	m g L	$\dfrac{1\text{m }(g,L)}{1{,}000{,}000{,}000 \text{ nm}(ng,nL)}$
μm μg μL	m g L	$\dfrac{1\text{m }(g,L)}{1{,}000{,}000\,\mu m(\mu g,\mu L)}$
mm mg mL	m g L	$\dfrac{1\text{m }(g,L)}{1{,}000 \text{ mm}(mg,mL)}$
cm cg cL	m g L	$\dfrac{1\text{m }(g,L)}{100 \text{ cm}(cg,cL)}$
km kg kL	m g L	$\dfrac{1{,}000\text{ m}(g,L)}{1\text{ km}(kg,kL)}$

From Unit	To Unit	Conversion Factor
m g L	pm pg pL	$\dfrac{1{,}000{,}000{,}000{,}000 \text{ pm }(pg,pL)}{1\text{m }(g,L)}$
m g L	nm ng nL	$\dfrac{1{,}000{,}000{,}000 \text{ nm }(ng,nL)}{1\text{m }(g,L)}$
m g L	μm μg μL	$\dfrac{1{,}000{,}000\,\mu m\ (\mu g,\mu L)}{1\,m\ (g,L)}$
m g L	mm mg mL	$\dfrac{1{,}000 \text{ mm }(mg,mL)}{1\text{m }(g,L)}$
m g L	cm cg cL	$\dfrac{100 \text{ cm }(cg,cL)}{1\text{m }(g,L)}$
m g L	km kg kL	$\dfrac{1\text{ km }(kg,kL)}{1{,}000\text{ m }(g,L)}$

Convert "from" "to"	Conversion Factor
feet to inches	$\dfrac{12\text{ in}}{1\text{ ft}}$
yards to feet	$\dfrac{3\text{ ft}}{1\text{ yd}}$
miles to feet	$\dfrac{5280\text{ ft}}{1\text{ mi}}$
pint to fl. ounces	$\dfrac{16\text{ fl. ounces}}{1\text{ pt}}$
quart to pints	$\dfrac{2\text{ pt}}{1\text{ qt}}$
gallon to quarts	$\dfrac{4\text{ qt}}{1\text{ gallon}}$
pounds to ounces	$\dfrac{16\text{ ounces}}{1\text{ lb}}$
tons to pounds	$\dfrac{2000\text{ lbs}}{1\text{ ton}}$

Convert "from" "to"	Conversion Factor
inches to feet	$\dfrac{1\text{ ft}}{12\text{ in}}$
feet to yards	$\dfrac{1\text{ yd}}{3\text{ ft}}$
feet to miles	$\dfrac{1\text{ mi}}{5280\text{ ft}}$
fl. ounces to pints	$\dfrac{1\text{ pt}}{16\text{ fl. ounces}}$
pints to quarts	$\dfrac{1\text{ qt}}{2\text{ pt}}$
quarts to gallons	$\dfrac{1\text{ gallon}}{4\text{ qt}}$
ounces to pounds	$\dfrac{1\text{ lb}}{16\text{ ounces}}$
pounds to tons	$\dfrac{1\text{ ton}}{2000\text{ lbs}}$

Convert "from" "to"	Conversion Factor
cm to inch	$\dfrac{1\ \text{in}}{2.54\ \text{cm}}$
meter to feet	$\dfrac{3.28\ \text{feet}}{1\ \text{m}}$
meter to yard	$\dfrac{1\ \text{yd}}{0.9144\ \text{m}}$
mL to fl. ounce	$\dfrac{1\ \text{fl. ounce}}{29.574\ \text{mL}}$
gram to ounce	$\dfrac{1\ \text{ounce}}{28.35\ \text{g}}$
kg to lb	$\dfrac{2.2046\ \text{lb}}{1\ \text{kg}}$
km to mile	$\dfrac{1\ \text{mile}}{1.6093\ \text{km}}$
m to mile	$\dfrac{1\ \text{mi}}{1{,}609.3\ \text{m}}$
L to quart	$\dfrac{1.06\ \text{quarts}}{1\ \text{L}}$
L to gallon	$\dfrac{0.265\ \text{gallon}}{1\ \text{L}}$
L to cubic inches	$\dfrac{61\ \text{in}^3}{1\ \text{L}}$
mL to cubic cm	$\dfrac{1\,\text{cm}^3}{1\,\text{m L}}$
L to cubic cm	$\dfrac{1{,}000\,\text{cm}^3}{1\,\text{L}}$
cubic inch to cubic cm	$\dfrac{16.387\,\text{cm}^3}{1\,\text{in}^3}$

Convert "from" "to"	Conversion Factor
inch to cm	$\dfrac{2.54\ \text{cm}}{1\ \text{in}}$
feet to meter	$\dfrac{1\ \text{m}}{3.28\ \text{feet}}$
yard to meter	$\dfrac{0.9144\ \text{m}}{1\ \text{yd}}$
fl. ounce to mL	$\dfrac{29.574\ \text{mL}}{1\ \text{fl. ounce}}$
ounce to gram	$\dfrac{28.35\ \text{g}}{1\ \text{ounce}}$
lb to kg	$\dfrac{1\ \text{kg}}{2.2046\ \text{lb}}$
mile to km	$\dfrac{1.6093\ \text{km}}{1\ \text{mile}}$
mile to m	$\dfrac{1{,}609.3\ \text{m}}{1\ \text{mi}}$
quart to L	$\dfrac{1\ \text{L}}{1.06\ \text{quarts}}$
gallon to L	$\dfrac{1\ \text{L}}{0.265\ \text{gallon}}$
cubic inches to L	$\dfrac{1\ \text{L}}{61\ \text{in}^3}$
cubic cm to ml	$\dfrac{1\,\text{m L}}{1\,\text{cm}^3}$
cubic cm to L	$\dfrac{1\,\text{L}}{1{.}000\,\text{cm}^3}$
cubic cm to cubic inch	$\dfrac{1\,\text{in}^3}{16.387\,\text{cm}^3}$

Conversion	Formulas
°F to °C	$^{\text{o}}\text{C}=\dfrac{\left(^{\text{o}}\text{F}-32\right)}{1.8}\qquad \text{or} \qquad ^{\text{o}}\text{C}=\left(\dfrac{5}{9}\right)\left(^{\text{o}}\text{F}-32\right)$
°C to °F	$^{\text{o}}\text{F}=1.8\left(^{\text{o}}\text{C}\right)+32 \quad \text{or} \quad ^{\text{o}}\text{F}=\left(\dfrac{9}{5}\right)\left(^{\text{o}}\text{C}\right)+32$
°C to K	$\text{K} = {}^{\text{O}}\text{C} + 273.15$
K to °C	$^{\text{O}}\text{C} = \text{K} - 273.15$

Convert "from" "to"	Conversion Factor
in^2 to ft^2	$\dfrac{1\,\text{ft}^2}{144\,\text{in}^2}$
mm^2 to cm^2	$\dfrac{1\,\text{cm}^2}{100\,\text{mm}^2}$
in^3 to ft^3	$\dfrac{1\,\text{ft}^3}{1{,}728\,\text{in}^3}$
mm^3 to cm^3	$\dfrac{1\,\text{cm}^3}{1{,}000\,\text{mm}^3}$
cm^3 to m^3	$\dfrac{1\,\text{m}^3}{1{,}000{,}000\,\text{cm}^3}$
cm^3 to L	$\dfrac{1\,\text{L}}{1{,}000\,\text{cm}^3}$
in^3 to gal	$\dfrac{1\,\text{gal}}{231\,\text{in}^3}$
cm^3 to gal	$\dfrac{1\,\text{gal}}{3{,}785.412\,\text{cm}^3}$
in^3 to L	$\dfrac{1\,\text{L}}{61.0237\,\text{in}^3}$

Convert "from" "to"	Conversion Factor
ft^2 to in^2	$\dfrac{144\,\text{in}^2}{1\,\text{ft}^2}$
cm^2 to mm^2	$\dfrac{100\,\text{mm}^2}{1\,\text{cm}^2}$
ft^3 to in^3	$\dfrac{1{,}728\,\text{in}^3}{1\,\text{ft}^3}$
cm^3 to mm^3	$\dfrac{1{,}000\,\text{mm}^3}{1\,\text{cm}^3}$
m^3 to cm^3	$\dfrac{1{,}000{,}000\,\text{cm}^3}{1\,\text{m}^3}$
L to cm^3	$\dfrac{1{,}000\,\text{cm}^3}{1\,\text{L}}$
gal to in^3	$\dfrac{231\,\text{in}^3}{1\,\text{gal}}$
gal to cm^3	$\dfrac{3{,}785.412\,\text{cm}^3}{1\,\text{gal}}$
L to in^3	$\dfrac{61.0237\,\text{in}^3}{1\,\text{L}}$

Convert "from" "to"	Conversion Factor
seconds to minutes	$\dfrac{1\,\text{min}}{60\,\text{sec}}$
minutes to hours	$\dfrac{1\,\text{hour}}{60\,\text{min}}$

Convert "from" "to"	Conversion Factor
minutes to seconds	$\dfrac{60\,\text{sec}}{1\,\text{min}}$
hours to minutes	$\dfrac{60\,\text{min}}{1\,\text{hour}}$

Made in the USA
Middletown, DE
21 August 2019